17°00′
17°00′
REPUBLIC OF VIETNAM BUFFER ZONE
VIETNAM TERRITORIAL WATERS ADIZ
GULF OF
VỊNH BẮC
01
02
17
50
21
VIETNAM TERRITORIAL WATER
VIETNAM WESTERN BOUNDARY ADIZ
CUA VIET
Hà Lợi Trung
Triệu Phong
TRIỆU PHONG
QUANG TRI
QUANG TRI / LA VANG / 21 / s
HAI LANG
Hải Lăng
MY CHANH / 12 / s
Phong Điền
PHO TRACH / 11 / s
Quảng Điền
Hương Điền
PHA TAM GIANG
ĐẦM THANH LAM
Hương Trà
THI XA HUE
HUE
Phú Vang
Hương Thủy
HUE / PHU BAI / 56
Nam Hòa
VAN XA / 11 / s
YD7
16°30′
107°30′
TA LUONG
ABANDONED TA BAT / 15 / s
NOT USABLE A LUOI / 34 / s
DONG CA PUY
NUI BAI CAY TAT
NUI MAY NHA
DONG BA RANG
DONG MANG CHAN
Hàm Hòa (abandoned)
NUI MO TAU
Thôn Dương Hòa (abandoned)
GENERALLY FOREST COVERED

MSGT Donn R Proven USMCR (Ret)
3/1/97

USMC/VIETNAM HELICOPTER PILOTS & AIRCREW REUNION

TURNER PUBLISHING COMPANY
Paducah, Kentucky

TURNER PUBLISHING COMPANY
Publishers of Military History
412 Broadway, P.O. Box 3101
Paducah, KY 42002-3101
Phone: 502-443-0121

USMC/Vietnam Helicopter Pilots &
Aircrew History Book Staff:
Publisher: R. A. Herman
Editor: A.M. Leahy
Chief Historical Writer: J.Van Nortwick
Editorial Assistant: M.E. Salter
Historical Aircraft: J.L. Shanahan

Turner Publishing Company Staff:
Chief Editor: Robert J. Martin
Designer: Ina F. Morse

Copyright © 1996 Turner Publishing Company. All rights reserved. Additional copies may be purchased directly from Turner Publishing Company.

Library of Congress Catalog Card No. 95-60324
ISBN: 1-56311-190-X
Limited Edition. Printed in the U.S.A.

This book or any part thereof may not be reproduced without the written consent of Turner Publishing Company.

Photo: HMM-364 "Purple Foxes" aircraft maintenance area at Phu Bai, 13 Sept 1968 (Photo credit: HB Staff)

Table of Contents

Foreword	4
Dedication	5
Acknowledgements	6
Publisher's Message	7
A Chronology of Marine Helicopters in Vietnam, 1962-1975	10
Vietnam Helos	19
Special Stories	25
Glossary of Terms	49
Veteran's Biographies	51
Index	120

FOREWORD

Archie J. Clapp, South Vietnam-1962

"First in"

Preparations for the operation had been cloaked in secrecy from the outset. For one thing, it wasn't until the LPH flagship, USS Princeton, had set sail from its last "friendly port" that the embarked Marines were even told where they were headed.

After an uneventful transit, the 7th Fleet Ready Group advanced toward the foreign shore under cover of darkness. Then, soon after first light on April 15, 1962, the helicopters of HMM-362 were launched.

Now designated Task Unit 79.3.5 and code-named SHU-FLY, the squadron proceeded to its classified objective: a former Japanese airstrip in the heart of South Vietnam's Mekong Delta. Located near the town of Soc Trang in Ba Xuyen Province this airfield would be the site of the first Marine "tactical landing" since the Korean War.

Upon our initial touch-down in Vietnam, following a stealthy approach, we received our first, but certainly not last, taste of the contradictory nature of our presence in that troubled country. Instead of being confronted by a hostile force, as the security-oriented prelude would have suggested, MajGen John Condon, CG, 1st MAW, and a few members of his staff nonchalantly awaited our arrival. But that was just the beginning of that day's "education."

Rather than returning to the ship for another load right away, which was my original plan, I shut down my UH-34 to report to my boss. With a smile, he said, "Let's go meet the Ba Xuyen Province Chief, Archie." It became apparent that amenities were at least as important in this "war" as my helping to get the squadron completely established ashore by dark. Then, on the short jeep ride to town, my eyes were opened wider still. Along the way were several banners proclaiming "Welcome Task Unit 79.3.5." So much for my secret mission!

This episode marked the beginning of the lengthy Marine Corps presence in the Vietnam War. And, to me, it exemplified the ambivalence concerning our role there. As illustrated here, we could be called upon to comport ourselves in accordance with accepted military doctrine and procedures, and at the same time ensure that no social or operational transgressions occurred.

An example of the latter requirement can be found in the Rules of Engagement for our deployment. Our primary mission was to land Vietnamese troops virtually on top of their Viet Cong enemy. Yet, the rules stated that we were not to fire our weapons (such as they were) until the other fellow (whoever that might be) had definitely shot at us first. And in the beginning, the only aircraft armament we had were hand-held M-3A .45 cal. "grease guns." Furthermore, we weren't allowed to call in supporting air strikes. That had to be done by a Vietnamese officer, who may or may not have understood what needed to be done.

Still, despite the Alice in Wonderland nature of our operational orders, the super bunch of Marines I was privileged to have working with me— veterans and rookies alike— learned to cope, and to cope very well indeed. Since there was no precedent for what we were expected to do, improvisation was the order of the day.

We devised tactics and techniques, and adapted equipment, to meet the day to day needs of situations we encountered. While we were at it, we endeavored to record our trials and errors in order to spare squadrons following us from having to "re-invent some of the necessary wheels."

While serving in subsequent Marine Corps and JCS positions, I was able to keep fairly well abreast of Vietnam helicopter operations which followed, from the time my squadron was relieved until that final evac mission from the roof of the American embassy in Saigon some thirteen years later. And, since I had been in on the creation of the deployment, I must admit that I had a sense of shared pride during that entire time.

Furthermore, from the reports I saw, I suggest that everyone who served in a Marine helicopter unit in Vietnam is more than entitled to feel that same shared pride. I believe that's what this book is about.

Semper Fi!

Colonel Archie J. Clapp USMC(RET)

DEDICATION

Roger A. Herman, President

    The idea for a USMC/Vietnam Helicopter Pilots & Aircrew Reunion was first conceived in 1985. At that time we didn't have an official name or any idea that we would eventually form into a veteran's non-profit organization. What began with a few former Vietnam-era Marine helicopter pilots getting together for a one time event soon turned into many former Marines wanting to get together again on a regular basis. Word spread and it wasn't long before the organization was to include former crewchiefs, gunners, corpsmen, flight surgeons, chaplains and other support personnel as well. Qualification to join was simple ... to be one of us today, you had to have been one of us back then.

    Having a sense of humor is a must to belong to our group, just as it was necessary in our ranks during the Vietnam War. It's what got a lot of us through combat during a very controversial and difficult time. As we get together again at our biannual reunions it is a time to re-tell war stories and revive the camaraderie that was born amongst us so long ago. We were Marines, and always will be. There is a special bond among our members, and it extends to our families as well. Most of us bring our wives and other family members to participate in these get togethers. We have a lot of fun with the closest friends that we will ever have in life. In addition to the good times and laughter, we also take time out to remember those friends of ours who didn't return from Vietnam. They paid the ultimate price. I think they would be glad to know that today their buddies are staying in touch and remembering them fondly during these reunions. They were just like us, and but for a lucky break here and there any of us could have been the ones not to return home. And it is to those who didn't come back that we proudly dedicate this book. They were the bravest of the brave, and they were our friends. We will never forget them.

    Now, all of our adventures and friendships with our fallen brothers will be forever be remembered in this publication. It's a tribute that I have no doubt they would have made to us had the roles been reversed, and we were the ones not to come home. That's what being a combat Marine helicopter pilot and aircrewman was all about during the Vietnam War... we got the job done and we took care of each other until the final mission was flown. We knew we could always count on each other. That will never change.

                                               Semper Fidelis, Marines.

                                               Roger A. Herman
                                               President

ACKNOWLEDGEMENTS

A. Michael Leahy

Many members of the USMC/Vietnam Helicopter Pilots & Aircrew Reunion had a strong hand in production of this volume. It is not possible to individually name them all. However, certain members have made major contributions and deserve special mention.

John Van Nortwick tackled the brunt of the boilerplate history of Marine helicopter outfits in Vietnam from the initial landing of HMM-362 in Soctrang, 1962, until the pullout from Saigon in 1975. His bone-tiring efforts to contact individuals for meaningful information and to assemble all of the operational history merits the highest praise. Gene Salter ably assisted John and also provided very personal recollections of individual flight actions. Danny Crawford, Head, Reference Section, Historical Branch of the Marine Corps Headquarters History and Museums Division, examined the historical manuscript and pronounced the major events and actions accurate for publication. A very special thanks to John T. Dyer, Curator, Marine Corps Art Collection at the History and Museums Division, for providing virtually all of the transparencies used to produce the art in this book.

Archie Clapp provided an insightful foreword to the book from his perspective as the first commanding officer of a Marine helicopter squadron assigned to duty in-country. Jim Shanahan wrote several interesting summaries which described the most-utilized helicopters during the Vietnam War, i.e., the CH-37, UH-34, CH-46, CH-53, UH-1E and the AH-1. His fresh, unstereotyped descriptions of the helos are in a class by themselves. Kellan Kyllo provided many thoughtful poems from his book *WHERE LIGHT IS AS DARKNESS*, while poet LCpl Vieira wrote of many soul-searching experiences from back in the cabin. Norm LaFountaine and Roger Herman provided the initial thrust and RPM to commence production of the history book with Turner Publishing. Without detailed fact checking and proofreading by J. D. Barber and Roger Herman, we never would have made it.

Finally, most grateful appreciation goes once again to the individual members who chose to be an integral part of the history by providing their individual stories, biographies and photographs to form the basis of this book. When our kids, grandchildren and others want to know about Marine helicopter involvement during the "down and dirty" Vietnam war, this book will stand as a monumental testimony to the brief moments of terror and the long hours of backbreaking work to "keep the faith" with America and to ourselves.

A. Michael Leahy
Editor and Combat Artist

Publisher's Message

Dave Turner, President - Turner Publishing Company

It is a tremendous privilege to publish the proud history of the USMC/Vietnam Helicopter Pilots & Aircrew. From the first Marine Corps helicopters to serve in Vietnam in 1962 until the lift off of the last mission flown by an HMM-165 CH-46 crew thirteen years and fifteen days later on 30 April 1975, your story in Marine military history was etched in stone.

Contained in these pages is a record of the events spanning 1962-1975. However, it is more than facts and figures. It is the account of men and machine developing the ultimate test of character and performance during combat. Marines are known to be the first. First to land on the beach, first to fight, and first to develop new Marine tactics in helicopter warfare.

The men and the machines passed the test having been forged in the furnace of armed conflict in the steaming jungles of Southeast Asia. You, Marines, have left a legacy for those who will follow in your footsteps. You have shown the courage to try new things, to adventure where others would not.

A special thank you to Mr. Roger Herman, President, and to Mr. Michael Leahy, Editor and Combat Artist, who worked closely with Mr. Robert J. Martin, our Chief Editor and a fellow Vietnam Veteran, to help write, compile, and edit the manuscript. Most of all, thank you to the Marine Veterans who sent in their biographies and photographs. I thank all of you for your sacrifice in defending our nation in Vietnam.

My only brother, S/Sgt. John H. Turner, was Killed-In-Action in Vietnam during his second tour of duty. My oldest son is a Marine and currently serves in the Reserves. For those families and friends who have lost loved ones or have a Marine in their family, I hope that this book will help you remember the FEW, the PROUD, THE MARINES ...

Sincerely yours,

Dave Turner
President

HMM-362 was the first USMC helicopter squadron into Vietnam, landing at Soc Trang, in 1962. This photograph shows the aircrew members of HMM-362 which was also the last UH-34D squadron in Vietnam on Aug. 22, 1969. It was located at MAG-36, Phu Bai. Pictured from left, front row: Gordon, Marquey, Medina, Herman, Nazareth, McCallum, Silva, unidentified, unidentified, unidentified, St. Pierre, McDade. Second row: Culver, unidentified, unidentified, Hardesty, Duda, Sigman, Barnes (behind Sigman), Thurman, unidentified, Interrante, Diaz (behind Interrante), unidentified. Third row: Stragal, La Croix, unidentified, unidentified, Elwood, Lorisey, Elder, Daniels, Koral, Ryan, Hansen, unidentified, unidentified. (Courtesy of John Sigman)

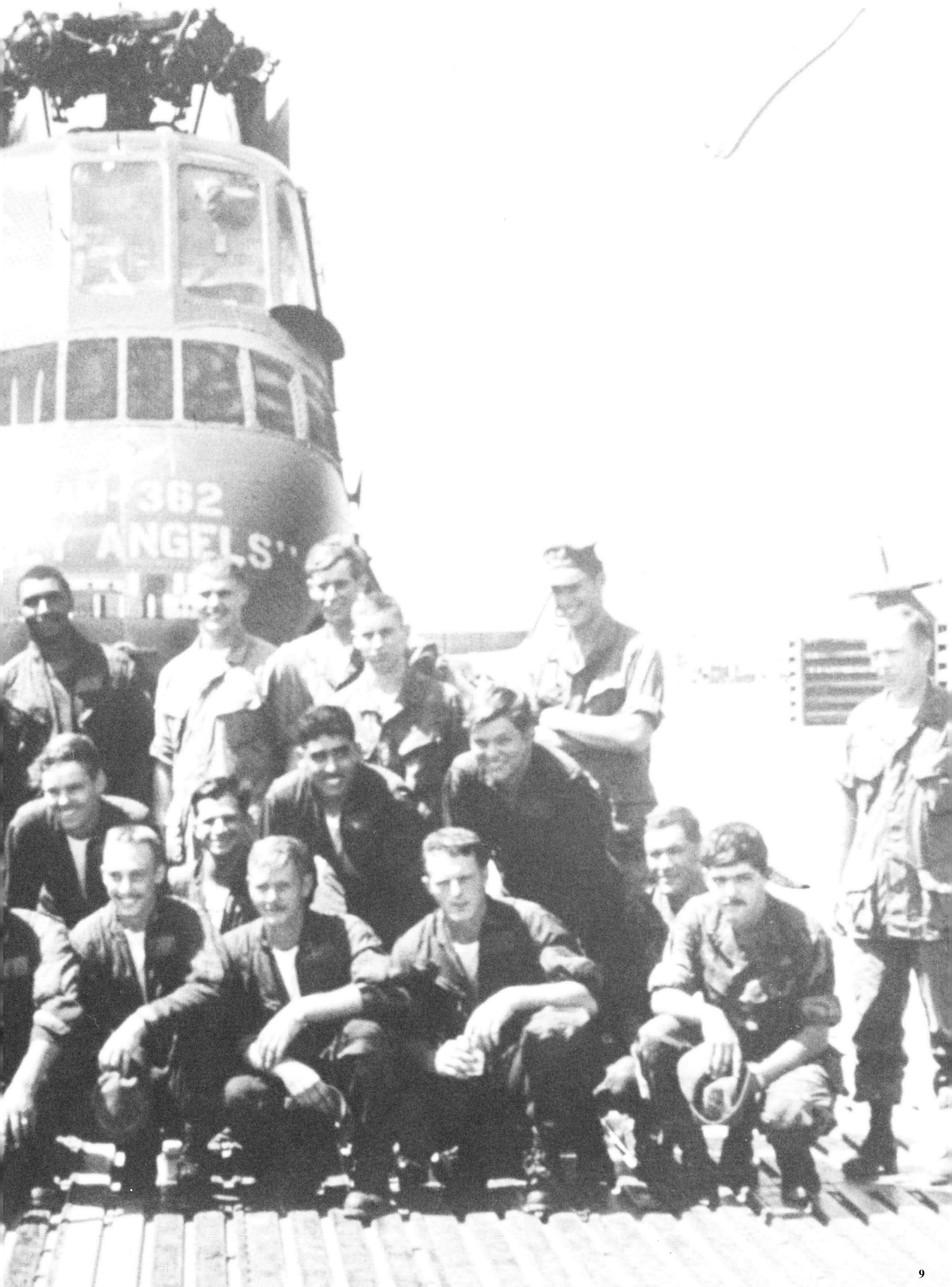
362
ANGELS

A Chronology of Marine Helicopters in Vietnam 1962-1975

The first Marine Corps helicopters to serve in Vietnam arrived just four months after the first American helicopters were deployed in-country.

On 15 April 1962, a UH-34D crew from LtCol Archie Clapp's HMM-362 touched down on a World War II Japanese fighter strip 3 miles from Soc Trang, southwest of Saigon, in the Mekong Delta. They were first in. Thirteen years and fifteen days later, on 30 April 1975, an HMM-165 CH-46D crew lifted the last Americans out of Vietnam, from the rooftop LZ of the American Embassy in Saigon. They were last out. WHAT HAPPENED TO ALL OF US IN BETWEEN IS WHAT FOLLOWS——.

1962

Operation Shu-Fly was activated on 15 April 1962 with the deployment of HMM-362 into Soc Trang. HMM-362, assisted by HMM-261 was ashore by mid-afternoon and ready to accept missions the following day. The first helicopter-borne assault with ARVN troops was conducted 6 days later. Two days following that, HMM-362 suffered the first combat damage to a UH-34D during Operation Nightingale when a bullet pierced an oil line in its engine compartment.

During May, 1962, HMM-362 flew its first night medevac. The Eagle Flight tactic, developed by HMM-362, was first employed on 18 June. An Eagle Flight employed troop-loaded H-34s orbiting a tactical area to engage escaping VC. The first joint USMC, US Army, VNAF assault mission took place in July.

The policy of rotating squadrons into Vietnam every four months commenced in August when HMM-163 relieved HMM-362. HMM-163 suffered their first battle damage 18 days later. In September, three UH-34Ds were hit by small arms fire, and a crew chief was wounded, becoming the first Marine helicopter aircrew casualty of the war.

In response to a MACV request for more capable aircraft and instrument qualified pilots in mountainous I Corps, HMM-163 moved north to Da Nang on 20 September, thus setting the stage for the Marine buildup to follow in the next 3 years. This squadron provided the majority of the helicopter support in I Corps. The first fatalities in Marine helicopters occured in October 1962, when a UH-34D crashed due to mechanical failure, killing 7 of the 8 aboard.

1963

In January, HMM-162 relieved HMM-163. With squadron rotations continuing three times a year for the next two years, half of the Marine Corps' squadrons received invaluable combat experience prior to the large scale deployments that started in 1965. HMM-162 conducted a major lift of 300 ARVN into three LZs 15 miles west of Da Nang. The area would be revisited many times in the next 10 years.

April 1963 saw the tempo of operations pick up with the advent of clear weather. This also caused the loss rate to climb, with many instances of aircraft being hit and crewmembers wounded. Army gunships from the Da Nang Utility Tactical Company regularly escorted the Marine 34s. On 27 April, a year after the arrival of Marine helicopters into Vietnam, the first loss occurred of an aircraft that was not recovered, and directly attributed to enemy action.

HMM-261 relieved HMM-162 in June, continuing the building of combat experience in the helicopter community. During August, HMM-261 conducted a major retrograde operation lifting 1300 ARVN from LZs near the Laotian border to Thuong Duc, SW of Da Nang.

In October 1963, HMM-361 relieved HMM-261. Soon afterwards, the squadron lost 2 aircraft while on a SAR mission and 10 aircrew were killed. The ensuing recovery operation lasted 3 days and required the insertion of Marine and ARVN security forces numbering over 150 men. These and other actions caused the Marines to develop procedures to perform quick engine changes, "QECs," in the field.

1964

MACV announced, in January, that all Marines would be withdrawn from RVN during the first half of 1964 as part of a plan to end direct US participation in the war. Wonder what happened? In the meantime, the missions continued. HMM-361 conducted a critical retrograde of a 200 man CIDG unit under heavy VC pressure.

During February, HMM-361 was relieved by HMM-364, which was informed that they would be the last squadron in country and to assume the mission of training VNAF helicopter pilots in the UH-34. Their other missions continued. The action started to pick up and many of HMM-364's UH-34s and their escorting ARMY UH-1B gunships were hit.

By April, the VC were using many new tactics to lure helicopters into ambushes, such as popping smoke near LZs and waiting until after the gunships had made their runs, to fire on transport helos. The most costly and bitterly opposed helicopter operation to date, Sure Wind 202, took place on 27 April. One VNAF and two Marine H-34s were lost, and 17 of the 21 committed were hit.

MACV directed that HMM-364 transfer their 34s to the VNAF in May, and prepare to depart Vietnam by 30 June. Another fully equipped squadron would be ordered in. So much for the withdrawal ordered 6 months earlier. HMM-162 arrived back in country in June. HMM-364 personnel departed, their H-34s now flown by the VNAF.

HMM-162's operations were expanded to continuously maintain a two aircraft section at Quang Tri or Khe Sanh to perform SAR NORTH in support of operations in Laos and North Vietnam. During July, HMM-162's H-34s, working out of Khe Sanh, first reinforced the besieged Special Forces camp at Nam Dong, then 2 weeks later, helped evacuate the camp, as it became untenable once more. HMM-162 conducted a major lift of the 2nd ARVN Div. from Kham Duc to an LZ 24 miles NW in September.

In October, HMM-365 relieved HMM-162, and soon suffered its first battle casualties, two WIA, during a medevac 10 miles SW of Tam Ky. Operation Shu-Fly was redesignated Marine Unit Vietnam (MUV) in December.

L. to R.: LTCOL Joseph Pultorak, COL Charles B. Armstrong and COL McDonald D. Tweed in front of Marine Helicopter HMM 361 sign. COL Tweed was the Commanding Officer. (Courtesy of M.D. Tweed)

HMM-162 receiving Combat Aircrew Wings at Da Nang Vietnam. (Courtesy of Zin Warford)

1965

HMM-163 returned for a second in-country tour, relieving HMM-365 in January. HMM-162 returned to Da Nang, joining HMM-163 in March in response to increasing helicopter requirements. Sadly, HMM-163 made the cover of Life Magazine (16 Apr 65) when it showed YP-13's copilot, hit by gunfire. Crewmen worked feverishly and futily in an attempt to save his life over the Que Son Valley.

During March, the 9th MEB landed at Da Nang. The MUV was redesignated MAG-16 as added helicopter resources began to arrive in-country. The 3rd MEB landed at Chu Lai in May. VMO-2 arrived in-country with the first UH-1E gunships. HMM-161 came in-country and moved into Phu Bai airfield just south of Hue. HMM-162 relocated out of country.

In July, HMM-161 moved from Phu Bai to Qui Nhon in II Corps to support the SLF BLT which came ashore to secure the port area during the arrival of the 1st Cavalry Division. HMM-163 went aboard ship as the SLF squadron. MAG-16, with HMM-261, HMM-361, and VMO-2, moved from Da Nang to the newly completed Marble Mountain Air Facility in August.

Operation Starlight was conducted on the Van Tuong Peninsula 15 miles south of Chu Lai by RLT 7 and the SLF, and supported by MAG-16. This was the first significant contact with major VC units. The tactical success there led many to believe that US forces would achieve a quick victory and hence, we would all soon be home. The rumor among MAG-36 Marines onboard the USS *Princeton* enroute from California was that we would turn around and head back.

September brought MAG-36 from California with HMM-362, HMM-363, and HMM-364 (UH-34D squadrons), H&MS-36 with sorely needed CH-37Cs for downed aircraft retrieval, and VMO-6 with more UH-1E gunships and O-1's. All the MAG-36 squadrons moved into the newly constructed Ky Ha base near Chu Lai except HMM-363 which remained at Da Nang temporarily.

While at Da Nang, HMM-363 lifted the first Marines into Hill 55 south of Da Nang, which remained a Marine CP for many years to come. HMM-363 deployed to Qui Nhon in October in direct support of the Army's 101st Airborne, which relieved the Marine BLT securing the port. HMM-161 returned to Phu Bai. HMM-261 relieved HMM-163 as the SLF squadron.

On 27 October, shortly after midnight, explosions hit MMAF. An estimated force of 90 VC/NVA launched a well-planned, coordinated attack on MMAF. Three or four enemy teams were involved. They attacked the H&MS hangar area, the MAG-16 bunker area, and the maintenance and administrative tents of the squadrons. Afterwards they began a methodical attack on each helicopter. The sappers destroyed 20 helicopters and damaged another 30 that evening.

Shortly after HMM-363's arrival in Qui Nhon, LtCol George Kew, CO, brought back a "one-legged" H-34 having left the other landing gear strut in an Army LZ. Later that night in the club, he made the following statement: "You all know that the 1820 engines are a critical supply item. Main Landing Gear Assemblies are not. I made the decision NOT to overboost that engine, even if I bent the gear. Any questions?" The reply was loud and unequivocal: "NO, SIR!!"

During November, the district headquarters at Hiep Duc, in western Quang Tin Province, a place we would come to know well, was overrun. MAG-16 and MAG-36 helos lifted in the ARVN counter-attack force, taking hits in 20 of the 30 UH-34s involved. Hiep Duc was retaken, only to be aban-

"Tiger Airlines" CH-46s of HMM-362 in their revetments at ProvMag-39 digs at Quang Tri, 16 Sept. 1968. (Photo credit: HB Staff)

BGen Frank Garrettson, Assistant 3rd Division Commander, extends a thumbs up to emplaced artillery personnel from a VMO-6 UH-1E helicopter at LZ Nanking, NW of Khe Sanh, 5 Oct. 1968. (Photo credit: HB Staff)

"Take five" at Phu Bai. Mike Royer and Ed Bosler. (Courtesy of Robert Hachtel)

LTCOL Mac Tweed, Commanding Officer, Marine Medium Helicopter Squadron 361 briefs flight crews on enemy held territory in the Republic of Vietnam near Ky Ha, on Oct. 23, 1966. This was just prior to making the rescue of a recon team trapped behind enemy lines. (Courtesy of M.D. Tweed)

Checking out a UH-34 wheel assembly while hovering. (Courtesy of E.J. Coady)

Things Go Better with Háy uóng. Huh? This was taken outside of Da Nang. (Courtesy of Gene Wesolowski)

doned later by the ARVN, as undefendable. Operation Harvest Moon was launched in the Phouc Valley west of Highway One between Da Nang and Chu Lai in December. Both MAG-16 and MAG-36 plus HMM-363 up from Qui Nhon and HMM-261 from the SLF support Task Force Delta and the 5th ARVN Regiment swept this familiar ground for VC and NVA units.

Seven squadrons were now in country, plus one aboard ship, all with UH-34Ds. These squadrons, plus a detachment of CH-37Cs: and two VMO squadrons with UH-1Es and O-1Cs made up Marine helicopter assets in Viet Nam.

Late in 1965, a pilot from VMO-2 was flying alone, heading north from Chu Lai to Marble Mountain Air Facility (MMAF) along the beach in a UH-1E gunship. Near Hoi An, he observed a lone American spread-eagled on the beach, his hands and feet tied to stakes. Numerous VC were in the area and appeared to be preparing to beat the American. The pilot immediately rolled in, firing his six M-60s and scattered the VC. Soon joined by a second UH-1E gunship flown by Capt Bob Stoffey, the pilot landed his gunship on the beach next to the American, who turned out to be a captured Marine. He left the ship, armed only with his .38 pistol, freed the Marine and hustled him aboard the UH-1E. Stoffey kept up relentless and effective covering fire while the VMO-2 pilot, his crew, and the Marine escaped. (note: we couldn't identify the VMO-2 pilot)

1966

January kicked off with Operation Double Eagle consisting of a two battalion amphibious operation south of Quang Ngai supported by MAG-36 and a heliborne assault by the SLF BLT. A similar US Army operation south of the Marines in Binh Dinh province was carried out at the same time. HMM-363 moved furthur south to Tuy Hoa, again in support of the 101st Airborne which was opening another airfield at Phu Cat. In February, HMM-363 rejoined MAG-36 at Chu Lai, after five months in support of US Army and Korean operations at Qui Nhon, Tuy Hoa, and Phu Cat in II Corps.

Capt John Van Nortwick of HMM-363 was out on a mission wearing an XXL infantry flak vest because his regular vest was "borrowed" while he was away from the squadron. He had to wear the big XXL vest with the right and left flaps overlapped because it was too large. He took a round square in the middle of his chest. The round penetrated the outer flap and bounced off the inner one. XXL flak jackets suddenly became a high fashion item.

HMM-364 moved aboard ship as the SLF squadron. HMM-163 rotated out of country to re-equip with CH-46s. Northwest of Phu Bai, HMM-161 conducted a night assault with a Marine battalion going to the relief of an ARVN battalion under attack. HMM-164 joined MAG-16 bringing the first CH-46As into-country in March. The Special Forces camp in the A Shau Valley, near the Laotian border, was attacked by 3 NVA battalions and the resulting defense, air support and, finally, evacuation of the camp was costly as three Marine UH-34s were lost.

In April, numerous 3rd Marine Division operations commenced south of Da Nang in an attempt to clear this area. All were heavily supported by MAG-16 and MAG-36. These operations went on for several months.

The first significant contact with an NVA division occured north of Cam Lo in the DMZ. Task Force Delta, initially consisting of three Marine battalions, grew to six battalions, and five Army of the Republic of Vietnam (ARVN) battalions. They were in close contact for over a month in what came to be known as Operation Hastings. Again, MAG-16 and MAG-36 were fully committed in support. It was costly as five H-46s were lost the first day. Some MAG-16 units, including VMO-2 moved to Dong Ha to support Hastings.

Operation Prairie, now reduced to three battalions, followed Hastings in the DMZ area during August. The SLF BLT, airlifted by HMM-363, joined this continuing operation, later during the month. Operations continued as helicopter squadrons came in country to relieve those going out for refitting and retraining. The line up of squadrons at the end of the year included HMMs 163, 164, 263, 265, 361, 362, 363, and 364 plus VMOs 2 and 6.

1967

As the year began, a detachment of HMH-463 joined MAG-16 with CH-53s relieving H&MS-36's CH-37Cs. Regardless of its eccentricities, the "Deuce" was the Marines' "Heavy Lifter" since its introduction to the Corps 10 years earlier. Few helicopters ever looked so "formidable." Nevertheless, the arrival of the CH-53As ushered in a new era of turbine-powered heavy lift capability. Operation Prairie, begun in 1966, continued along the DMZ. The Third Marine Division moved north from Da Nang to Hue; the First Marine Division moved up from Chu Lai to Da Nang, and the operating areas of Marine helos moved similarly. Operation Stone commenced 12 miles south of Da Nang in February, in an area that was and would be cleared many times with support from MAG-16. In March, two SLFs were formed, Alpha and Bravo, each with an HMM assigned. Initially these were HMM-363 with UH-34s and HMM-164 with CH-46s.

During April a Marine rifle company stumbled onto a large NVA contingent near Khe Sanh, uncovering an enemy intent to attack in force. MAG-16 helos lifted in a reinforcing Marine regiment and intense fighting continued well into May. Additional helicopter strength arrived with the balance of HMH-463 (CH-53As), and HMM-165 (CH-46Ds). Helicopter flexibility improved in July when VMO-2 and VMO-6 were reorganized to form VMO-2 (O-1E's) and HML-167 (UH-1Es) in MAG-16 at Marble Mountain, and similarly equipped VMO-6 and HML-367 in MAG-36 at Ky Ha.

Capt Steve Pless of VMO-6, flying gunship escort out of Ky Ha for a medevac package, was enroute back into Ky Ha when he heard an Army Chinook (CH-47) pilot on the "guard channel," indicating that he had landed in an unsecure LZ with mechanical problems. He then came under VC attack, and lifted off, leaving some of his passengers in the LZ who immediately came under heavy VC fire. Pless was overhead and saw the VC capture the six Americans who were stranded there. He rolled in with rockets and machine guns which caused the VC to scatter. Pless then landed, and together with his copilot and crew, they left the helicopter to assist the survivors aboard his UH-1E. As they scrambled aboard, the VC resumed their attack on the helo crew and its new found passengers. Overloaded, Pless jettisoned his rocket pods, and nursed his helicopter out of the LZ. Pless was awarded the Medal of Honor for his actions, while his remaining crew members received Navy Crosses for their actions that day.

During August, the 1st Marine Division again went into the Tam Ky-Hiep Duc-Thang Binh triangle south of Da Nang. MAG-36 went in as well, together with the SLF Alpha squadron, HMM-362. The seemingly media-inspired "Siege of Con Thien," a few clicks south of the DMZ began in September. Marine helos provided massive sup-

The Rockpile with a helicopter pad on top. (Courtesy of Robert Hachtel)

Marble Mountain, ready for take off. (Courtesy of M.L. Poleski)

Looking out for "Indians." (Courtesy of M.L. Poleski)

port to Marines dug in there. The "siege" evaporated.

In November, changes to the helicopter order of battle took place as MAG-36 moved from Chu Lai to Phu Bai to be nearer the 3rd Marine Division. HMM-263 rotated to the Continental United States (CONUS) to re-equip and re-train. HMM-262 returned in-country, equipped with CH-46Ds. By the following month, two full strength CH-53 squadrons were in country, HMH-463, MAG-16 at Marble Mountain, and HMH-462, MAG-36 at Phu Bai. By year's end, seven HMMs, and two each: HMHs, HMLs and VMOs were now in-country.

1968

This pivotal year found 13 of the USMC's 24 helicopter squadrons in-country. They were: HMLs 167 and 367; HMMs 161, 163, 165, 262, 265, 361, 362, 363, and 364; and HMHs 462 and 463. Additionally VMOs 2 and 6 continued to serve. With the arrival of US Army units in southern I Corps, Marine ground forces were now deployed primarily around Da Nang and near the DMZ in the north. Marine helicopters also operated primarily in those areas. During February, the much publicized Tet offensive took place and Marine helo squadrons were fully involved in many actions including the battle for Hue City, the defense of Khe Sanh, and many other operations throughout I Corps.

"The Siege" of Khe Sanh continued into March with much of the logistic support provided by "super gaggles" of CH-46s hauling external loads under IFR conditions from Dong Ha into the battalion LZ's in the hills north of the combat base. Losses of helicopters were high with at least 17 being destroyed. Crew losses were equally high. Khe Sanh was abandoned in April.

In order to provide improved response to 3rd Marine Division units along the DMZ, PROVMAG-39 was established at Quang Tri with attached units HMM-161, HMM-262 and VMO-6. MAG-36 at Phu Bai retained HMM-362, HMM-363, HMH-462 and HML-367.

In June, squadrons continued to rotate in and out of country. HMMs 164 and 364 came in with CH-46Ds. HMM-361 went out with their UH-34Ds to refit as a CH-53 squadron. The first fixed-wing OV-10As arrived in-country in July, and were assigned to VMO-2 at Marble Mountain. The OV-10s were a remarkable addition to the observation role. With their capability to carry rockets and machine guns, they also served in a "light attack" role, although that role was not the OV-10s "MOS." Maj Michael Leahy accompanied Capt Gene Kimmel of VMO-2 in an OV-10 on a mission to support a Marine rifle company SW of Da Nang, on a sweep of a paddy/village area. The remarkable aircraft lazed just a few hundred feet overhead until called upon by the company commander to attack tree lines and other entrenchments in the company's path. If it weren't war, it could have been a symphony. Capt Kimmel and his copilot were killed when they were shot down a few weeks later.

August and September were surprisingly quiet in the Marine areas of operations (AOs), although many Marine search and destroy, cordon and patrol operations required helo support from MAG-16, MAG-36 and the SLF squadrons, HMMs 265 and 362. During mid-September, 24 CH-46s, primarily from Prov/MAG-39 at Quang Tri landed 1700 troops along the south side of Ben Hai River, in the middle of the DMZ, in a pincer movement. The LZ was honeycombed with well-used trails leading from North Vietnam to the south.

In October, the Thuong Duc Special Forces camp, which controlled the route from the A Shau Valley into Quang Nam Province, southwest of Da

1stLt James O. Atkinson Jr. holding a piece of shrapnel from the ammo dump at the Dong Ha Airfield which blew up Sept. 3, 1967 as a result of enemy shelling. (Courtesy of J.O. Atkinson)

2ndLt Bill Ring, Khe Sanh, 1967.

Viet Cong battle flag captured off the top of Marble Mountain, Aug. 23, 1968 by UH-1E crew of HML-167: Capt John Henry Key, WIA, Aug. 26, 1968; Capt Arthur L. Stockburger, KIA Aug. 26, 1968; Cpl John B. Becker, KIA, Aug. 26, 1968; Cpl Michael E. Williams. (Courtesy of John Henry Key)

Nang came under heavy NVA pressure. A classic relief operation, Operation Maui Peak, took place with a ground column fixing the NVA in place while MAG-16 helos lifted in one Marine and two ARVN battalions behind them to inflict heavy NVA casualties and relieve the camp.

Operation Meade River continued into December in the familiar "Dodge City" triangle, south of Da Nang, supported by MAG-16.

1969

As the year began, HMM-263 returned to MAG-16 with CH-46Ds. MAG-16 was comprised of squadrons HML-167, HMM-165, HMM-263, HMM-364, HMH-463, and VMO-2. Operations kicked off with Operation Taylor Common in the "Arizona territory," northwest of An Hoa, supported by MAG-16. Both SLFs participated in Operation Bold Mariner on the Batangan Peninsula, last visited during Operation Starlight in 1965. HMMs 362 and 165 flew in support. To the north, MAG-36 and PROVMAG-39 provided helicopter-borne logistic support to Operation Dewey Canyon, which took place in northwestern-most I Corps.

In April, VMO-2 received its first AH-1Gs. Major operations took place along the DMZ, in the Da Krong Valley, and on Charlie Ridge, southwest of Da Nang. These ops continued into June, and all were supported by the three helicopter groups.

An HMM-263 medevac effort brought out 27 WIA from a hot LZ just SW of Thuong Duc following the ambush of a 7th Marines patrol. The medevac crews were cautioned not to bring out a deceased Marine with a live RPG round in his chest. VMO-2's newly arrived Cobras provided very effective close-in fire support. During May, Operation Pipestone Canyon cleared the Go Noi island area south of Da Nang.

The lineup of helicopter groups and squadrons in July, just prior to the beginning of troop withdrawals: Provisional Marine Air Group (PROVMAG)-39 at Quang Tri with HMM-161, HMM-262, and VMO-6; MAG-36 at Phu Bai with HML-367, HMM-265, HMM-362, and HMH-462; and MAG-16 at Marble Mountain with HML-167, HMM-165, HMM-263, HMM-365, HMH-463, and VMO-2. HMM-164 was the sole SLF squadron, when SLF Bravo was deactivated.

Maj John Van Nortwick, HMM-263 XO, tells of a poignant reward following a night medevac. After launch on a routine flight, he landed to load a WIA Marine, who was a booby trap victim. He then took off for Da Nang to land at the NSA hospital. The Marine was offloaded and carried under the nose bubble of the CH-46. As he passed under

Khe Sanh Combat Base. (Courtesy of Robert Hachtel)

CH-53s from HMH-463 are overhauled in the MAG-16 maintenance area at MMAF. November 1967.

Ugly Angels logo on a sign at HMM-362's squadron area, Phu Bai. (Courtesy of Robert Hachtel)

UH-1E from VMO-6 departs from Camp Carroll, December 1968. (Courtesy of Joseph A. Syslo)

the nose, he looked up, grinned, gave a big thumbs up and could be heard to say: "Thanks for coming down and saving my life!" Only after that did John see that the Marine's right leg was missing from the knee down. To this day he cannot tell this story without some tears.

In August the first major troop withdrawal was announced. HMM-165 and HMM-362 redeployed. HML-367 joined MAG-16 from MAG-36. HMM-362's departure was noteworthy in that they took the last UH-34s out of country. The "Dogs" had seen service continuously since April 1962 and had logged nearly a million combat sorties. HMH-361 with CH-53s returned in-country to replace HMM-362's UH-34s. Operation Durham Peak brought many aircrews back into the Que Son Mountains and the Que Son Valley, which was first entered during Operation Harvest Moon in 1965. The following tells of some pilots' attitudes when the prospect of going back to the Que Son AO arose:

Location: MMAF Officers Chow Hall; Time: 0430.

Marine Major #1: "Let's get going, I need to get down to the flight line."

Marine Major #2: "What's the hurry? We just sat down. We don't need to brief for an hour yet."

Marine Major #1: "I need to get down there to barf at least two more times!"

Marine helos supported a major joint US Army-US Marine operation in Hiep Duc once again. The final SLF landing was conducted by HMM-265 in September.

Support of the First Reconnaissance Battalion was a continuing mission for the HMM squadrons of MAG-16. An excerpt from STARS AND STRIPES described one mission for an H-46 crew from HMM-263. "... a recon team, operating in the Que Sons made contact with the NVA, and in the fire fight that followed, suffered two WIA while killing three NVA. The team requested an emergency evac. Upon arrival over the area, the helo crew saw that the recon team was located approximately 200 yards from the only LZ. A hoist extraction was ruled out because of the wounded. Covered by smoke laid down by VMO-2 Cobras, the 46 spiraled in and backed up to a narrow ridge with just the rear main gear on the ground. As the pilot held this precarious position, Sgt Lemuel Cook, gunner, and LCpl David Penn, crewchief, left the aircraft to help evacuate the two wounded Marines. Advancing under enemy fire, they were halfway to the recon team's position when they met the team running down the trail. "We told them we would bring down their wounded," said Penn. They found the two wounded Marines with the recon team leader and a Navy corpsman. Cook and the recon team leader carried one of the wounded down the trail to the aircraft while Penn and the corpsman followed with the other. The corpsman soon became exhausted and Penn attempted to carry the man by himself. Soon Cook returned and helped Penn get this last wounded aboard the 46. With heavy fire support from four Cobras and two OV-10's, the CH-46 was able to spiral safely upward from the LZ under heavy ground fire. There were no further casualties. The extraction had taken 20 minutes.

More withdrawals occurred in November as MAG-36 with VMO-6, HMM-164, and HMH-462 redeployed. PROVMAG-39 was deactivated. HMM-161 and HMM-262 moved from Quang Tri to Phu Bai, joining MAG-16. HML-367 re-equipped with the AH-1Gs of VMO-2, and VMO-2 moved from MAG-16 to MAG-11 at the Da Nang air base. MAG-16 remained the only helicopter air group in-country after the second redeployment.

During December, a most unusual and heroic

Convoy Patrol, Hwy. 1, Quang Tri Province, HMM-262. (Courtesy of Peter G. Sawczyn)

USS Tripoli (LPH-10). (Courtesy of J.B. Wegener)

PFC Mike Clausen, HMM-263 crew chief, was presented the Medal of Honor by President Nixon. (Courtesy of R.M. Clausen)

Aboard USS Tripoli (LPH-10), HMM-165. (Courtesy of Mike Wolter)

D. Moore, V. Pilck, and G. Armstrong, Quang Tri. (Courtesy of Robert Hachtel)

Sgt Flores, HMM-262, "Ham Shot" while at Marble Mountain, June 1970. (Courtesy of Samuel Flores, Jr.)

1970

HMH-361 redeployed as part of the third redeployment of US forces in January. The Marine TAOR shrank to essentially Quang Nam province. Full employment of Marine helicopters didn't lessen up, though. ARVN units moving into old USMC areas still required Marine helo aupport from MAG-16.

On 31 January, Pvt Mike Clausen, an HMM-263 crewchief, was the embodiment of the highest ideals of Marine helicopter aircrews. He participated in a rescue mission to extract elements of a platoon which had inadvertantly entered a minefield while attacking enemy positions. Clausen skillfully guided the pilot, LtCol Walt Ledbetter, the squadron CO, to a landing in an area cleared by one of several mine explosions. With 11 Marines wounded, 1 dead, and the remaining 8 holding their positions for fear of detonating other mines, Clausen quickly leaped from the CH-46 and, in the face of enemy fire, moved across the extrememly dangerous, mine-laden area to assist in carrying casualties to the waiting helicopter. Despite the ever-present threat of additional mine explosions, he continued his valiant efforts, leaving the comparatively safe area of the helicopter on six separate occasions to carry out his rescue efforts. At one point, the pilot gave him a direct order to remain in the helicopter. Clausen had removed his hardhat and was unable to hear the order. He continued to bring wounded aboard. On one occasion, a mine detonated near him, killing the Navy corpsman, and wounding three more Marines. Clausen did not falter. Later, PFC Clausen was presented the Medal of Honor. The pilot, Col Ledbetter, was awarded the Navy Cross.

In March, a few formalized operations, Imperial Lake and Catawba Falls were initiated in familiar areas such as the Que Sons and on Charlie Ridge. However, increased emphasis was placed on "quick reaction packages," variously known as Pacifier, Kingfisher and Sparrowhawk. These packages consisted of four to six CH-46s, two to four AH-1s and one UH-1 on standby to be used against targets of opportunity in the last months of the war. HMM-161 redeployed in August. Marine support was then focused on the ARVN Operation Hoang Dieu, a final systematic search of every hamlet in Quang Nam province.

medevac took place. While on alert at LZ Baldy, two 46s and two Cobras were launched to pickup a critically wounded Marine located on Hill 845 in zero/zero weather. After many unsuccessful attempts to locate the LZ, the 46s climbed out of the weather to wait for a break in the overcast. The HML-367 Cobra flight leader, Capt Roger Henry requested permission to attempt an approach up the mountain side to the LZ. The request was granted and Capt. Henry and his co-pilot, 1st Lt Dave Cummings, proceeded upward through the ravines, just above the canopy, much of the time at low airspeeds. They were guided by the Marines in the LZ who were listening for the sound of the Cobra's engine. Three hours later, after five attempts, they found the LZ, and landed on the edge of a bomb crater. Without hesitation, Cummings climbed out of his seat; got the wounded Marine strapped in; then straddled himself atop the starboard rocket pod facing backward, and gave Henry a thumbs up to launch. After a seemingly endless climb and then descent out of the weather, a safe landing was made at a nearby aid station LZ, with Cummings still clinging to the rocket pod, a little messed up by the elements, but otherwise OK. This was incredible airmanship and true heroism. The wounded Marine survived. Years later, LtCol Dave Cummings suffered a fatal heart attack.

1971

The year started with significantly reduced operations for both Marine ground and aviation units. Although ARVN operations continued to increase, the overall need for helicopter support steadily decreased. There were, however, exceptions. Operation Upshur Stream was one last sweep of that old nemesis, Charlie Ridge.

During February, the ARVN kicked off their massive multi-division incursion 40 miles into Laos from a point west of Khe Sanh. Known as Lam Son 719, the operation had only minor American involvement. However, one element of this support was the daily use of from two to eight CH-53Ds of HMH-463 to lift heavy equipment and artillery for ARVN forces. The 53s, escorted by the AH-1Gs or newly arrived AH-1Js of HML-367, moved massive amounts of tonnage of material in and out of Laos. One CH-53 was lost in a hot LZ near Tchepone. LtCol Chuck Pittman, Maj Mike Wasco and their crew spent a few anxious hours in a water-filled bomb crater until an Army UH-1 snuck in to get them out.

HMM-364 and HMM-263 redeployed in April. The last Marine operation, Scott Orchard, was conducted west of An Hoa with minimal contact. In

HMH-463 CH-53 crewmen refuel an aircraft at the quick refuel facility at Phu Bai while enroute to Con Thien and Khe Sanh, November 1967.

May, HMM-262 redeployed. On 7 May, all combat by Marines ended. On 26 May, helicopter operations ceased. MAG-16 with HMH-463, HML-167 and HML-367, stood down and redeployed. We were out of Vietnam— or were we?

1972

During March, the North Vietnamese launched the Easter offensive. The ARVN Division on the DMZ was overrun and the NVA marched south towards Hue.

The South Vietnamese Marine Division held the advancing NVA just north of Hue in April. Two composite squadrons, HMM-165, and HMH-462 joined 9th MEB to support Vietnamese Marines, conducting four major assault landings along the Quang Tri coast.

In June, HMA-369 with AH-1Js commenced "hunter-killer" operations against small boat traffic off the North Vietnamese coast from three LPDs in the Tonkin Gulf.

Vietnamese Marines retook Quang Tri and the NVA invasion stalled as the monsoons arrived in September. HMM-165 and HMH-462 returned to other missions. HML-369 continued their LPD-launched hunter killer ops off the North Vietnamese coast.

1973

In January, HML-369's very unique and successful "hunter-killer" operations were terminated to prepare for the upcoming Operation Endsweep.

HMH-463, HMM-165, HMM-164, and HML-167 joined Task Force (TF-78) in February, and moved into Haiphong Harbor for Endsweep, the clearing of US-laid mines. This was in accordance with the provisions of the Paris Peace talks.

Many of the Marines in these squadrons had served multiple tours in-country fighting the North Vietnamese and questioned that they would now in effect be working in cooperation with their former enemies. However, two factors overode. One, the success of Endsweep was directly linked to the release of US POWs held in North Vietnam, and two, Marines obey the orders given them. However, the operation was unusual as it seemed that the daily flight schedule was based on the progress of the Paris peace talks.

One rainy night, the HMM-164 SAR crew became firemen as they made many IFR trips into Lach Huyen carrying fire fighting gear to the crew of one of the Navy's wooden MSOs which was on fire. Despite their efforts, the ship burned to the waterline. The crew was rescued safely.

The following day, bad luck dogged Operation Endsweep as HMH-463's XO, Maj Bill Smith, took his CH-53 "submarine" YH-11, to the bottom of Haiphong Harbor. Bill and the crew got out with minor injuries and swam to the surface. Bill said that he was down there looking for a missing rotor pitch link.

Endsweep was completed when HMH-463, HMMs 165 and 164, and HML-167 were reassigned other missions in July.

1974

Marine helicopter activity recorded few significant events this year.

1975

In January, the NVA commenced probing attacks on many major South Vietnamese cities. NVA attacks multiplied and overran the ARVN during March. The continuing defeat of ARVN units throughout the country caused planning for the humanitarian evacuation of Saigon to accelerate. During April, the 9th MAB with PROVMAG-39 consisting of HMM-165, HMH-462, HMH-463, HML-367, and HMA-369, was activated for the evacuation of Saigon. At 1420 on 29 April, Operation Frequent Wind commenced.

A total of thirty four CH-53s, twenty seven CH-46s, six UH-1Es and eight AH-1Js were committed and flew 682 sorties between first light on 29 April and 0825 on 30 April. At 0458, Capt Gene Berry of HMM-165 lifted the U.S. Ambassador to safety. The last personnel evacuated, Marines of the combined security force, were lifted out at 0825 on 30 April. A total of 6,968 persons were evacuated. Flight crews averaged 13 hours in the air, and Capt Berry was high timer with 18.3 hours.

AND WE WERE FINALLY ALL THE WAY OUT ...

Contributors: Head Writer: John Van Nortwick, assisted by Gene Salter, Jim Shanahan, Roger Herman, Joe Novak, Bob Stoffey, Alan Barbour and Ted Read.

Vietnam Helos
HR2S/CH-37C

Conceived in the early 1950s, the HR2S-1 (Deuce) was supposed to be the helicopter that the CH-53 became. Its basic mission was beachhead supply, ship-to-shore and operations within the beachhead. It was not conceived as an assault vehicle. The war scenario in effect when it was designed envisioned amphibious landings into areas either denuded of opposition by nuclear weapons or in which minimal opposition existed or could be mustered.

As initially planned, it was to be powered by two turbine engines, one on each side of the fuselage. But in the 1950s these engines were in short supply and priority was given to fighter aircraft, so two Pratt & Whitney R-2800 piston engines were substituted. They powered the machine very well and, after a pilot learned how to use the twist grip throttle, were quite responsive to varying power/rpm needs throughout the full flight regime.

In its day, the HR2S was a quantum jump in size and complexity over any other helicopter the Corps had ever had. It was nowhere near as sprightly as any of it predecessors. There was an art to anticipating what the aircraft required in order for the pilot to get it to do the maneuver he next wanted and when he wanted it done. Until the twist grip throttle was mastered, maneuvers near the ground could be thrilling indeed. On the other hand the machine was stable, had an excellent ASE (Auto-

A US Marine Corps CH-46A Sea Knight helicopter is viewed from another Sea Knight during a combat mission near Da Nang in South Vietnam. (Courtesy of J.A. Cutri)

HR2S/CH-37C. (Courtesy of A.M. Leahy)

matic Stabilization Equipment) system, was a joy to fly under actual instrument conditions, and could lift and precisely deliver loads (both internal and external) never previously considered possible. It pioneered the automatic positioning of the rotor head and folding of main rotor blades. Many a crew chief, first mech, and helo maintenance man has gone to his shipboard bed and in his prayers thanked God for this system. It could only be appreciated by those who have had to manually fold rotor blades over 30 feet long at night in the wind and rain without dropping them on a rolling, pitching, slippery flight deck.

Though it never was glorified, this helo did everything desired of it and more. It was the basic platform for proving (and improving) most of the components in the CH-53. Until the arrival of the CH-53 in RVN, it did all the heavy helo work there, little or none of which was of a dashing, headline grabbing, dagger-in-the-teeth nature. In RVN it was plagued with constant supply problems as could be expected of an out-of-production machine of which only a limited number were ever produced. Many were cannibalized in order to keep the remainder flying, their carcasses winding up on the junk pile at Marble Mountain when there were no useful parts left. But, until the CH-53 came along, the Deuce had no equal.

HUS/UH-34D

This machine was a hit from the day the first one rolled off the production line. Relative to its predecessor helos, its performance justified its looks. It was responsive, agile, well-powered, and forgiving. It required few hours of maintenance per hour of flight time. For its time it was the ideal helicopter in reliability and performance.

Prior to the commitment of helos to combat in RVN there was considerable wonder among the aircrews as to the ability of helos to survive direct enemy fire, there being scant practical experience in this matter. The initial commitment of H-34s to assault and medevac operations proved that helos were more difficult to hit than imagined and that they could absorb a lot of damage and still fly home. That so many damaged H-34s made it home is not so much a criticism of the VC gunners... they put a lot of holes in helos and scared the hell out of a lot of aircrews... but a testament to the aircraft and its crews, frightened or not.

In RVN there was literally no mission to which the H-34 was not assigned, from "Holy Helo" (delivering chaplains to conduct services) to emergency recon team extractions. No matter the mission or conditions, the aircraft and crews found some way to get the job done. Missions in the mountains west of the coastal plain were especially challenging since

Greg Armstrong and Ben Cascio of HMM-362. (Courtesy of Robert Hachtel)

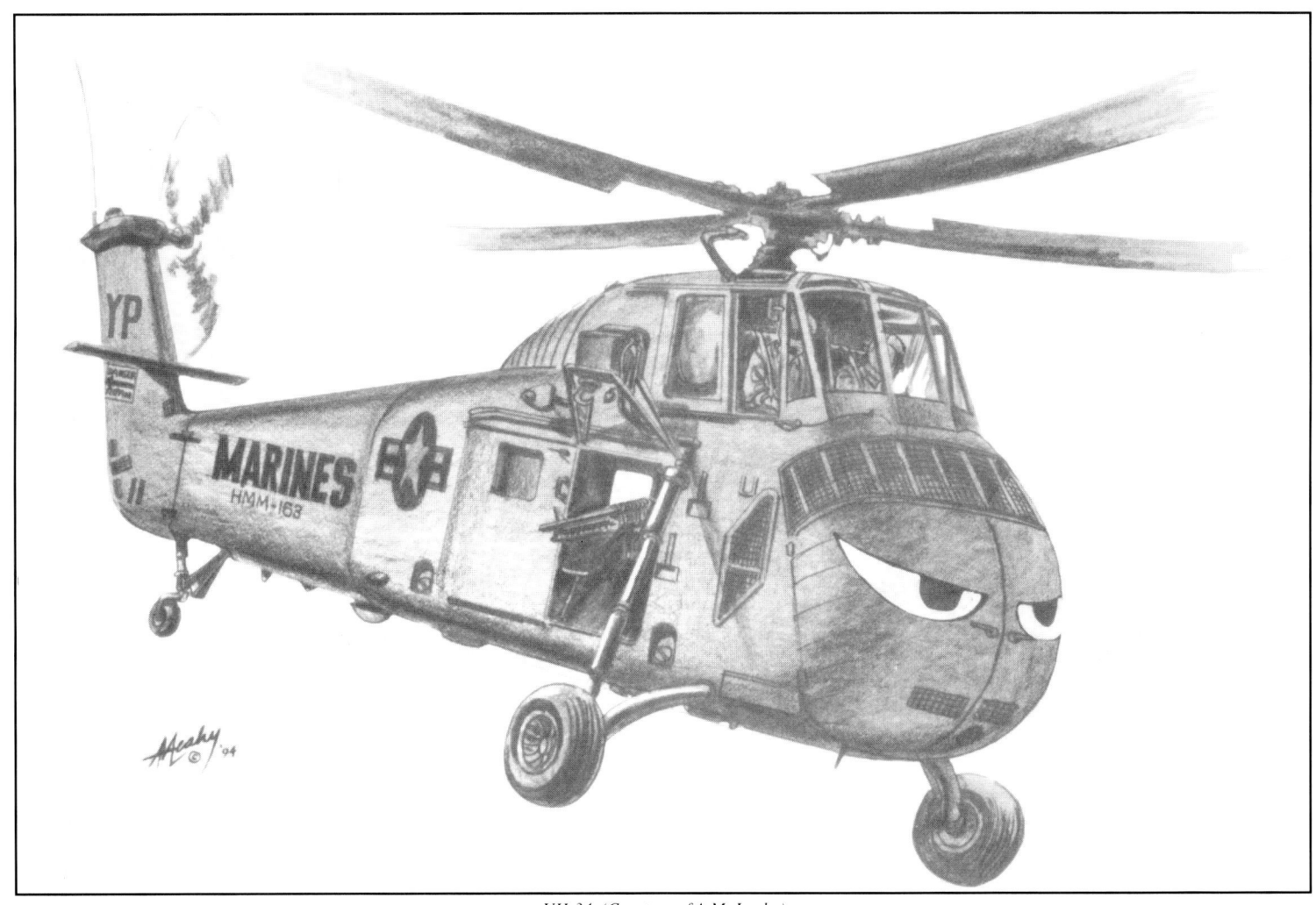

UH-34. (Courtesy of A.M. Leahy)

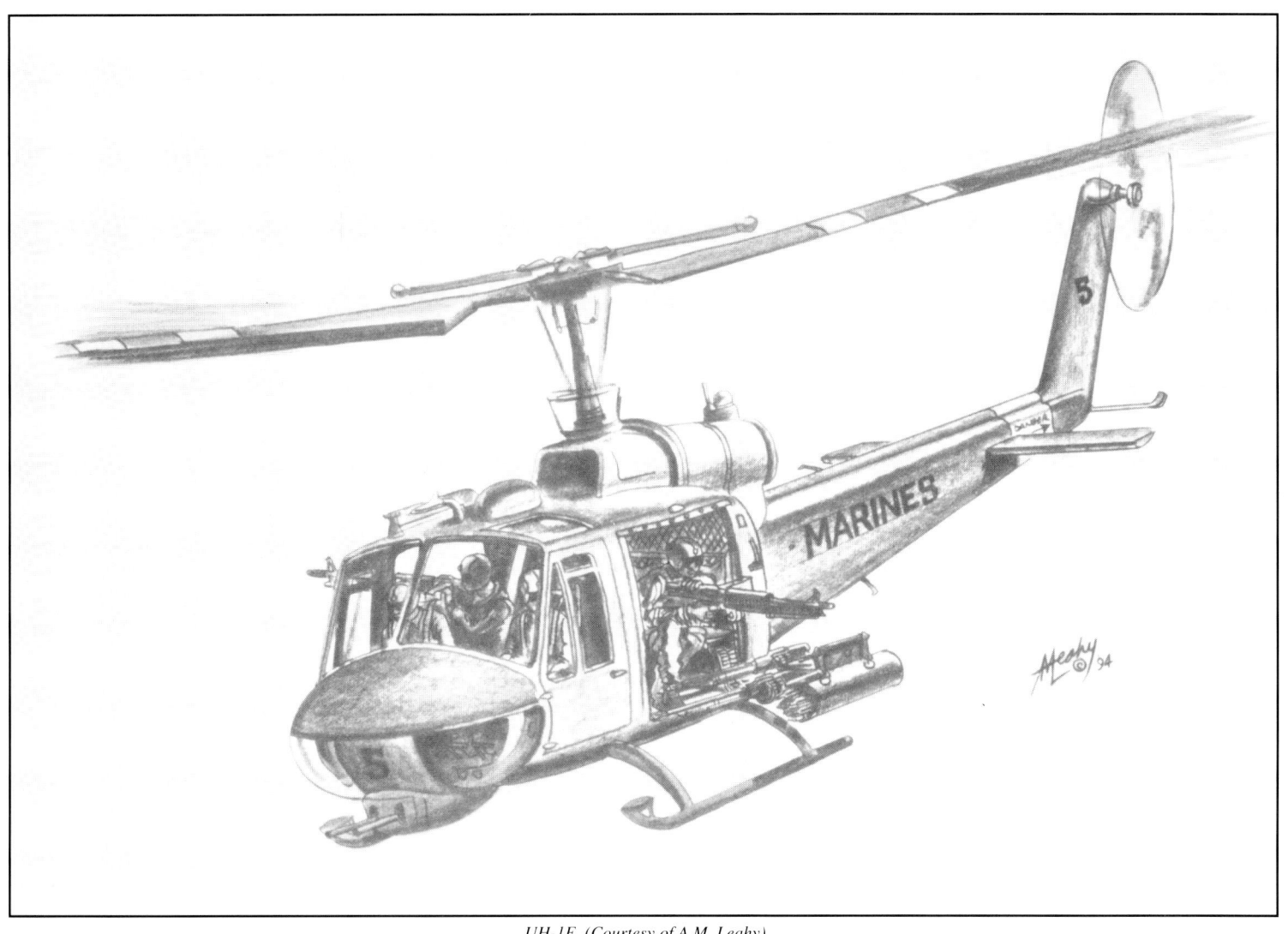

UH-1E. (Courtesy of A.M. Leahy)

SSgt Richard H. Hall on M-60 machine gun in the door of H-34 helicopter. (Courtesy of Richard H. Hall)

UH-IE

The UH-IE was an Army UH-1B to which minor modifications were applied to make it suitable for use by the Marines. Of these mods, chief were a different electrical system and radios, a rotor brake, and an aluminum fuselage. It was the first turbine powered helicopter to enter service in the FMF (Feb '64) and was procured as a replacement for the HOK and as an assault support helicopter.

The UH-1E was soon to be at the center of a controversy over doctrine within the Corps regarding armed helicopters. The issue was brought to a head by unique limitations imposed in prosecuting the war in RVN. It included the increased success of the VC in shooting up unescorted transport helos in LZs, and the increasing reliance of Marines on armed Army helos as escorts during operations into hot LZs.

In order to control costs, Bell had built the initial order of the UH-1E exactly like the Army UH-1Bs except for changes specified by the Corps. As a result, all the featues required to fully arm the Army version were built into the Marine UH-1Es. Thus, once a decision to use the UH-1E for armed helicopter support was made, this cost control measure greatly expedited the conversion of a portion of the UH-1E fleet into gunships.

There were never sufficient Marine gunships in RVN to fulfill all the missions requested of them. Medevac escort, recon team inserts and extractions, assault operations, direct fire support of engaged Marine ground units, TACA duties, and convoy escort, as well as other missions flying 'slicks" had VMO aircrews overcommitted throughout their entire tours. If ever the Marine Corps got its money's worth out of any aircraft and aircrews, it was these.

rotor and engine performance deteriorated quickly with altitude. Under these conditions an H-34 with any appreciable payload was operating at the edge of the envelope.

For many reasons, the most rewarding mission to the aircrews was probably medevac. It provided a lifesaving service to fellow Marines, often when the LZs were hot. There were a lot of nuances to doing this mission correctly; land with the helo between the medevac and the source of enemy fire so as to shield those carrying the casualty to the chopper; land as close to the medevac as possible to reduce the carrying task and exposure of the grunts to the VC. In the event of enemy fire have the escorting gunship blow up the general part of the world from whence the fire came. Flying the UH-34 "Dog" in RVN changed the meaning of the phrase, "Dog days." They were a joy to fly. H-34 pilots and aircrews hold a special place in their hearts for the "Dog."

21

Few summaries of the UH-1Es in RVN ever spent much space on its non-gunship missions such as reconnaissance, VIP and passenger transport, command & control, and medevac, in part, because gunship missions were more exciting and these other missions were frequently assigned to other helos.

CH-46

For as far back as anyone could remember, the FMF had always been near totally equipped with Sikorsky helicopters (there had been a few HTL-4s in Korea and the HOKs were stateside in the late 1950s). There was shock and disbelief when it was announced that the replacement for the UH-34D would be the Boeing Vertol CH-46 and not the competing Sikorsky design. Nicknamed "The Frog" due to its posture and appearance on the ground, the CH-46 was anything but frog-like in the air. As the new primary assault helo it brought a greatly increased payload, cruising air speed, and ease of loading/unloading people and cargo while retaining all of the UH-34's virtues.

As quickly as aircrews and maintenance people could be adequately trained, "Frogs" were deployed to RVN replacing the UH-34s in-country. All went smoothly until a series of catastrophic accidents in 1967 resulted in the H-46s being grounded. They flew only on emergency operations in RVN which could not be flown by other aircraft. Many of these accidents involved the chopper coming apart in flight, and it is easy to imagine how aircrews felt when dispatched on flights before the cause of these accidents were determined and corrective actions taken..Even so, not one needed sortie was missed.

While the Frog distinguished itself throughout its service in RVN, surely the battles around Khe Sanh and its outposts were high points. In generally bad weather and from bases near the coastline, CH-46s shuttled throughout the mountainous Khe Sanh area. They constantly moved medevacs, people, and supplies as the tactical situation required, often under IFR conditions, all the while being strenuously opposed by North Vietnamese Army regulars... all pretense of indigenous VCs with home-made weapons had long since vanished. Operations around Khe Sanh took place over a prolonged period, severely taxing the endurance of aircrews and maintenance people. That the NVA finally packed up and withdrew was due in no small part to the CH-46, her steadfast aircrews, and those who supported both.

First delivered to the FMF in 1964, the Frog was still serving over 30 years later... the Corps' primary assault chopper.

CH-53

In many ways the CH-53 is the ideal helicopter which all the early Marine Corps vertical envelopment planners had in mind when initially conceiving requirements to move large amounts of troops or supplies. It has power to spare; passengers and cargo are easily accommodated in large quantities; it is stable, maneuverable, long range, all-weather, and the cockpit is actually comfortable. Everything on the helo except the seat belts are hydraulically boosted.

First delivered to the FMF in 1966, the H-53 made it to RVN just as the final few Deuces were breathing their last gasps. Once there, the H-53 began establishing records as people and cargo haulers and became familiar all over I Corps wherever there was heavy hauling to be done.

Although they worked out of fire bases, outposts, and CPs in the field from their first days in RVN, their first known use of carrying troops on an assault occurred during the 1972 Easter Invasion by North Vietnam. Once the NVA had been stopped north of Hue along the My Chan River, plans got underway to use 9th Marine Amphibious Force helicopters carrying RVN Marines in offensive operations to drive the NVA out of the territory they had just taken. Because of the strength of the NVA near the LZ, a strong initial assault force with the capability for rapid subsequent buildup was required, and it exceeded what could be accomplished with CH-46s only. Accordingly permission was obtained to use CH-53s loaded with RVN Marines in the assault waves. The planning figure was 58 RVN Marines sitting on the cargo bay deck, front to back, per CH-53,and that figure was met or exceeded. One CH-53 was lost during the initial operation but with only minimum injuries to crew and embarked troops.

Similar operations succeeded in forcing the NVA northward along the coastal plain until the province capital of Quang Tri was liberated. During the last such operation a CH-53 carrying troops in the assault became the first Marine helo known to be shot down by a guided missile. Its starboard engine was blown off by a heat-seeking SA-7. It continued in flight on the port engine until it reached the LZ where it crash landed, with its final mission accomplished and its aircrew becoming unanticipated reinforcements for the assault troops

GUNSHIPS

As early as spring of 1949 the Corps had conceived of using helos firing rockets in an anti-tank role, and by 1951 HMX-1 had tested the mounting and firing of machine guns and 2.75 inch rockets from an HTL-4.

The instability and limited development of armed helicopters was not stopped, neither did it become a front burner project. It was to be a long and winding road from these early efforts to the Corps' first fully capable and deadly gunship.

Contrary to opinions which became popular among early gunship advocates, there were valid,

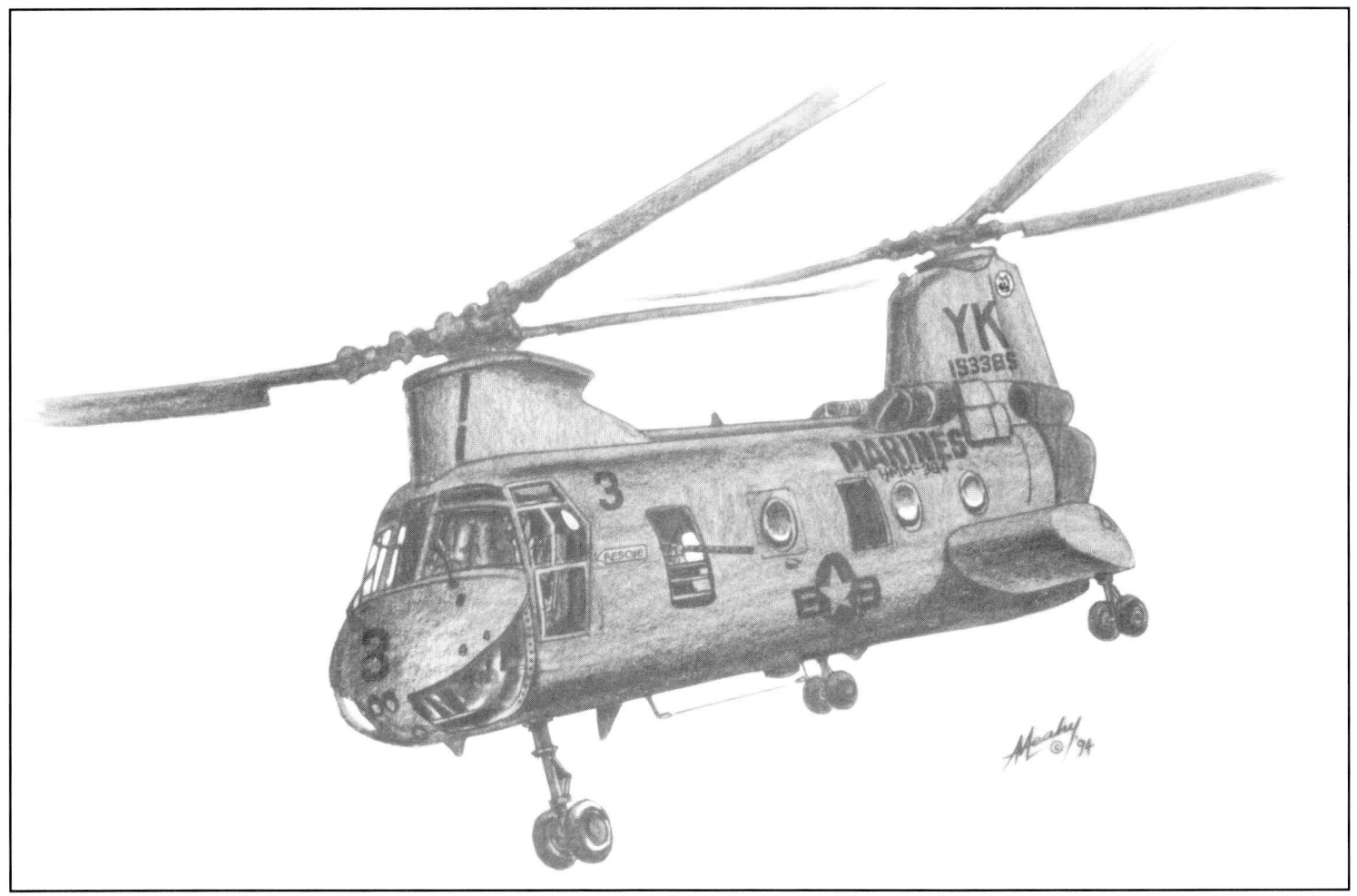

CH-46. (Courtesy of A.M. Leahy)

practical reasons for this delay. Instability and limited lift capability were engineering problems more easily solved than other issues. The combination of budget limitations and force structure provided CMC with more difficult choices. The budget limited the number of squadrons and airframes. If you want two squadrons of gunships, give up two squadrons of attack aircraft. The Corps had to be prepared to respond to a variety of threats all over the world, and the inter-related issues of force structure and doctrine were based on this. It was not at all obvious that swapping attack aircraft for the gunships which could be developed at that time would be a smart action to take.

The Vietnam war provided both impetus and solution to gunship development. Budget constraints relaxed which resulted in the availability of funds for development and hardware. By 1963 combat damage and losses of H-34s grew to the point that Army UH-IB gunships were called in to escort them on most missions. By 1964 even this escort was not adequate and the Corps developed a kit of two rocket pods and two machine guns to convert some H-34s to gunships, a mission for which this model helo proved unsuited. A late 1963 proposal to arm the UH-1E had run aground on controversy over using helos as attack aircraft. In late 1964 this problem was resolved by specifying that arming the UH-1E was for self-defense. A month later, after the program was underway, it was expanded to include escort missions. By such devious means the UH-IE became a superb gunship, so the gunship concept was slipped into the FMF, there to remain.

In comparison with today's gunships the UH-1E was primitive. On each side it carried a rocket pod and two fixed M-60 machine guns. The crew chief and gunner each had a swivel mounted M-60 to provide fire abeam of the helo. In moments of great excitement the copilot could, and occasionally did, fire his revolver out the left cockpit window. The gunsight was a swing-up, swing-down rod on the end of which was a metal ring containing a pair of wires which crossed in the center. This was lined up with a grease pencil dot on the windscreen. The dot showed where, more or less, the forward fired ordnance would hit the target. Simple, basic, and deadly, the UH-I E gunship proved the concept beyond any doubt and paved the way for the AH-I gunship series which followed.

It is sort of fitting that even as the Flash Gordon fighters of today can trace their origins to WWI pilots shooting pistols, rifles and throwing bricks at each other, so also can the current gunships. They can eat up a whole column of tanks for breakfast. Gunships trace their lineage to copilots with pistols, firing out the side windows and gunsights based on grease pencil marks.

HMM-364 port machine gunner in a CH-46 with his .50 cal Browning. Enroute to Quang Tri from Phu Bai, 14 Sept. 1968. (Photo credit: HB Staff)

CH-53. (Courtesy of A.M. Leahy)

Medevac crewchief from an H-34 helicopter flying out of MMAF carrying wounded Vietnamese child to the aircraft after release from his parents. The location is the Hoi An runway south of Da Nang. (R.E. Hendrie)

Special Stories

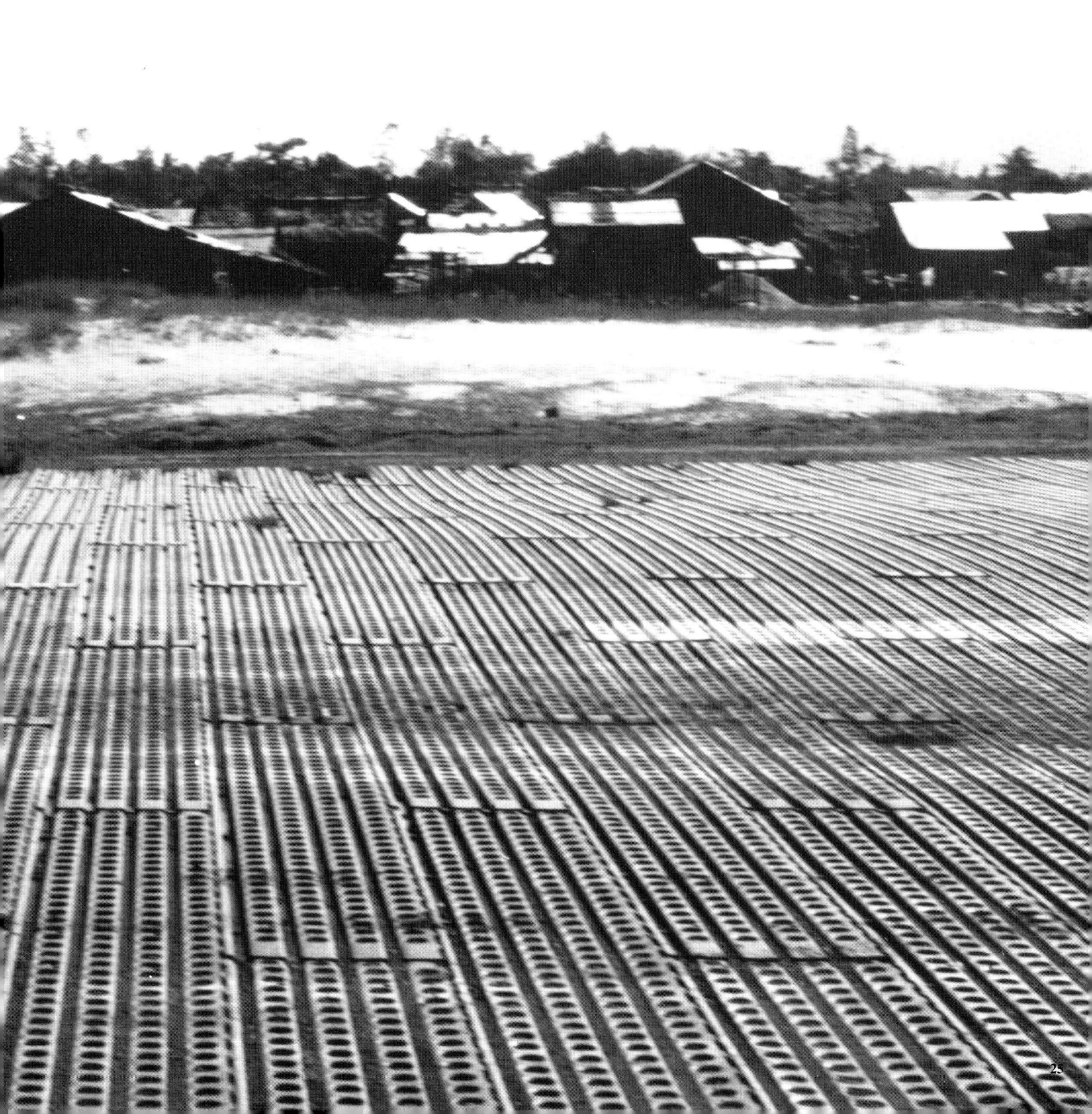

COMBAT EXTRACT
Gene Salter

It's another of those incidents that sticks in your mind no matter how long you live. It frightens you so much that when it's over you lock it away in a closet of your memory so that no one, not even you, remembers what happened. Or, maybe, you don't want others to know just how frightened you were. Anyway, the emergency extract of a recon team from the side of a mountain during the Viet Nam war, is one of those experiences that is hard for me to forget. The memory of it was thrown back at me when I saw my helicopter in Mike Leahy's painting called *VIETNAM: HOT RECON TEAM EXTRACT*.

Although this happened many years ago it still frightens me. I do admit that war is exciting and makes the adrenaline flow. Don't get me wrong, I do not like war nor do I want to see it again— or for my children to see anymore war or for my childrens' children to see war—but it does make you appreciate being alive. During the war the fear and excitement forces you to develop a narrow mindset. You need to develop a narrow mindset to survive war. My mindset consisted of taking off in a helicopter, flying to the "zone," picking up the wounded Marine/recon team, stopping off at the hospital/recon team HQ and returning to the airfield. Then, a cold beer when the day is completed. Take off, pick up, drop off, cold beer. That seemed to be the extent of my whole world. Anything else was a stinking panic, a frightening nightmare with sweating, bleeding, bullets, enemy, and the smell of death surrounding me. It was better to keep a narrow mindset.

During 1967, Mike Leahy was a Marine Corps Reserve Major who had been recalled to active duty as a combat artist. According to Mike's journal, November 11th was one of those days he had elected to fly with me as the port machine gunner on an H-34 helicopter. I was the operations officer for HMM-363, stationed at Marble Mountain Air Facility (just out of Da Nang, South Vietnam). Mike and I had known each other for years, as helicopter pilots at New River, NC, and at Quantico, VA, where we both had been selected as helicopter pilots for the President of the United States. In 'Nam, Mike went along to observe how we did our jobs, and to record it in the form of drawings and paintings after the flights. He said he felt confident of getting back safe and sound when he flew a mission with me.

My flight log shows I flew two missions that day (every pilot has his own personal record of each flight he has ever flown, well, almost all of them). And I remember that day because of the screw ups, the "frictions of war" as Clausewitz calls it. The flight schedule had to be rearranged; other people who had flown the night before, or were on "stand by" the prior night and should be resting were pressed into the schedule as replacements for those of us who had a "priority mission." We had to fly some stateside-style colonels around the area so they could "observe." We didn't have a real VIP helo in the squadron so I grabbed what was available, had the crewchief set up the regular canvas and aluminum frame seats, and took off for the "VIP area orientation." I left my assistant operations officer, Capt Joe Clark, with instructions to give me a call on the squadron radio net if we received any more unscheduled missions or a greater than usual number of medevacs. Joe, who had flown the night before and should have been resting, knew that I'd dump the VIPs in a "New York second" if there was any build up in the medevac load.

The weather was beautiful (pilots call it "CAVU" i.e., Ceiling and Visibility Unlimited.) The blue of the South China Sea was on my left as I lifted off from the Air Facility and headed south toward the huge Marble Mountains, standing like sentinels on the flat beach area just south of the Air Facility. It was a good VIP sightseeing and picture taking area. We flew on down to Hoi An (quaint Viet Nam native village good for more pictures). On the trip down you could see the mountains off to the west. They were a light green near the tops and a darker green as your eyes followed the line down the side of the mountain to where it finally touched the flatlands and the rice paddies, villages, and roads. The mountains looked like the tourist photos that might have been touched up for commercial purposes. They were that beautiful, and scenic to the eye of the uninitiated, but the hostility and death was there for the unwary. Without a war, the area around Da Nang truly is a lovely sight.

As we were making a big circle around Hoi An we received a message from Joe Clark that a 2nd Division recon team was in a "shit sandwich" and needed an emergency extract. I told the crewchief to pass the word to our VIPs that we had to cut the tour short and return to MMAF to make ready for an emergency mission (should make good cocktail hour conversation when they get back to the States). I immediately called for ground transportation for the VIPs and then on the separate FM radio net I instructed the squadron to refuel and prepare another chopper for the recon team extract (that meant to get rid of any seats or other superfluous gear onboard and refuel the chopper with a combat minimum fuel load— front tank only).

Captain Clark relayed to me, on the squadron freq, the position of the recon team taking small arms fire from a pursuing enemy. I told him to arrange for a couple of UH-1E gunships to accompany me in my attempt to extract our Marines. Our Marines, recon teams, troopers - we had lots of impersonal, generic names for some very young individuals, men who saw their duty as "just doing the job."

I'm sitting here thinking about the time one of the recon teams were required to stay on patrol for a couple of extra days and I had to resupply them. One case of "C"s (rations) and a five gallon can of water. Enough for six men for one afternoon, night, and next morning while they quietly watched an empty valley for any possible enemy movement. I decided to add a case of canned beer to their rations (very illegal).

After locating them on the back side of their lookout position, I approached from the side away from any possible enemy observation and the crewchief lowered the "C" rations, beer and water via the rescue hoist because there was no place to land. As they saw the two cases of "rations" being lowered, their radio man came on the FM with an angry "We said one case of " C "s! We don't want to have to haul an extra case of —— hello, That's OK." We picked them up the next day and when we let them off at the recon compound near Da Nang, they ran out with a big smile and gave us a big thumbs up.

Then, there was the morning we had one of those "Oh-Dark-Thirty " briefings and an FNG (F——— New Guy) briefer asked if we wanted the area "prepped" prior to inserting the recon team into its initial position. To "prep " a position means to prepare the position either by dropping bombs, napalm, or rockets from the air or firing into the position with artillery, or both. When attempting a clandestine insertion, drawing attention to the exact location of your insertion is the last thing you want either as the pilot flying the helicopter or the five or six individuals who make up the reconnaissance team. (We looked at each other— we being the aircrew and the recon team.)

Then I spoke up, "Yeah, we want you to nuke the area! "

"Let's keep this on a professional scale", the briefing officer told me.

"I'm being professional, AND, as serious as I can," I answered him.

"Nuke it! Nuke it!" broke out from everyone at the briefing.

"No nuke, no prep,", said the briefer. I told the FNG where to get off, and we got up and walked out.

Yeah, I knew those young men on that mountain top, and they knew me. I always liked to know the people I was working with, and I wanted them to know me. Hadn't General Morgan familiarized himself to his recruits the night prior to the "Battle of Cowpens"? Didn't Rommel visit with the troops before he sent them into combat? (Not that I expected to become a general, but, you learn from the winners.) I always felt that if the busy generals could find time to visit with the troops, I didn't see why a Marine Major couldn't work in a short visit with the men who trusted him to take them into combat and get them out if the going got too tough.

I was the pilot they worked with, the one who came for them in an emergency or after a patrol or when they needed a medevac bird. I knew them when they were wearing camouflage utilities and "make up" on their faces. I recognized them in their dress uniforms. And, I remember the day they invited me to their compound and presented me with the big red, white and blue banner that read "F—COMMUNISM." In smaller print it read "For additional copies, contact the Daughters of the American Revolution." (I've still got that banner. It's on the wall of my workshop. My wife still won't let me bring it into the house). Yeah, I know that team, or most of them. And I was going to do everything in my power to get them back safely.

The recon team must train together and work together. It's after they have worked together for a while - on missions - that they become a team. They learn who is the best planner, the best weapons mechanic, the strongest, etc. Any weakness must be discovered. Once they have worked together, committed to each other and to their tasks, they learn to depend on each other. They have shared the risks, the hardships and dangers as a team. They have learned to depend on each other, and accept the shortcomings and failures of each individual. They know they must depend on each other as a team, and the team is more important than any error one or the other might commit. Bad habits must be corrected, pride and jealously must be shoved aside. Each accomplishment is a shared team achievement.

I dropped off the VIPs (What the hell, they were just doing their job). They told the crew chief 'thanks' and gave me a "thumbs up" as I taxied toward the squadron parking area. I left the helicopter and let the copilot shut it down and I shifted to the new chopper waiting for me. And here comes big Mike Leahy, my gunner, all six foot two and too many pounds of him.

A different copilot, a different crewchief , me and Mike. With the rotors turning, we began to taxi out for takeoff while we radioed the tower and the Huey gunships, as I was strapping in. Then we were airborne into a CAVU sky. Thank goodness for small favors.

I see this on paper and I know you ask why I had to be the one to go on this mission? Why couldn't I have assigned it to one of the other squadron pilots? Was I so egotistical that I thought I was the only one who could carry out this mission? At the moment it was a decision based on knowledge of the other required missions being flown by the squadron, the availability of the helicopters in a "up" flying status, and my familiarity with the recon team and the area they were working in. I was the most experienced combat pilot available and was flying the least essential mission. It was one of those decisions all too frequent in combat, that had to be made at that very moment and I made it.

On the way to the "hot zone" the recon team was operating in, I contacted them for a head count; an estimate of the number of enemy pursuing them; and the direction they were traveling. All of this information was necessary for me to plan my extraction of the team. I knew they were "humping it" when I didn't get an immediate response to my call. (I didn't like to be bothered with radio calls when I was trying to evade the enemy either). I continued toward the area when a second H-34 joined on me. I could see that I was going to need all of the help I could get on this mission. If everything works out when you make these snap decisions, they are called "bold" decisions. If everything winds up in a mess, it was a "stupid" decision. Sort of like the old saying about victory has many fathers but defeat is an orphan. I had made the decision and now I had to live with it.

We needed more than one chopper, for if something happened to me, he would be able to finish the mission. About then the recon team came up on FM with the information that I had requested, namely "eight men," "a hell-of-a-lot," and "away." I suggested that "away" was not a direction I could home in on. Then I spotted them- or rather I spotted the automatic weapons fire of the enemy and from their position I could ascertain the probable direction of the recon team.

I turned the chopper so that we could cross behind our own troops and in front of the enemy. We began firing our machine guns at an aggressive enemy in pursuit of the exhausted recon team. There was no place in the immediate area to land my helicopter and expect to make an extraction. However, the helicopter covering fire gave our recon Marines time to get a little further away and it kept the enemy troops from concentrating their firepower on us- so long as we could keep them pinned down. I made a pass on the enemy with the machine guns clattering and the crewchief's M-79 grenade launcher "thump-thumping." Then the second H-34 followed with his arsenal, then the UH-1E hit them a good lick with machine guns and rockets. And like the radioman said, there were a "hell-of-a-lot of 'em."

It was very obvious that we would have to hold up the enemy troops for a much longer period of time if we were going to be able to allow the recon team to get far enough away from the enemy so we could pick them up without getting the hell shot out of all of us. They were more or less following a path along the crest of a hilltop (mountain?). (Why in hell do Marines always travel along the crest of a mountain?) I know you can see to either side of the mountain, but so can "they" see you from either side "they" are on.

In this case, it was easier (and faster) traveling for the recon team. There was tall grass, scrub brush, and uneven footing on either side of the path. But, there was no place to land a helicopter without being shot out of this world and into the next. So, I had to look for a suitable extract location.

I remembered seeing a place a couple of clicks ahead, where I could hover aways down the side of the mountain without being exposed to any direct enemy fire. If the team could reach that spot, it would be a snap picking them up. Any enemy fire would be over the top of the helicopter because I would be part-ways down the side of the mountain and they couldn't see me- I hoped! But we would need refueling and additional ammo to hold out 'til the recon team could "hump it" that far.

I directed the UH-1E gunship to stay on station while the other H-34 and I dropped down the side of the mountain to the small airfield at An Hoa to refuel (the UH-1E had a full load while we carried only the full front tank). I relayed to the recon team what we intended to do so they wouldn't think we were abandoning them.

When we received an acknowledgment from the recon team leader we cut away from the firefight and left it to the UH-1E gunship to stall the enemy pursuers while the two H-34s refueled and, hopefully, found some additional ammo. (The UH-1E had ten times the firepower of the H-34. It carried six machine guns and a covey of rockets. The H-34 was configured as a troop transport/medevac ship with two 7.62 M-60 machine guns for self defense. The M-79 grenade launcher was something the crewchief had "picked up" in his travels.) The Huey was able to slow down the pursuers by making feint dives in one direction and immediately turning to shoot in another direction.

I dropped down to the An Hoa emergency airfield. The other H-34 followed. Lucky for us no one else was refueling so we had the pumps to ourselves. As I pulled up near the refueling hose, I saw the crewchief jump out of the helicopter and remove the cap on the forward fuel cell. At the same time I saw big Mike leave the cabin. I knew he had gone to the supply tent setup nearby for a couple of cans of 7.62 ammo as I watched him disappear out of my sight. It wasn't 'til later that I discovered he was hassled trying to obtain ammo. Picture this, I'm refueled and pull up my helicopter so that the second H-34 can refuel at the single hose and nozzle emergency fuel stop. And across the blacktop trots Mike in flight suit, flack vest, side arm, and a whole case of ammo. I won't say he was running. Maybe he wasn't trotting. Maybe when you have a too-heavy case of ammo in your arms it's called "humping it'. The crewchief jumped out, gave him a hand with his load, and we were on our way. Mike had slimmed down some. He would sweat off a few more pounds before we returned to Marble Mountain that day.

Not 'til that afternoon at An Hoa's outlying airfield did I realize how I desperately wanted to climb back up to that mountain top to get to some Marines who needed me. We launched and I immediately contacted the recon team leader. They were running out of mountain top. They didn't have that much more room to maneuver in before they came to our agreed on pickup point. The other H-34 joined the gunship in harassing the enemy and pinning them down, or at least, slowing them down.

"Recon team leader, this is the Red Lion. Do you copy?" I called on the recon team's radio net. "I've got you, Red Lion," he came back. "Keep going in the same direction you are headed now", I said, "You will find a path leading down the side of the mountain. I'll be waiting." "We are moving out!" he screamed back at me.

We circled the area once, made a pass at the NVA enemy troops and sprayed them with as much 7.62 as we could and made a run for our rendezvous point to take aboard our Marine recon team. I found a slight projection from the side of the mountain where I could rest my right main landing wheel while I hovered. It was low enough on the side of the mountain where the enemy couldn't get a good shot at me from the crest of the mountain while I loaded the recon team. The mountain sloped just enough so that no one was in danger of being decapitated by the rotor blades (if they ducked down) as they approached the helicopter.

It appeared I was in the ideal spot to extract a recon team from a "hot zone." All I had to do was hover there while the recon team scrambled down the slight distance, protected from enemy fire, and let the gunship and the H-34 "hold them off at the pass." Wouldn't you know somebody would find a way to mess it up!

The enemy pursuers didn't stay on the path running along the crest of the mountain. (All the plans of mice and ..., frictions of war,... etc.) They saw that the part of mountain that our recon team was on had played out. They left the crest for the tall grass and scrub on either side of the mountain to out-flank the recon team and to get a better shot at them and the helicopter.

"They're taking to the tall grass!" I heard the gunship pilot yell.

It was my first indication of the enemy's intention to outflank me and it contributed considerably to the "pucker factor" I was now operating under. Then I saw the first of the recon team slide over the crest and dive for the cabin door. And I heard the "ping" of the first bullet to hit the tail section of the helicopter.

"Somebody's got me in their sights," I told the gunships trying to protect me. We wouldn't be much good to the recon team if I was hit and tumbled down the side of the mountain losing a helicopter and a helicopter crew.

Then I heard a rocket explode behind me. "I got him!" shouted the UH-1E driver.

"WHERE LIGHT IS AS DARKNESS"
Vietnam Poems by Kellan Kyllo
New Sweden Press
(used with permission of the author)

PEOPLE WITHOUT DIMENSION
The most descriptive profanity
ever used,
traced an outline
around
an absence of feelings.

BURNED MEMORIES
On the side of a bare hill
white phosphorous
exploded
hot
and killed
lasting odors of death into men.
Faces of the dead
stared forgotten years,
old times,
at those who carried them away,
at the world.

LESSONS OF HISTORY
The terribleness of war
written since the beginning of time
in words
no one has ever
been able to understand.

CLOSE CALLS DIDN'T COUNT
Mountain ridges
with dark forests to the sky,
bullets
through the helicopter,
between the gunner's feet.

WAR'S BIGGEST SURPRISE
Death took a few every day,
without
telling anyone ahead of time
who
they would be.

PITCH AND VOLUME OF VIETNAM
His friends
killed in that war
hurt
the inside of his head
every day.

"Can't you guys move any faster?" I demanded of the recon team as two more Marines ducked the rotor blades and headed for the "safety" of the helicopter.

"One man is carrying a case of "C's", the radio came back.

"Drop them and get your asses in here!" I was somewhat vexed.

"Can't. The enemy will get them."

"If they are that important to you", I said, "Why don't you stay there and guard them while the rest of us get the hell out of here!"

They opted to go with me.

YANKEE PAPA EXTRACT— A GRUNT'S PERSPECTIVE

Mick Carey

I remember now a cold, foggy morning northwest of Khe Sanh. I was the team leader of a five man recon insert. We were to penetrate north and west to get intel on suspected improvements to the Ho Chi Minh trail that would allow heavy truck and tracked vehicle traffic. We were also supposed to check on some new seismograph detectors that were not sending back signals. Part of the patrol orders were to avoid contact, although G-2 (Intelligence section) advised us that they did not have any indication of any enemy troops in any strength in the area. The insert went smoothly on the first day.

Just about sunset we walked right up to the perimeter of an NVA regimental size base camp, complete with field hospital and more NVA troopers than I had ever imagined could be in one place. How we got so close without detection remains a mystery to me to this day...pure dumb luck. We knew we had hit on something VERY big and I made the decision to mark the coordinates on my map and stay put, in place, until near dawn, then bug back out to our extract LZ as fast as we could. Getting away from the camp was a lot different than getting in... the bad guys weren't just sitting around sunning themselves; they had very aggressive patrol activity going on. We shot a straight azimuth for the extract LZ and "dee dee'd" outta there!

Four of my people were carrying M-14s with about twelve mags each and I was toting a shotgun. We were in NO position to slug it out, even briefly, with the number of NVA troops in the area. We made contact almost immediately on our way back; it was virtually unavoidable. We were moving as fast and as quietly as we could, but they were like hornets with their nest stirred up. Thinking back, it is easy for me to understand just how concerned they were that anyone who saw what we saw not be able to call it in. They were all over us and a running fire fight commenced. My radio operator got on the emergency freq and broke radio silence and started yelling for any air in the vicinity. A flight of Army "slicks" with Cobra escorts advised us that they were about 30 clicks to our south, but that they could not come down to the deck because of the fog. We kept yelling on the radio about the shit that we were in, running our butts off, trying to avoid running smack into the NVA and at the same time I was trying to get the camp coordinates out over the air in case we didn't make it to the LZ. The NVA were so close we could hear them shouting orders. They were busting "beau coup" caps in our general direction and finally, the inevitable happened ... we adjusted our direction southwest and ran right into about a platoon size NVA patrol.

I think they might have been more surprised than we were because we literally ran right through them with everybody on full auto. Three of my people took hits but kept going. My RTO took a grazing hit across his forehead that literally removed both of his eyebrows and he was blinded. Two of us took him by the arms and we kept going. I started yelling into the handset ... can't even remember anything I was saying, but someone up there heard me.

I heard a very southern-accented voice telling me to try for the LZ, that they would try to get into it too. I didn't even ask who they were ... didn't give a shit, either! Whoever he was, he was American, and I needed some more Americans around me real bad!

We broke through the triple canopy and there was a ridge line about 50 meters directly in front of us. We were doing everything we could to get to the ridge, when out of the fog I saw the most beautiful sight I have ever seen in my life. It was a big, green, ugly, beautiful UH-34 with the word MARINES on it, coming into a hover with his right wheel just touching the ridge line. Another '34 was sweeping to our immediate rear and the door gunner was burning out a barrel with continuous M-60 machine gun fire.

We pushed everyone on board and I was last to get on. When I turned to dump my remaining five rounds, I saw about 50 NVA hell-bent for us and that chopper. Just a solid line of muzzle flashes. I dove in the door just as the pilot literally dropped the bird off the ridge to his left to get out of their kill zone as fast as possible. It felt like a roller coaster nightmare, but it sure felt good!! The door gunner took a grazing neck wound and somehow broke his arm. That 'ol '34 had a bunch of holes in it and made a lot of very un-helicopter-like noises all the way back. I gave the copilot my marked up map and he was calling in arty and air before we even got half way home.

I never had the chance to meet or thank that crew. Five grunts owe them their lives, and we may never know how many lives were saved by hitting that area with arty and air before the Ninth Marines moved in on the ground. I seem to recall that part of his call sign was "Yankee Papa". I found out later that the flight of two that came in for us were on a routine re-supply when they copied our transmission to the Army but couldn't reach us on the radio. They dumped their loads and came in for us. The pilot didn't even see the ridge until the door gunner told him he was about 50 feet from it!! AND, they had no gun escorts!!

I still get very emotional when I think about the courage those Marine aircrews displayed on that cold, foggy day so long ago. I cry to myself when I think about the love of one Marine for another and the sacrifice those men were willing to make for five very scared, about to die grunts. I can't really find the words to express how I truly feel about those men that day. That kind of stuff only happens to Marines!! It was too foggy for the Army guys and that was it for us...we would have died. DAMN! But those insane, brave, Marine "airdales" came right into the shit and hauled us out ... Thank you from the bottom of my heart! I keep remembering and I keep crying, but that's okay, because it is a grateful, cleansing crying. We were restricted and cheated ... NEVER DEFEATED!

THE GUNNER'S LOT

A. M. Leahy

During my brief experiences as port .30 cal. machine gunner in VMO-2's UH-1Es, HMM-363's UH-34s and starboard 50 cal. machine gunner in HMM-161's CH-46s, several thoughts occurred to me during my 1967 and 1968 assignments to III MAF as a combat artist serving in I Corps.

As a Marine buck sergeant helicopter crew chief, naval aviator and Marine helicopter pilot prior to Vietnam, I served as a Major during the Vietnam war, the executive officer for the Marine Corps' Combat Art program. After arriving in Vietnam in

Gunner's trial, October, 1967. (Courtesy of A.M. Leahy)

September 1967, I went out with my M-14 to join ground Marines in Combined Action Program (CAP) units, then to Dong Ha, Khe Sanh and then on to Operation Medina, where I joined up with Charlie Co., 1/1, in the Hai Lang Forest, SSW of Quang Tri.

Afterwards, I joined up with several helicopter and fixed wing outfits. So as not to take up precious weight aboard a helicopter in order to accomplish my combat art duties, I decided to man a machine gun, thereby getting a really meaningful look at helicopter operations. What I didn't realize was that my good old Marine Corps helo aircrew and pilot buddies (?) would often put me in the lead aircraft on several missions and operations!

I quickly learned that one doesn't just sit there and pull a trigger. There was much more to the "gunner's lot," in actuality. You MUST become an integral part of that helicopter crew, or the whole thing doesn't work. As a UH-1E gunner, I was responsible for twin fixed M-60s on the port side, and a whole 19 pod of 2.75 rockets, in addition to the flex M-60 on a swivel mount. Next to my feet were 250 round boxes of ammo, an M-79 grenade launcher and its peculiar ammo in addition to my trusted M-14.

My first firing mission during November,1967, was in a VMO-2 gunship escorting medevac UH-34s. With my old friend, TSgt Alford as crewchief and starboard gunner, watching over me like a hawk, I passed my first test well, as a helo machine gunner. Shortly afterward, VMO-2 was assigned to Operation Essex, in Antenna Valley, SW of Da Nang. I flew in the lead UH-1E, which was loaded with Willy Peter 2.75 rockets to mark targets in a ville for the A-4s and F-4s to bomb and strafe.

Having the rockets leaving the pods right near my left knee was a thrill I'll never forget. Especially when the hot caps, which burned off the rear of the rockets flew all over my nomex flight suit, burning neat little holes in it, and into the flesh of my legs as well! The next thing that happened startled me. I looked down at the 19 rocket pod, still more than half full. Four or five of the white colored rockets were creeping forward out of their tubes! I told the crew chief of my dilemma. He said: "Kick 'em back where they belong!" I told him I couldn't reach them without taking off my safety belt and sitting on the stoop. He said: "So what? Do what you have to do!" So I hung half out in space with one foot on the skid and I kicked the offending rockets back where they belonged. I was scared as hell.

Since I cleaned my own gun after firing missions, one evening I was tightening my gas cylinder lock screw with a combination wrench which had a screwdriver feature on the box end. The wrench slipped and my hand slid along the handle. The screw driver appendage wiped out the webbing between my right thumb and forefinger, resulting in my getting six stitches at sickbay. To this day, whenever anyone asks me about the resulting scar, I tell them: "Vietnam, I'd rather not talk about it."

I found out on another mission, how just about anything goes wrong at the worst possible time. We were returning to MMAF after completing a mission, when the gunship was diverted to the Hai Van pass. A truck convoy had been ambushed transiting the Pass and stalled, taking fire from up on the west slope. There was only one little area of trees up the slope and the enemy had to be using it for concealment. Our gunship commenced a shallow banked turn to the left in order to line up the machine guns and rockets for the attack.

While firing my M-60 until the UH-1E could line itself up to fire its other fixed machine guns and rockets, suddenly, my swivel gun jammed. But even worse, some of the brass and links had fallen into the chutes which fed the fixed machine guns! They jammed as well. No machine guns on the port side, as we were turning to port to line up the helo! I did the only thing I could, at the time. I grabbed my M-14; put the selector on full automatic and blasted away until the rockets left the pods!

CH-46s were standing down because of tail structure problems during fall, 1967. I decided to join up with HMM-363's Red Lions as Operation Essex was still in full swing, and leading into the follow on Operation Foster. The guys in VMO-2 had given me a souvenir custom-made 100 round ammo box as a remembrance of my short time with their squadron. I didn't realize why they gave me that gift when they heard I was headed for HMM-363's UH-34s. I found out soon enough, however. The port gunner leans out of a small portal with his M-60. If you used a standard 250 round ammo box on your M-60, your field of fire was severely constricted because the big box would bump into the sides of the enclosure! When I opted to use the 100 round box, I had enormously more flexibility, however, it doesn't take long to shoot up 100 rounds. It's a bitch to have to reach down for another box of ammo in the middle of a recon insertion or extract!

During the last part of Operation Essex, we made numerous troop insertions with plenty of firing. We also took WIAs back to a field hospital, on many occasions. As a fully functioning crew member of a UH-34, I had to load KIAs into our chopper for the doleful trip back to the Da Nang air base. It was one of the most moving situations in my life, when we loaded several full body bags into our helo. We were so loaded, that I had the upper part of a body bag in my lap, as I manned my M-60, upon takeoff. I could feel the trooper's head and upper torso, on my lap, still warm.

On Operation Foster, things in LtCol Frankie Allgood's lead helicopter bordered on the line between very busy to downright frantic! For all I know, I might have shot through my own rotor arc at one point! (Actually I did. The mechs back at MMAF recorded the bullet holes in the main rotor blades as "combat damage"). What's it like to be an effective gunner, firing away on every troop insertion along the Song Thu Bon? You get a good idea after you've cut your knees, kneeling in the brass and links littering the floor of the helicopter.

As we were beginning to turn a 180 to land another load of troops, we took several rounds. One of the rounds hit our hydraulic lines, knocking out the control servos. Several other bullets (along with mine) hit our rotor blades; another round hit my gun enclosure. A small piece of the bullet or shrapnel hit me in the neck but didn't penetrate, as the bullet buried itself in the ceiling near the main transmission.

Our UH-34 wobbled down to a forced landing somewhat away from the LZ. The crewchief checked the aircraft over quickly for evidence of fire. There was none. He then presented me with the bullet which had struck just a few inches from my head. He had dug it out of the ceiling. I sat down in my gunner's seat to reflect on what had just happened. I realized that there was brass and links ALL OVER the floor of the cabin. I then realized that my butt was extremely uncomfortable. I got up and

---

A Day And A Wake Up

Twenty-one hundred, when he got the call,
sadly four or five grunts down,
in a distant rice paddy.
Said the Corpsman: "Can't help 'em."
He's just prayin; kneelin',
and he thought about his ol' chopper
that keeps bucking and squeelin'
But he knew right away
he had to get em, you see
Hell, no hard decision man,
THAT COULD BE ME!

While he threw on his helmet
and tightened it on
His stomach kept squeezin'
and jumpin' around
He kept thinkin' "No problem
for sure it will pass"
while an inner voice yelled
"This flight is your last."
This suckers' too old, too shot-up
It can't fly."

As he smiled at his crew chief, his
right fist tightened, thumb up-high.
He went through the process
that started the motor
looked up out in front
saw the turn of the rotors.

His ears were wide open,
perched, his ears alert
for any strange sound, he
pulled up on the collective
she's off of the ground.

It was up in the clouds
the vibration began
Co-Pilot said "Sir?"
he waved back with is hand.

Don't worry 'bout nothin'
she's just talkin,' you see
Says them grunts on the ground
COULD EASILY BE ME!

But he thought real hard
as he lowered her down
Better take my chances
with fire from the ground.

Just when he thought
his odds were too low
"Look! Straight ahead,
a marker —
Let's go!"

He set the craft down
First left then right tire
And wouldn't you know it
Here comes the ground fire.

'Bout twenty or so enemy
we ain't gettin' out
when just as he thought it
He heard Marines shout!

"Go men! Move it up!
Move them wounded!
Cover that chopper!"
"Open ya sixty! Keep 'em down!
The bastards ain't gonna stop 'er."

Three troops and the corpsman
moved the wounded all intact
While two squads of grunts,
held and pushed 'em back.

The pilots chest swelled
in incredible pride
He saluted a corporal
and got set for the ride.

When the blades created
a typhoon on the ground
From the troops or the enemy
He heard not a sound...

Now back in the sky,
He set off for the ship,
Thought of his wounded,
felt the motor sputter and skip.

As he thought of the guys
who were back on the ground.
His eyes puddled
His jaw tightened
His mouth stuck in a frown.

Now off of the coast
and out over the sea
He spotted the ship
Where he'd hoped it would be.

A huge white vessel
With a burning red cross
He thanked the good Lord,
That he didn't get lost.

And at that very moment
When he was sure he would make it
The engine started skipping
The fuel gauge?
Empty! Nothing in it.

No time for approval
from the ship
He could land
He knew this was it
The time was at hand

The voice in his ears
so loud and so close
"Sir, you've not yet been cleared to land.
On the good ship Repose."

He held onto that stick
with all that he had
And kept his eyes glued
on the landing pad.

About fifty foot left
and the engine just stalled
He calmly and quietly said
"Repose, hear my call."

"Your nurses, your docs and
your surgeons don't hoard
Were comin' in now
with five souls on board."

LCpl Vieira, grunt, RVN 66-67

looked at the two pieces of heavy metal gunner's armor on my seat, which I put there to protect my ass. A gunner's thought of getting shot in the butt is real and caused me much concern!

HMM-363 drew the short straw to extract a recon team along a ridge in Elephant Valley, one afternoon. Major Gene Salter, HMM-363 Operations Officer, led the flight. I was his port gunner. Low clouds obscured the ridges which precluded any chance for fixed-wing cover. After a few firing passes in the area, we flew back to An Hoa to refuel and re-arm. While the crewchief put gasoline into the tanks, I trotted over to a mini-ammo dump at the other side of the matting, and told the NCO I needed some cans of 7.62 ammo. He said he couldn't issue cans in broken lots, but only in crates, so I was out of luck. After I put on my dungaree cap, I *informed* him I'd take a *crate* of ammo since that was the case, which he was only too happy to provide. I put it up on my shoulder, and began to hump it back to my helo. I quickly realized that it was no little chore to carry that ammo at a trot, along with my heavy body armor. I was breathless by the time I got back to the ship. I crawled into the cabin and lay on the floor exhausted as Maj Salter bolted into the air and headed back to Elephant Valley.

The situation where the pickup point was located was grim. Each time we approached the landing point, there was a tremendous volume of fire from up on the slope. Salter uttered a few of his well known expletives and headed right back into that shit sandwich, resolutely determined to get the guys out.

Salter lurched onto the site and put the starboard main landing gear on a knoll, hovering at altitude. I sensed the blades beginning to cone as he lost engine and rotor rpm during the frantic effort to get the guys on board. We were taking a breathtaking volume of fire. I had just ripped my 100 round ammo box off, and was replacing it with a 250 rounder when Salter lurched off the knoll and dove down the side of the ridge, desperately trying to pick up some airspeed and regain some "turns." The other UH-34 and the gunship in our flight were making runs parallel to the ridge in a desperate attempt to provide covering rocket runs and machine gun fire which was successful. The recon troopers were really happy campers to have checked out of that situation.

During my second extended assignment to Vietnam in June of 1968, I flew with VMO-2's OV-10As SW of Da Nang. Afterwards, I participated in an operation on the ground just south of the Da Nang airfield during August. VC infantry and sappers were attempting to cross a river at the Cam Le bridge and infiltrate the airfield. Grunting with an M-14 is another story!

The weather was absolutely "blue bird" upon my arrival at PROVMAG-39, during September of 1968, By then, the CH-46s had re-proven themselves as the primary transport workhorses for the HMM squadrons, in-country. An old friend from my pilot days, LtCol "Tiny" Niesen, welcomed me aboard HMM-161, and ushered me into Hut #1 for my stay. I brought along a half gallon of Jack Daniels with me, to try to trade it off for a souvenir SKS communist rifle. When I produced the bourbon in Hut #1 for trade, the pilots said with one voice: "We don't have any SKS. Where do you think you're going with the Jack Daniels?" Well, so much for my big trading plans. (A month later, back at Da Nang, a beautiful SKS was delivered by my Quang Tri buddies to my quarters. Reputedly, it belonged to the top Quang Tri VC province chief Bui Thu who was ambushed by an ARVN patrol).

Maj Jim Bolton and Lt Bob Odom took me

Eversharp 1 Charlie, CH-46, RVN, October, 1968. (Courtesy of A.M. Leahy)

HMM-161 CH-46 night medevac over Mutter's Ridge RVN, October, 1968. (Courtesy of A.M. Leahy)

THE WALL

Hey man, guess what ?
your name is in town.

It's carved in bold letters
planted firm on the ground.

So many will see it
But sad as it is they won't know who you are,
where you were or what you did.

From what I can see
I'm the only one around
who knew you and loved you
'fore your name came to town.

Yeah many will see it
lookin' up, walkin' round
but won't remember you
after your name has left town.

An I wanna try to tell it
real relaxed and real calm
How it was on the day
in a bag you left Nam.

A few guys threw you,
in a stack of five others.
All of 'em; beautiful people
everyone of 'em brothers.

I didn't know you were
there in that stack on that day
when the choppers came in
and they took you away.

Then early that night-
long after that fight -
someone mentioned the fact.
That you were medevaced.
"How bad was he hit?"
I asked right way.
He just shook his head
"No, Man, K.I.A."

When I heard them words
I started to shake
My stomach got sick
All night, stayed awake.

It hasn't stopped yet;
these twenty-six years
Wrote about it in a poem.
"My eyes full o'tears."
They sent me home too
not long after you
The war ended without glory
The rest - a long story.

So here we are now
All those years; so much later
An I'm seein' you again
Man what could be greater?
But man it could never
ever be quite the same
All I get to do now
Is look at your name.

But just think, my good buddy -
take pride, feel real tall
Cause while the wall's here in town
Your name's most loved of all.

"Hei', dozo, buckatarni, meezo!"

Yeah, Yeah, Yeah.

LCpl Vieira, Grunt, RVN 66-67

out on a flight to check me out on firing the potent .50 cal. aircraft machine gun. We made a few passes near the east base of a distinctive mountain called "Razorback," which was a little east of the Rockpile, out along Rte 9. I quickly learned how to direct the fire of a mix of special purpose .50 cal. rounds. It included tracers, ball, incendiary and armor-piercing rounds. In forward flight, perpendicular to the aircraft's heading, if the starboard gunner trailed the target by aiming 6 inches behind it, the arcing rounds would virtually always find their mark. Incidental to checking me out on the .50, we had to complete an admin mission by picking up some troops at the base of the Rockpile. We landed with our other flight aircraft, piloted by Capt Buck Herron and Lt John Glenn. We sat there uncomfortably in the LZ, and waited almost too long for the troops to board the helos.

KER-WUMP- - -KER-WUMP! We were under mortar attack! One of the rounds landed between our 46 and the one behind us, spraying our craft with shrapnel. Maj Bolton immediately took off, engines and transmissions screaming as we clawed our way out of the area. Looking to my left, I saw that our crewchief, Cpl Beasley was hit in the upper left arm and was bleeding profusely. The port gunner, Cpl Aardema frantically tried to staunch the blood flow, with little success, because Beasley was trying to stand up and inspect the aircraft for serious damage. At that point, I pulled rank on Beasley and ordered him to stay put. I would check the aircraft over myself, for damage and leaks. Having done that, I lay down on the rear cargo ramp for the rest of the flight home, observing the engine and rear transmission areas the whole time.

On another occasion, Maj Bolton came up to me in the HMM-161 operations shack early one evening and told me that he had to swap an aircraft with another out in the field somewhere near Camp Carroll. It seems that the Search and Rescue helicopter had made an approach for landing, and someone in a tank with a big searchlight over the main gun, had turned on the searchlight to "aid" the pilot in his approach to landing. What he did was to BLIND the pilot who as a result made a hard landing; slightly damaging the SAR helo. Bolton said that if I went along, I could enjoy a breathtaking view of Dong Ha at night. Like a fool, I picked up my .50 and climbed aboard the waiting CH-46.

As we approached the Camp Carroll area, a call for an emergency medevac came in, from the infamous Mutter's Ridge area west of the Rockpile. We were the only helo airborne and in the area, so we were "it." We barreled westward to Mutter's with a stripped CH-46, which had only a "hoss collar" aboard, for any rescue purposes. When we arrived in the Mutter's Ridge area, the mountainous hills looked like a monstrous, roiled sea, frozen in place. A C-130 circled far overhead, but positioned behind us, dropping high powered flares, to illuminate the rescue area.

Bolton hovered over some dense canopy trees in a half hour long drama, trying to successfully bring up a few emergency cases, using only the hoss collar. Our two crewmen worked feverishly over the hell-hole, desperately trying to get the injured Marines aboard. I felt helpless, because I couldn't leave my gun for an instant to aid them. We were definitely in perilous country, and if I saw one flash of gunfire, it was my duty to return it instantly to keep us from sustaining serious damage. To complicate matters, the damned green running light under my gun barrel made my gun barrel look like a phosphorescent green popsicle, almost blinding me as it affected my night vision. Then it happened.

As one of the crew went down on the hoss collar to bring up the injured men, the flares from the C-130 died. What Maj Bolton used for a horizon to steady the hover of the CH-46 completely escapes me to this day. He maintained a successful hover in the pitch darkness until the medevacs were brought successfully aboard by the undaunted crewmen. It was the "hairiest' flight of my life. What made it so frustrating perhaps, was that I was not at the controls, but hovering over my .50 cal., which was equally as intense a feeling. On our way back to an aid station, I could see one of the crewmen stripping down to his skivvies in order to wrap his flight suit around one of the chilled medevac cases.

As LtCol Niesen remarked upon our return to the flight line that evening, as I wearily toted my .50 back to the armory: "No one said it was going to be easy!"

ANTENNA VALLEY BEANS
Len Mazon

February 16, 1967. On an overcast, raining, monsoon morning at approximately 7:00 a.m., our flight of two Marine Corps UH-34D helicopters took off from the outpost airfield at An Hoa, which was approximately 50 miles southwest of Da Nang, Republic of Vietnam.

Our mission was to re-supply a company of Marines from what I remember was "Bravo" Company of the 2nd Battalion of the 7th Marine Regiment. They were on a week long search and destroy sweep designed to flush out Viet Cong elements in the Antenna Valley about 20 miles west of An Hoa in a mountainous area which was the only coal mining region in South Vietnam.

Our squadron was the HMM-263 "Screaming Eagles" from Marble Mountain Airfield near Da Nang. We arrived the previous afternoon and slept the night in the USMC compound, after getting our briefing from the aviation liaison officer. It was monsoon season, cold and damp and it rained during most of the night.

Major Robert Dorfeld was our section leader with 1st Lt Mickey Walker as his co-pilot. Mickey had been a fellow MARCAD in my flight class at Pensacola and was a personal friend. I was the aircraft commander of the wing aircraft with 1st Lt Jon Baker as my co-pilot. Our helicopter's side number was EG-18, and it was freshly painted and refurbished after being recently reworked at the Naval Aviation overhaul facility in Shimewah, Japan. At the most, it had only 20-25 flight hours after coming back fresh and new to the exposure of mud, bullets and blood in Vietnam.

Major Dorfeld's aircraft was loaded with approximately 1,500 pounds of C-rations along with his crewchief and gunner, each with a swivel mounted M-60 machine gun mounted on the starboard and port side of the helicopter. My helicopter was loaded with about 800 pounds of C-rations for the field re-supply, and in addition to our crewchief and gunner, we carried a USN medical corpsman and two photographers from the 1st Marine Wing Headquarters. They didn't know it then, but they were about to get one wild ride worthy of many photos.

After taking off to the west and climbing uneventfully to about 2,000 feet, we leveled off staying just below a dark, grey, flat scud cloud deck. We flew west toward Laos along the river valley. As the terrain began sloping upward toward higher mountains and valleys we could no longer keep the altitude above ground we wanted. The cloud base remained solid at about 2,000 feet above sea level. We were now flying about 500 feet over the valleys with obscured mountains reaching into the clouds above us.

The lead aircraft was unable to plot a navigational steer to our rendezvous coordinates due to our low altitude and my UHF Navigational TACAN radio becoming completely inoperative. We were now out about 20 minutes and searching for B Company 2/7 by flying through the valleys between hill masses. To compound matters, Major Dorfeld's UHF communications radio quit suddenly, but we were able to communicate on our FM radio. We were navigationally blind and he was communications deaf. I tried repeatedly to make contact with the tactical communications center on our UHF radio and report that we hadn't been able to locate "Bravo" 2/7. I also tried reaching "B" 2/7 on the ground on both UHF and FM radio frequencies.

We were now flying a random zig-zag pattern through the valleys between the hills about 500 feet above ground level. Jon kept trying to make contact with ground troops but still no radio or visual contact.

Finally after 40 minutes of flight time we made contact with the company. They could hear the sound of our helicopter in the distance but could not see us. We told them to pop a smoke grenade and identify the color. It was yellow, but we couldn't see it. We continued flying a meandering course over the jungle valleys and over an occasional rice paddy clearing for another five minutes when suddenly the ground radio operator told us we just flew past them within a mile. We were just to the north of the ground troops with a hill mass between us.

As we approached a jungle clearing with rice paddies about 1/4 mile in diameter, Major Dorfeld radioed that we'd make a left hand 180 degree turn and sweep back over the area we just flew past. I ducked in close behind him on his port side and slightly above him as the cloud deck started obscuring flight vision. Just as he was about to start his turn, he tersely radioed that he was taking heavy ground fire. Almost immediately the front engine clamshell door area of our aircraft was pelted with small arms fire and the engine quit immediately as if someone had abruptly turned off the electrical magneto switches. "The damn VC were shooting at him, but they got us," were my instant thoughts. "They were never able to lead a flying object - dumb, ignorant North Vietnamese and VietCong"

I yelled over the intercom for Jon to back me up on the controls in case he or I were hit by ground fire, and I told the crew to strap in and open fire on the tree lines around the rice paddy as we were making an autorotation to the ground only 400 feet below us with our engine shut down.

Simultaneously with my instructions to my crew, I slammed down the blade pitch collective control with my left hand to flat pitch and punched the aircraft nose down with the cylic stick in my right hand to maintain forward airspeed.

While Jon was backing me up on the controls, he radioed Major Dorfeld of our situation. For a few anxious minutes we were very, very busy in the cockpit to say the least.

The autorotation glide and descent took less than a minute at most, but it felt like it was over in 10 seconds, it went so fast. It definitely was a "Bang-Bang" play as they say in baseball terms. Good, repetitious, emergency flight training procedures worked as advertised because instinctive reaction allowed me to make a perfect, dead-engine landing in a slightly flooded rice paddy as if I had made a controlled power landing from a hover. Jon quickly turned off the fuel control, the magnetos and the battery switch as we settled into the mud - the damn in-country ever-present mud!

While descending, we took hits in the cabin floor. Fuel was leaking into the cabin and our crewchief was hit in the foot by a round. Yet all the while our crewchief and gunner were firing away at the VC we could now see in the tree lines. Our trigger-quick gunner did drop one VC at the tree line clearing and one water buffalo in his excitement. There were definitely four very busy crew members trying to stay alive by using all the skills they'd been trained for.

As the aircraft came to rest we were taking sporadic small arms fire from the jungle, so I yelled for the crew to get out of the aircraft immediately and take both M-60 machine guns and as much ammo as they and our passengers could carry, then to follow my lead away from the aircraft about 200 feet to the east, and to get down behind a two-foot dike in the paddy. Because of my concern over the aircraft going up in flames due to fuel spillage after being hit by a tracer round, we didn't take the time to destroy the aircraft by activating a thermite grenade and leaving it in the cabin.

As we were running away from the aircraft to our shelter in the paddy, I spotted one Viet Cong who emerged from the tree line about 150 yards in front of us who raised his rifle to fire at me. I had my .38 caliber revolver in hand, took a running aim at him, and cranked off all six rounds I had in the cylinder. I was filled with survival adrenalin and wished that I had a .45 pistol and an M-3 "grease gun" with me. The .38 sounded like a pop gun and on the run all I could do was cause the VC to duck behind a tree.

All seven of us hit the ground behind the paddy dike and began shooting at the tree line as we set up the M-60 machine guns. Suddenly the VC fire stopped. We were in a position where small arms fire couldn't really hurt us - just keep us down. I vividly remember thinking about dispersal if the enemy began lobbing rifle grenades or mortar rounds on us.

The elapsed time since we exited the helicopter couldn't have been more than five to seven minutes, but it felt like an eternity. Major Dorfeld's aircraft had departed the area after his crew machine gunned the tree lines as we were running to the dike.

My next reaction was to organize our group into a small cohesive ground force and fight our way into the jungle so we could escape and evade the local VC, at least until B/2/7 could bail us out - that is, if they could find us.

All of a sudden back came Major Dorfeld in EG-11 flying low over the opposite tree line at the rear of us on a low, fast approach to land next to us and pick us up. He was in a position between a rock and a hard place. He didn't know where we were on the map coordinates; he couldn't find B/2/7, and because his UHF radio was out, he couldn't tell any agency what had happened. He had no choice but to come back, pick us up, and try to find our way out.

We laid down fire suppression toward the trees as he landed. Small arms fire began again from the jungle against his port side. I could actually hear his co-pilot, Mickey, yelling over the helicopter engine and blade noise for us to hurry as he was a sitting duck from where he was located in the cockpit. He was just firing and reloading his pistol as fast as he could.

Quickly, we made it to our rescuers' aircraft and climbed in. We were airborne before anyone could strap into a troop seat. The Major was hauling to get out of there! In addition to two mounted M-60s firing toward the VC, everyone else aboard was shooting out of the open windows in the cabin. We must have looked like an Al Capone gangster get-away.

As we pulled up and banked at least 60 degrees very sharply left at about 200 feet above the ground, the helicopter began to pitch violently. The crewchief yelled that we had been hit and lost flight controls. For a fleeting moment I felt we would roll over and auger into that damn oozing mud, and that would be the end of one very bad day. But without thinking, I yelled out as loud as I could, "Level out the turn!" I don't know if Major Dorfeld heard me or instinctively did it himself, because when he leveled out the maneuver the aircraft settled down into normal flight. We had taken a few hits but had not lost flight controls. We got into retreating blade stall, due to excessive air speed, one very scary, and often fatal experience.

Now we were in another pickle! The aircraft had an inoperable UHF communications radio, so we couldn't radio our predicament to an airborne or ground tactical controller. We couldn't climb high enough to pick up UHF navigation aids due to low overcast cloud cover. And on top of all that, the good Major was flat lost! No one really knew where we were for sure.

I prayed while we flew at tree-top level looking for an exit. Finally after about 10 excruciating minutes we found some cloud breaks where we were able to climb to 3,000 feet between cloud decks and finally lock onto the Da Nang TACAN for a navigational bearing back to the An Hoa air strip.

After a total elapsed time of about an hour and 40 minutes we landed safely, thanks to a great effort on the part of all crew members. My crewchief, although shot in the heel of his right boot, was not hurt except for a bruise. The bullet split the rear of the rubber heel and the leather rear of his boot. No one else suffered any wounds.

I've thought about this experience over the years and except for a few brief comments to my wife in letters, this is the first time I've recorded this escapade in a narrative form. I truly feel thankful and blessed for being able to survive this episode and a few others where I might not have come back to tell about it. Our green machines and our crews dodged more than a few bullets, mortars, and rockets, and we took more than our share of hits. I'm proud to have served with such an elite group of pilots and aircrews, especially on that real bad day in Antenna Valley, RVN.

By the way, "B" Co., 2/7 got into a hotly contested fire fight with the VC later that day, and the operation escalated into a week-long battle with more of the 7th Marines brought into the fray. They never found EG-18. But about a week later, a flight reconnaissance effort finally found the remains of that old bird. The VC had taken the C-rations and burned the helicopter completely except for the engine, rotor head, and rear tail pylon. I truly hoped then, and now, that those Antenna Valley Viet Cong all choked on all that "ham and lima beans!"

Finally, as a foot note, let me close with a direct participant's comment about Vietnam. Let it be known now and forever, that we, the Marines, and all the American forces had won all our war objectives - some many times over by the time we left. Too much innocent blood was spilled for us to ever forget the sacrifices of our fallen Americans.

"CHATTERBOX ONE ONE GOING DOWN OFF THE ROCKPILE"

Cpl "Bobby" Norcott

In the spring of 1969, HMM-262 was flying in support of the 26th Marines in northern I Corps. An unforgettable mission involved a routine "hop" to the Rockpile.

The rectangular landing pad on the "Rock" was not well suited for a CH-46. In order to safely set down, pilots enlisted the assistance of the flight crew. Gunners would hang out over their 50s and eyeball the stubwing gear, port and starboard, while the crew chief observed the nose gear. When properly aligned over the pad, the pilot set the bird down.

The hack driver on the mission that day was Lieutenant Tom "Mixer" Mix. Sadly, I don't recall the remaining crewmembers. I was manning the port gun position.

Grasping the barrel of my gun with my right hand and the bulkhead with my left, I craned my head out the hatch until I could see the stubwing gear. Winds buffeting the LZ compounded Lieutenant Mix's task of hovering the bird and adjusting its altitude as required.

Once all three gears were safely over the pad, Lieutenant Mix began to set the bird down. Without warning, there was a tremendous BANG followed immediately by a jarring vertical beat. I had been hanging out over my gun and received a good whack on the head as I attempted to jerk myself back into the aircraft. Vibrations were so bad that my vision blurred. A sick, whacking sound broadcast from the rotor blades was immediately a cause for concern.

While I struggled to hold on to my gun for stability, the lieutenant's desperate call came over the radio: "Mayday! Mayday! Mayday! Chatterbox One One going down off the Rockpile! Mayday! Mayday! Mayday!

It is the absolute truth that your life flashes before your eyes in moments of imminent peril. Mine did. Images of my mother, my grandmother and my fifth grade teacher overtook my consciousness. Thoughts of my dog, my bicycle and my bedroom back home ran through my mind. I prayed hard and regretted all the bad things I had ever done.

Thankfully, Lieutenant Tom "Mixer" Mix had it all together that day. Maintaining control of our would-be coffin, he began a steep, banking descent. I have no idea how long it took for us to reach the valley below. It seemed an eternity.

The landing was an abrupt but joyous one. After shut-down, we exited the aircraft to survey the damage. I felt as tired as if I had been lugging bricks all day. The energy exerted on the ride down had me physically exhausted. Once out on the ground, the problem was staring down at all of us. One of our rotor blades was missing most of its flying surface. All that remained was the spar! Evidently, a section of the corrugated roofing on the little hootch atop the Rockpile had blown off in all the turbulence, taking most of the rotor blade with it. The imbalance that resulted had caused the horrendous vertical beat.

When our situation was assessed, our wingman returned to the flight line and another blade was delivered to us in the field. We removed and replaced the damaged blade and flew back to Quang Tri, in one piece. Hats off to Lieutenant Mix for keeping a cool head and displaying admirable flying skills!

WHAT GOES AROUND... COMES AROUND

Robert A. Yaskovic

Around 1970 at Marble Mountain, Da Nang, RVN during an early launch, a CH-53D belonging to HMH-463 taxied outbound. The main rotors struck a fortified (concrete) hangar, sending shattered rotor blade fragments across the flight line. I witnessed this as I taxied toward the runway. Shrapnel went everywhere but as I later found out no one was injured. I continued with my mission.

At the O'Club that evening, during the normal "boring" movie the crowd was moved outside to watch a spectacular display as a quonset hut (hootch) went up in flames. Rounds exploding, the crashcrew stood back and let it burn to the ground. The excitement over, I returned to my "hootch" and went to shower and retire. Upon entering the shower hootch I encountered a fellow pilot, a first lieutenant (name still unknown) who stood robed in a towel, toothbrush in his right hand, mumbling to himself in front of a shaving mirror. I essentially, as best I can recall, stood back and after trying to assess the situation, queried as to what the problem was, and could I be of help? He told me . . "You know the '53 this morning?" "Yes" "Well it was me, and the XO said he's gonna have my ass for it. I went to the club to have a drink and forget about it and someone said a hootch was burning. I went to the door and it was my hootch on fire; my life was ruined, this was all I have left (pointing to his torso towel and waving a toothbrush.) I don't even have toothpaste!" At a loss for words, essentially dumbstruck, I grabbed his toothbrush and squeezed some paste onto it and said "Brush your teeth," and, "Here use my soap." After his shower I told him to bed down at my hootch but he declined; he said he had a place to stay.

Days later all of Marble Mountain knew that the XO of HMH-463 did, himself, strike a hangar. This time taking two lives of personnel in the transient operations shack. The word "went out" that the lieutenant was now out of risk.

Several years later, at an O'Club (in Futema, Okinawa, I believe) a drink was placed before me by the still unknown Marine who said, "Thanks—you gave me soap and toothpaste when I needed it the most." I still don't know his name and I'm sure he doesn't know mine, but for a moment we touched; he's a Marine and we both remembered.

FROM A DOG DRIVER
E. C. Southworth

A lot of luck and a little guile,
Leaves me around to reflect awhile,
My friend on this wall,
Who forfeited all,
Will always be owed this "live man's smile."

FIRST BLOOD
F.G. Wickersham III

It was a bright, sunny afternoon in the Da Nang TAOR. As I briefed with my helicopter aircraft commander (HAC), I wondered when I would be allowed to strike for HAC. I had arrived in RVN on November 5, 1967, and as I looked at the date on my "combat Seiko" I noted it was already January 15, 1968. Damn, this war was slipping by, and I hadn't even had a chance to fly my own airplane!

The brief droned on. It was to be another afternoon of "routine medevac." My plane commander was a Captain called the "Sugar Bear" by the other HACs in HMM-363. I just called him "sir" in that I was a lieutenant, a co-pilot and an FNG! The "Sugar Bear" had beat me "in country" by just a few weeks . . . with his 500 helo hours and a HAC check from Hawaii! That made him boss, but I still reckoned that I could fly circles around him. All this seniority stuff was sure a drag.

My mind snapped to the business at hand as the "Bear" asked me if I understood the brief. "Sure," I responded. A guy didn't have to be a rocket scientist to co-pilot a UH-34D. Fact was, I'd be lucky to touch the controls this afternoon. All I was along for was to move the mixture from rich to normal, and vice versa, when the HAC growled! So off we went to pre-flight, turn-up and do our checks.

The afternoon started off rather slowly. We didn't get a launch call until we had been on duty a couple of hours. Well, any day with flight time was going to be an okay day to me. We launched with the "Bear" looking serious and me still wondering what a guy had to do to get a HAC check. Then it struck me. I hadn't been bloodied yet. Hell, on the few missions I had flown where we reportedly took fire, I hadn't even heard it! I'd never been in a plane that had been hit. Yeah, that was it. As soon as I got myself bloodied they were sure to give me a crack at HAC.

By the notes on my old kneeboard . . . the one with the bullet hole in it . . . we launched at 1510

Sign at Air Force gate before going into town, Da Nang, 1964. (Courtesy of Richard H. Swoszowski)

HMM-365 hangar area, Da Nang, 1964. (Courtesy of Richard H. Swoszowski)

Hai Van Pass, northeast of Da Nang, February, 1970. Clouds crept up the mountain, "Socked in" within 20 seconds of taking the photo. (Courtesy of Peter G. Sawczyn)

that day. We made a few uneventful medevac pick-ups, got refueled and headed out again. DASC assigned us a mission Six Kilo, a priority medevac consisting of three Vietnamese civilians and one Marine WIA. We were to contact a unit called "Seven Hotel India" on 35.5 FM. He was located at grid Alpha Tango 914504. This was approximately 202 degrees at 14 nautical miles from the Da Nang TACAN.

We called for smoke to mark the zone when we were about two miles away. It was an easy LZ. Just open rice paddies with a village 50 meters to the south. There was another treeline to the east, but it was nearly 500 meters away. Not a factor. No enemy fire reported. No worries. Down we went in a rather lazy spiral from 2500 feet. "Geez, the Sugar Bear really is loose on his spirals," I thought.

The "Bear" got the UH-34 on the ground okay. I would have landed closer to the grunts . . . so they wouldn't have to carry the wounded across so much open paddy to get to our machine . . . but who cares what a co-pilot thinks. In less than a minute we were

33

in and out. So much for my silent critique of the "Bear's" approach. With the wounded on board, we headed for "Charlie Med" in Da Nang . . . the roar of our R-1820 engine disturbing an otherwise "peaceful" afternoon.

Lifting out of Charlie Med, we checked in with Da Nang DASC. They had another mission, Six Mike. It was south of Liberty Bridge. "Bear" pulled on about 27 inches of MAP and headed us down range. I told DASC that we were "ready to copy." DASC read me the mission as, "Priority, one US WIA, Alpha Tango 914504, 202 at 14 nm, contact Seven Mike India on 35.5." There was no noticeable acknowledgement of the mission from the "Bear." He seemed too intent on maintaining his turns at 2500 RPM and not over-boosting as he pulled on more power. I "rogered" the DASC, and pointed out the coordinates on my map to the "Bear". Still no reaction.

Obviously, "Bear" knew it was the same LZ we had just been in. I could feel him getting tense. Co-pilots sense these things in their HACs. We were roaring along at 85 knots . . . trying for 90 . . . when "Bear" called 7 Mike India. He didn't ask for a zone brief or a smoke. All he said was, "Is the zone cold?" There was a rather long pause from the lad on the other end of the radio. Finally he responded, "My CO says the zone is cold."

When a medevac pilot hears a Lance Corporal radio operator tell him that the skipper "says the zone is cold," it is a sure tip-off! "Cold zone, my horse's behind," I thought as I gave an extra tug to my shoulder straps. I secured my pencil inside the little holder attached along the side of my kneeboard. It was something I normally did only at the end of a flight. Somehow it seemed appropriate.

I stood by to move the mixture to rich on the "Bear's" nod. I reckon I really expected him to say something dramatic. He didn't. The "Bear" didn't say a word . . . not even a "hang on, here we go!" It didn't matter. I was already "dramatic" enough.

It is a well-known fact that things appear to move in slow motion when you come under extreme stress or distress. That's the way it was for me that day. In what was actually just a few seconds, I mentally rehearsed likely emergency procedures, wondered if I would ever get to go to Australia for R&R and prayed that I wouldn't freeze up if we did take serious fire. My thoughts ended abruptly as "Bear" nodded for the mixture.

We were offset from the LZ to the west just a bit. I rammed the mixture forward and hunched over to cover the controls as "Bear" split the needles. He rolled the aircraft up to about 35 degrees angle of bank to the right and pushed the nose down hard. I saw 100 knots on the airspeed indictor as we began spiraling down from 2,500 feet. We dropped through 800 feet wrapped up in a damn tight spiral and pushing near 120 knots. "Sierra Hotel approach," I thought to myself. "Geez, I didn't know Sugar Bear could fly this hot!" Then it dawned on me . .. he really couldn't. "Damn, the Bear's going to crash us into the flipping zone!"

In UH-34s one of the iron rules for co-pilots was, "Don't touch the controls until you are told—or the HAC is shot!" Despite being scared to death by the "Bear's" kamikaze spiral, I was determined to die like a good co-pilot. So I just sat there watching the altimeter spin through 500 feet and waiting for the impact.

It never came! At 350 feet "Bear" hauled back on the stick like the former F-8 driver that he was. The nose pitched up and we did an aerodynamically impossible maneuver for a UH-34 . . . pulled Gs! God and Sikorsky were with us as we leveled out at 50 feet and screamed across the open paddies toward the LZ. Now came the next problem for the "Bear" . . . how to get us stopped.

Most of the lads who flew UH-34s used a side

CH-46 (HMM-262) on final approach to landing pad. (center right edge of photo) Pad only accommodates the two rear wheels of CH-46. Quang Tri Province, October 1969. (Courtesy of Peter G. Sawczyn)

flare to get the "Dog" stopped after a hot spiral. The "Bear" apparently wasn't up to another non-standard maneuver after nearly losing it on the spiral down. So, he opted for a NATOPS quick stop. Problem was, he commenced his maneuver while we were still going 100 knots. The best I can recall, we were about 35 degrees nose up, 30 feet AGL and 95 knots when we zoomed through the designated LZ. Time for a wave-off for most folks . . . but not for the "Bear!"

The "Bear" held his nose high, quick-stop attitude until our indicated airspeed passed through about 20 knots. Then he rocked level, pulled on some power and established us in a 20-foot hover. All pretty amazing considering we were now about 100 meters away from the zone! I could hardly wait to see what he would do next. "Sugar Bear" didn't disappoint me . . . not for a moment. There we were in a hover, over 100 meters outside the LZ in the middle of open rice paddies, when he began to air taxi backwards!

It was then that the enemy quit laughing at us and went to work. I remember hearing what sounded like heavy raindrops pelting the left side of the aircraft. I wondered if that was what it would sound like if you were taking hits. Then the engine quit. "Bear" reflexively pulled collective, but we dropped hard into the dry rice paddy. For a split second I thought he had shut the engine down himself. I glanced at the mixture lever. It was still forward in the rich position. "Bear" looked at me as if to say, "What the hell did you do?"

We both got our answer as the crewchief yelled over the ICS, "We're taking hits!" In a blur of action driven more from fear than practice, I popped my harness' quick release buckle, unsnapped my kneeboard, tossed it up on the glare shield in front of me and thrust myself out through my port side cockpit window. Suddenly my head snapped back into the middle of the cockpit. "Damn, I forgot to unplug my helmet's radio cord!" I cursed myself. I frantically groped for my cord, vaguely aware that "Sugar Bear" was still thrashing about on his side of the cockpit.

As I popped myself free of the entangling radio cord, bullets crashed through the windshield on my side of the aircraft. My kneeboard, lying on the glare shield, took a round. Another burst through the plexiglass near my face, wounding me slightly. With a sudden lunge, I propelled myself head-first out of the port side window and crashed to safety in the hard dirt of the dry paddy. I crawled rapidly under the belly of the aircraft to the starboard side cargo bay hatch and yelled, "Get the guns out." It was an unnecessary command. The crewchief and his gunner were well ahead of me. In an instant they both fell to the ground around me with their M-60s and ammo cans. Without a word they crawled forward to a paddy dike and began to set up their guns. It was at this time that I realized that "Sugar Bear" was not out of the aircraft.

Fearful that I had left him wounded in the cockpit, I looked up toward the starboard side cockpit window. There was the "Bear" hung up in the window between the windshield frame and his starboard side armor plate. His head and one leg were in the cockpit and his bottom and the other leg were out. It reminded me of Winnie the Pooh being stuck in the hole to Rabbit's house. If it hadn't been so damn serious I would have laughed. "You okay?" I yelled, "Yeah," Sugar Bear responded as he extricated himself and plopped to the ground beside me.

Bolstered by the comical sight of the "Bear" hung up in the window, my courage and composure returned. "I'm off to help the crewchief with his gun," I hollered to the "Bear" over the steady chatter of our gunner's now unlimbered M-60. As I crawled toward the crewchief's position rounds continued to impact into both the ground around us and our stoic old dog. "Sierra Hotel," I thought to myself, "I'm bloodied today for sure ... now they're certain to give me a HAC check!"

KOOL-AID

(Unknown)

Everyone (it seems) sent us packets of Kool-Aid. Since none of us drank the stuff we collected it in a plastic bucket, and over the course of four months accumulated three or four gallons of the mixed crystals. Our portable water supply tank at Quang Tri sat on a tower immediately behind our hootch. Periodically one or more of the former hootch members would be selected to swim in the Colonel's branch water." However, the CO's biggest surprise came when he tried to shower with Howdy Doody's favorite beverage. Fortunately I was on R&R in Sydney and couldn't be implicated. Takes a long time and a lot of booze to properly dispose of 2,000 gallons of contaminated "branch." This is no S___!

My Most Unusual Mission
Gene Massey

On December 22, 1968, my crew and I were on night medevac standby with HMM-161 at MAG-39 in Quang Tri. The weather was bad, the ceiling being about 300 to 400 feet, and visibility at 1/2 - 1 mile. One of those nights when you don't want to be called. About midnight we were called out to pick up an injured Marine who was part of a recon team on an observation mission just south of the DMZ, and about six miles east of the Laotian border. All we knew was that he was injured and the team's position had been compromised, something you just love to hear. That meant that the bad guys would probably be trying to get to them before we did, which, combined with the bad weather and mountains, didn't sound like something you wanted to get involved with just before Christmas.

Our flight of two CH-46 helicopters had just lifted off when our compass started acting up, so we asked for radar vectors to the area. We climbed to about 10,000 feet to get above the cloud cover, and a C-130 flare ship on station, having made contact with the recon team and directed us over the valley where they were located. Fortunately, the cloud cover had become broken with some holes here and there, so we decided to attempt the extraction as flares were dropped by the C-130. It was quite a ride as we worked our way through the flare-lit clouds to the valley floor while staying clear of the mountains. Once we could see the valley we searched for the team, who could hear us but couldn't see us because of the dark and the poor weather conditions. They finally directed us close enough that we could see their light signal. The team was in a bomb crater, so we maintained a hover with the tail of the helicopter next to the crater and lowered the ramp. Concentrating on holding our hover and waiting the one or two minutes we thought were needed to load the team, we couldn't help but wonder if any unfriendlies were making their way toward us. It seemed as though it was taking a long time to load six Marines so I called back to the crewchief to find out what was delaying us. That's when we got the story that the injured Marine had been attacked by a tiger. As it turns out, the tiger had stalked the team to the crater, and as one of the team reported, "Suddenly I heard somebody scream, and then somebody else was yelling, 'It's a tiger!'" Another one of the Marines said, "I jumped up and saw the tiger on my partner. All I could think about was to get the tiger away from him. I jumped at the tiger and the cat jerked his head and jumped into a bomb crater 10 yards away, still holding his prey." The Marines quickly followed the tiger to the bomb crater and opened fire. When hit, the tiger released the young Marine, who quickly received first aid from the other team members. They called for a medevac, and that's where we came in, and that's what was taking so long. Large dead tigers aren't that easy to load into a hovering helicopter. After politely telling the crewchief to tell anyone who wanted a ride back to Quang Tri that they had better get on pronto, all were loaded aboard and we worked our way up on top of the clouds and were vectored back home to Quang Tri where we made an IFR approach. We delivered the wounded Marine, the recon team and the dead tiger to the hospital. The next day we were called by the 3rd Recon Battalion to see the tiger they had killed. It was nine feet from head to tail, and weighed in at 400 pounds. Our pictures taken with this man-eating tiger, we returned to our squadron with a great story about the most unusual medevac mission we had ever performed. It's my understanding that the mauled Marine was sent home and later received the gift of a large tiger rug from his recon unit. This most unusual mission was successfully completed due to the professionalism of my flight crew, the air controllers, and the crew of the C-130 flare ship. If any of you involved read this story, give me a call. I'm in the USMC/Vietnam Helicopter Pilots & Aircrew Directory. Semper Fi.

(Wish I could remember all the crew's names. My co-pilot was Lt. Terry Powell. The A/C of the other helicopter was Capt. Tim Smith.)

Cattlecall and Chatterbox
Michael M. McElwee

In May 1968, I deployed via LPH with HMM-161 to Vietnam. We arrived off the coast of Quang Tri Province on the 18th and flew our CH-46 "Super Ds" off the USS *Princeton*, taking them to the Quang Tri Air Base. Our squadron CO was Lieutenant Colonel Paul W. "Tiny" Niesen. Based on his performance in this capacity, he was later selected as the recipient of the Alfred A. Cunningham Award for being the Marine Aviator of the Year for 1968. I flew with him as his copilot on the day he won the Silver Star, on July 17, 1968.

On this particular day there was to be a large-scale tactical troop lift into LZs south and southwest of Con Thien in north-central Quang Tri Province, the northernmost province in I Corps. HMM-161 (call-sign "Cattlecall"), along with our sister Quang Tri Squadron, HMM-262 ("Chatterbox"), were to provide the troop-carrying CH-46s. It's possible that some HMM-364 aircraft from Phu Bai also participated, but I don't recall for sure. There was a mission briefing the evening before, as was often the case with big tactical troop lifts. I was assigned to flying with Captain Gene "Buck" Massey on the first lifts, which were to insert Marines to act as a blocking force along the area where "Helo Valley" and "Mutter's Ridge" run out into the plains of "Leatherneck Square." We were then to pick up a second Marine unit (unfortunately, I don't recall what battalions were involved, or where we picked them up) and lift them to LZs further west up in "Helo-Valley," where they were to make contact with the NVA that infested this area (a major infiltration area from across the DMZ, just a few "clicks" north), and drive them toward the blocking force. Concurrent with the insertion of the blocking force, Colonel Niesen took a flight of two aircraft up into the valley and inserted a seven-man recon team into an LZ named "Falcon." These recon Marines were to prepare the LZ to receive the troops of the maneuver element. All these events went off without any untoward happenings, at least from my perspective, because Captain Massey and I (flying BuNo 153388) and our flight of four (the typical tactical element with fairly large LZs) made several pickups and uneventful drops of the blocking force troops.

Back in the Southeast Asia hut that served as our ready room, some of us lieutenants/co-pilots were standing around discussing the merits of going to chow in the mess hall versus digging into some C-rations before we started the second phase of the troop lift. The ODO's phone rang over in the corner and Colonel Niesen was called up to it. He hung up and turned to the group and hollered, "Listen up!" Satisfied that he had everyone's attention he stated, "The recon team I just inserted is in big trouble and has to come out . . . I need somebody to fly with me right now." I imagine everyone's immediate thought was the same as mine, i.e., what about the guy that you just flew with to insert the team? I don't recall who that individual was, but apparently he wasn't a player this time. No one stepped forward immediately as Colonel Niesen's eyes swept the room. But for whatever reason, they settled on me and he asked "Mac, do you want to go?" "Yes, sir," I lied, for after two months of flying in Northern I Corps, we all knew that a recon team in trouble could only mean what we euphemistically called a "s— sandwich."

We quickly manned-up aircraft BuNo 154006 and were on our way as a flight of two 46s toward LZ Falcon, accompanied by two UH-1E gunships from VMO-6. It was a typical Quang Tri summer day – clear, stifling hot, and unbearably humid. Whether hot or not, after two months "in-country" I was pretty "salty" and had taken to wearing the sleeves of my flight suit rolled up (which would offer no protection if a fire developed in the aircraft). The only thing I remember Colonel Niesen saying to me enroute to the LZ, was, "Mac, roll those damned sleeves down!"

There was a FAC(A) in an O-1 "Birdog" orbiting over the recon team's position and we came up and checked in with him on his FM. He stated that the team was under heavy fire from the NVA and had taken casualties. Talking to the recon team, he told them that their pick-up bird was inbound and told them to get ready to pop a smoke grenade. When the recon Marine keyed his mike to acknowledge, the heavy small arms fire and screaming of the wounded team members could be clearly heard in the background. Now when a recon team was just "in trouble" they usually whispered into the

Marine outpost, as seen from a HMM-262 helicopter, May 1970. (Courtesy of Samuel Flores, Jr.)

35

hand-held microphone of their PRC-42 radio. That meant that the enemy hadn't found them yet but was so close to them that they might be heard if they spoke in a normal voice. Once the situation had deteriorated to the status of the aforementioned "s— sandwich," then they screamed into the radio - a normal human reaction, because if you have to scream just to hear your own voice above all the racket of small arms fire and explosions, then obviously you have to scream to be heard over the radio. I'm sure adrenalin had something to do with it as well. Colonel Niesen then broke in, talking directly to the recon team. He asked them to hold the smoke and just state their position in relation to where they had been dropped off. The Marine on the radio hollered that they were all in the big bomb crater just east of the LZ, but had all been hit and couldn't hold out much longer. By this time we were shuddering along at maximum forward airspeed for a '46 at about 1,500 feet above the ground and just clearing the second ridgeline southeast of LZ Falcon. Without further ado, Lieutenant Colonel Niesen nosed over into a shallow dive, heading directly toward where he remembered the LZ to be, stated on FM, "Cattlecall __, is going in, dash two, you orbit to the east and keep us in sight." At this point I believe we began to outstrip our escorting gunships because we arrived over the LZ before anyone could lay down any suppressive fire. As we transitioned to a no-hover landing, I could see the recon Marines sprawled in the big bomb crater that drifted by my side of the aircraft almost right below us. I could see a lot of blood on them and only a couple of faces turned up toward us as we drifted over them. We had already started to notice the peculiar hollow "tock" sound of rounds hitting us. Our door gunners must have seen some of the enemy muzzle flashes by now because I heard the heavier, pounding staccato of our own .50 calibers opening up. We had just sat down in the LZ, facing approximately westward, when the area off to our left erupted in a flashing line of rapid explosions. That apparently was a result of the gunships catching up to us and salvoing 2.75" rockets into the NVA positions. That seemed to even up the odds for this particular fight for a few moments so I ventured a peek around the side wing of my armored seat to see that the recon team, with the the help of our crewchief, was crawling or being dragged onto the tail loading ramp. As I turned around the windscreen right in front of us suddenly went opaque as a burst of automatic weapons fire came through it with a tremendous bang. Out of the corner of my eye I noticed Colonel Niesen jerk rather violently in his seat (I'm sure I did the same out of shock and surprise) and I hollered, "Are you OK, sir?" In my fright, I was forgetting to the use the intercom because Colonel Niesen responded to the crewchief who was on the intercom, yelling, "We got 'em! Let's go, let's go!," and answered my question as well, by pulling up on the collective pitch and lifting us up rapidly off the ground. The gunships were around for a second pass by this time, and in a matter of seconds we were climbing like crazy as Colonel Niesen gradually rotated the nose down and we accelerated around in a climbing right-hand turn. I could see about four bullet holes through the left side and center section of the windscreen. They had apparently come from about our 10 o'clock position. Colonel Niesen had taken a few plexiglass splinters, but I was unscathed. We discovered later that some of the rounds had impacted the bulkhead just above his head and had hit the left-hand wings of each of our armored seats. That probably accounted for the terrific bang we heard and the way his body jerked. (The right pilot seat had a fairly minimal inboard wing piece, which often caused an aircraft commander to elect to occupy the left copilot seat when faced with flying potentially "hot" missions.) In any case, the NVA soldier sure seemed to know where the drivers sat in the H-46.

Colonel Niesen eased back off the power a bit and we were heading for the medical facility at Dong Ha. In the cockpit access "tunnel" between the two seats there suddenly appeared the camouflage grease-paint and dirt-encrusted, sweat-streaked face of one of the recon Marines. That was a wild-eyed, but grateful, look if there ever was one! He began pounding us on the shoulder, and seemed to be laughing and crying at the same time. We tried to acknowledge him, but I had to try to keep him off the CO so he could continue flying. The crewchief came up and managed to get him back into the cabin.

After dropping the recon team off for treatment at Dong Ha, we returned to Quang Tri. Undaunted by the fierce resistance that the recon team had encountered, the powers-that-be had, in the meantime, decided that after extensive LZ preparation by artillery and fixed-wing air strikes, the H-46s would proceed with the lift of Marines into LZ Falcon. This time, flying BuNo 154012, Colonel Niesen, with me as his copilot, would lead the transports into the LZ. With a full load of combat troops (15 per aircraft) about 16 of us orbited a few clicks south down by Camp Carroll until the last fixed-wing strike was on its way in. We then proceeded toward the LZ, letting down and making a dog-leg turn to the west to land, four at a time. I was gratified to note that the landscape had been somewhat rearranged in the interval since I'd last been here, and we took no fire on approach or while in the LZ. As we climbed out to the west at about 300 feet above the ground, the NVA opened up on us from a ridgeline on the right side. The aircraft received a fusillade of heavy hits and lurched nose down as if suddenly the aft rotor was the only one providing lift. This turned out to have been several 12.7mm rounds that entered the so-called "control closet" just behind the right pilot's seat. They destroyed the Stability Augmentation System (SAS) and took out various and sundry linear and collective trim actuators, and other things for which I can no longer recall the terminology. At that moment I truly felt that we were going in out of control and that it would all be over in seconds - a much higher "pucker factor" than I recall experiencing when we were sitting ducks in the LZ the first time. However, by some miracle of superhuman flying skill, the CO managed to regain control. This time we really had no intercom or radios either, so we had to shout at each other and were not able to tell anyone else in our flight that we were in serious trouble. As the controls were getting more and more sluggish, we staggered over a couple of ridgelines and landed in a level area along a stream just a few clicks northeast of the Rockpile. All five of us got out of the aircraft, amazed that no one was hurt or hit. Within a few minutes we were joined by another H-46 from HMM-262, which set down just behind us with extensive combat damage. The pilot of this aircraft, I believe, was a Major Buck Crowdis. He sticks in my mind because both he and the other pilot were wearing flak jackets under their bullet bouncers. It occurred to me that this might not be a bad idea and I adopted it for some future missions. In any case, our crewchief and gunners counted over 60 bullet holes in 154012, a couple having passed just under the CO's knees to wipe out the center console where our radios were. Someone eventually came in and lifted us out of there and got us back to Quang Tri.

I was nearing the end of my time as just a copilot. My log book shows that I flew my first mission as an aircraft commander on September 7. In those first months I learned a lot from people like Colonel Niesen and several of the second tour Captains who had come back with 161. Names like Scott Sharp, Pete Angle and Tom Warring, stick in my mind. It's my regret that I don't recall the names of the other crewmembers that flew with us that day. Our crewchiefs and gunners won everyone's respect for their courage and dedication. My log book also shows that I only flew with Colonel Niesen on one other occasion, also in September. In November he gave up the squadron and went on to be XO of our group, Provisional MAG-39.

Later, in 1974, as a Colonel, he was the 2d MAW G-4 when I was a member of HMA-269 at MCAS New River. One day he was in our squadron area and he made a point of looking me up. My then CO, Lieutenant Colonel Huey P.L. Miller, later told me that Colonel Niesen had told him that I was one of the coolest heads under fire that he had ever seen. I certainly would not have said that I deserved that description, but it's my guess he might have thought that because I had asked him if he was OK that day that the cockpit seemed to blast apart all around us. Paul W. Niesen died in 1990 and was buried in the Barrancas National Cemetery, Pensacola, FL. I attended the service.

SURF'S UP
Paul Gregoire

One day, shortly after our arrival, the squadron was flagged for a major combat assault with the 101st Airborne into a hot area north of Qui Nhon. From that mission they were to proceed to another location for another assault later that day. It was a maximum effort and all A/C's were scheduled. John Murphy and I weren't scheduled because we hadn't flown our squadron check rides yet.

About two hours after the squadron had left, the operations duty officer came to my hootch and told me that I had to fly an emergency mission. When I got to Operations I found out that the 101st had run into a buzz saw and was running out of ammo. Because we were the only A/C's available, Murphy and I were going to fly the mission, check ride or no. Due to my many hours of squadron seniority I was designated section leader and Murph flew my wing. Our copilots were brand new. Mine was on his first mission. We had the machines loaded to about 1,000 pounds over max gross. We had to use the runway for a rolling takeoff because we were much too heavy to hover. We staggered off the ground and headed north to find out what kind of s— sandwich we had been handed.

When we reached the landing zone we found that the situation had not improved. The 101st was still trying to fight their way off the LZ and from the way the tracers were flying we could see why they needed more ammo. From our vantage point at 3,000 feet above the zone it looked like the bad guys had it surrounded. I decided to make a tight, spiraling autorotation into the LZ to minimize our exposure as much as possible. I told Murph what I was going to do and expected him to fall back into a trail position and follow me down. Directly above the zone I slowed to about 40 knots, dropped the collective pitch lever into autorotation and rolled into a 60 degree banked turn to the right. The VSI pegged at 4,000 FPM almost immediately and we headed for the deck.

Just at that moment I glanced over my right shoulder and got a hell of a shock. Instead of being in a trail position, Murphy had closed in on my right and was turning inside my turn. Our blades were overlapped by about 20 feet and I could clearly see the grin on his face and the terror on his copilot's. He stayed there through the entire descent until we turned onto a very short final when he finally moved away for the landing. It was at this point that I noticed that the paddy was flooded.

Let me set the scene. We're rolling out of a tight spiraling approach, overloaded and preparing for a semi-controlled crash into a flooded rice paddy. Tracers are flying all around us, the gunships are firing into the tree lines and I'm expecting to start taking hits at any moment. My co-pilot is terrified and I'm a little tense. As we slam into the paddy at about 50 knots a huge rooster tail of muddy water is thrown into the air. At that point, amidst all the noise, tracers, water and general pandemonium, Murph keys his mike and announces to all, "Surf's up!" That was probably the best tension breaker I've ever heard. Needless to say, we lived through that day and some worse. From that time on though, John was known in the squadron as "Murph the Surf."

Murph, I wish I could have been there in person to help you celebrate your retirement but I made the decision to stay away. You've always been a bad influence and have always brought out the worst in me. I was a perfect gentleman at our recent helicopter reunion primarily because you didn't come. The maids were glad too! (Explain that one!) Even though I couldn't be there I thought about you. If I were a drinking man I would even have drunk to your health.

All the best Murph. You're one of a kind.

OPERATION BARROOM
H.R. Scott

I was a maintenance officer in HMH-463 near the end of my tour when I got a suspicious call from operations. The skipper wanted me to fly wingman on a mission with him that was supposed to be a "good deal." His call did very little to put me at ease since I'd learned not to trust his "good deals" and it happened to be April Fools Day. I shouldn't have been surprised at my cargo considering that the last "good deal" had been delivering a load of toilet paper into Khe Sanh during the Siege, but when he told me I was going to be hauling an elephant — well, I figured it was another example of the skipper's sense of humor.

Operation Barroom, named after the old joke about an elephant fart, was actually an attempt by Army Special Forces to save a sawmill in the village of Tra Bong. The sawmill was the sole support of almost 200 Montagnard families who lived in the village. They had harvested the lumber nearest the mill but were unable to go any farther because the terrain was too rough even for machinery to carry the lumber back to the mill. The Montagnards had been abused by the Viet Cong for years and it was hoped that providing a way to save the sawmill would win the Montagnard support in fighting the Viet Cong. The elephants seemed to provide the perfect answer since they would have no problems navigating the terrain and could easily haul the lumber. The problem was in getting the elephants to them. Tra Bong was "Indian country" and the only way in without a major offensive was by helicopter.

There were a number of obstacles to the transportation of the animals besides just the helicopter lift. The elephants had to be tranquilized for the trip and finding a tranquilizer that was strong enough to knock out an elephant long enough for the duration of the trip became a major problem. After extensive research an experimental tranquilizer called M-99, which was made by a British pharmaceutical firm, was recommended for the job.

Unfortunately, in order for the drug to be commandeered for the operation it was necessary for Special Forces to apply for a Narcotics Import License. It took six weeks to cut through the red tape and get a license for the drug to be imported from England. In the meantime, wind of the operation had gotten to the British Society for the Prevention of Cruelty to Animals who were holding rallies against the operation on the steps of the American Embassy in England.

Six months later, the Heavy Haulers of HMH-463 got a "frag" order from Wing to deliver elephants, two each, grey in color, to Tra Bong. The elephants were to be transported by Air Force C-130 to Chu Lai and then air lifted the last 18 miles of their journey by helicopter. The elephants were tranquilized by veterinarians wearing gas masks (it turned out the fumes of M-99 were instantly fatal) then the animals were strapped down onto large pallets. Once strapped down a cargo net was placed over them to make sure they were secure.

Compared to other sling lifts, this was a pretty light load for a 53. However, as you might have guessed, an elephant is not the most aero-dynamic freight in the world. When an external load starts swinging on a helicopter you've got two choices, pickle the load or get it under control. I had to drop a UH-1E I was hauling out of Con Thien because I couldn't stop the swing.

We made the trip at about 50 knots to prevent swing as we wouldn't be too popular if we had to splat an elephant on a mountain somewhere. Another precaution was a 50-caliber machine gun that was continually trained on the animals. If the tranquilizers proved not to be strong enough, the animals would have to be killed in order to prevent them from bringing the helicopters down.

HMH-463's skipper, Lt. Col. Sadowski, flew the first lift, the one with all the press around. I carried the second elephant into Tra Bong on what turned out to be a rather routine flight. The elephants were picked up and then delivered to the saw mill without incident.

Despite the rather heavy press coverage, there was very little play of the story back home. Martin Luther King had just been killed and a humorous story about hauling elephants in Vietnam just didn't seem to rate special attention. Ironically, despite all of the efforts between Army Special Forces, the Marines, the Air Force, and numerous other agencies, the story had the usual 'Nam ending. The only USMC elephant lift was all for nothing.

The elephants were young and untrained, and much too small to move the large logs. They constantly broke loose and ran through the village wreaking havoc and destroying the village gardens. Eventually, the elephants were shot and eaten.

THE PHU BAI "O" CLUB
Robert A. McClellan, "MacAttack"

John "Nasty" App has always been noted for his quick wit and repartee. There have been a few individuals who maintained that at times his mouth may overload his anal orifice, but it certainly was a rare occurrence.

It must have been late summer or early fall of 1966 at Phu Bai when one of those rare times did happen. We were with HMM-161, operating as a detached squadron in direct support of the 4th Marine Regiment and doing whatever else came our way.

We had a small base which included the usual clubs for enlisted, sergeants, SNCOs, and officers. During off duty hours, squadron pilots and officers could often be found at the club as there was usually ice and a working fan. We were not flying this particular day, taking a break from our administrative duties at the 'O' club, when we had visitors.

A large special forces captain entered the club without stopping inside the door to place his weapons and cover on the pegs provided for that purpose. Some members of his team accompanied him. Ever alert, Nasty greeted the special forces officer and proceeded to point out our local protocol regarding covers and weapons in the club, "You can't come in here like that!"

"What did you say, lieutenant?" the visitor replied, as he unholstered his .45 auto and placed the muzzle of the already cocked weapon perpendicular to Nasty's temple.

Nasty quickly realized that this man had apparently undergone a great deal of stress recently, or perhaps just did not put up with minor B.S. as a matter of principle. He didn't bother to decide which applied as he smoothly responded, "Glad to have you here, can I buy you a drink?"

"That's what I thought you said."

Of course, there were others observing these events who were determined that this breach of protocol (not to mention safety and common sense) would not be tolerated by a sister service. Those pilots started individual maneuvers, all directed to a common purpose as if they were on a mission requiring radio silence. Shortly after the sun went down, all of the visiting special forces personnel had been carried unconscious from the Phu Bai 'O' club. Not a one could resist the temptation to try

Marble Mountain "O" Club, December 31, 1968, Pussy Cat A Go-Go Show. (Courtesy of Michael Wolter)

their physical skills at the demanding game of Carrier Quals. If the ramp strike doesn't get you, the hook skip will.

Fox Hunt
R. M. Talen

All throughout the afternoon we had heard the confusion on the radio. Not only was there the monotonous sound of the pilots in the spotter planes calling out the hits and misses, but there seemed also to be a lot of ground chatter. I was leading a flight that was involved in something else and my duties for the day required my full attention rather than something "hairy" I would never be a part of.

In the early evening, the radio chatter was incessant and I occasionally saw the glint of diving planes and bursting bombs a few miles to the northeast, not far from Dong Ha. I will never understand why, but I started feeling a tension that came with those missions that were especially trying or dangerous. My senses seemed to become more acute and I could think clearer and with more objectivity. I could feel something terribly wrong and I didn't have the slightest clue as to what it was.

After refueling and nearing shutdown, we could hear ground troopers talking with the spotter planes. There was some sort of "wild-ass" operation going on with a lot of contact with the enemy. Fighting had forced the withdrawal of our forces, and evidently someone had been left behind. People everywhere were looking for an unarmed radioman and the confusion over the radios indicated that he was in serious trouble.

Shutting down, I told the crew to go ahead and pick up some quick chow because we might be called on to get involved in whatever it was that was taking place. I walked over toward operations, but someone came running up to me before I got there. The message was brief and to the point . . . I was wanted in operations right away for some sort of mission.

When I hastened through the door, the serious looks on everyone's face validated my previous feelings of anxiety. Operations talked to the spotter plane and the ground forces, obtaining enough information for a quick brief. The man everyone was looking for could occasionally be seen by the spotter planes. The "grunt" was completely separated from his unit, without a weapon, and both the enemy and the friendlies were trying hard to reach him. The co-pilot went to round up our crew but they were already at the bird getting it ready for the mission. It always amazed me how the crewchief could almost "bond" with his bird to where he had a sort of instinct about what was going to happen next. It was as if there was some sort of unwritten pact between man and machine, working together to make it through one more day in that stupid-ass war.

We were up, strapped in, and turning within minutes. I had no brief other than what I already had gotten at the bunker until I was to check in with the spotter. It was dark as we climbed out and I could see explosions and tracers to the northeast. Adding to the confusion, a low fog had moved in and it was keeping the friendlies from finding their man. Someone sounded more adamant than the others on their radios, and as I got closer, I realized someone on the radio was very much afraid. Terror was evident. Checking "Highboy" flight in with control, I was now linked up with a UH-1E which took over the mission. The spotter stayed high to try to coordinate any possible use of fixed wing or naval gunfire. A "dragonship" aircraft was on station to help keep the bad guys at bay with suppressive gunfire. The fixed wing couldn't come down because of the fog, and flares were being dropped to help light up the battlefield. There would be an intense flash of light followed by the flare slowly making its way down only to be swallowed by the fog. Once into the fog, the flares gave off eerie, subdued light with no movement, shape or substance to help us in our efforts.

The UH-1E tried to get more information but radio chatter was unbelievable. Everyone was concerned with getting this lone individual out of his predicament. We learned the man was called "Fox" and we could occasionally hear him respond to his parent unit as they tried to pinpoint his location. Our helo lead broke in and tried to locate Fox by turning on all his lights so Fox could see him and give directions. Fox would go off the air for a time and we wondered what was happening. Then the radios would sound like static and Fox would be up once again, transmitting breathlessly and incoherently as we tried to find him among the enemy below. The UH-1E lead called with a plan. He would lead me down below the fog layer with his lights on and mine off, serving two purposes: one to get Fox to focus on giving us some kind of directions so we could find him, and two: enemy fire would probably target him and allow me to slip in without drawing fire. We dropped down to only a few feet over the ground. The UH-1E's lights were just in front of me and I stayed close behind so I could pick up on Fox once he spotted us. The ground was dimly lit by several flares making their way through the thick fog and I caught a shape flash by my right window. I had just missed a spent flare and parachute and I wondered how many more were out there, waiting for us to fly into them. I called Huey lead and he clicked his response while continuing to get Fox to give us adequate directions. Suddenly and unexpectedly, tracers flashed in front of us. They were incredibly close and I was confused because it wasn't ground fire. Then we realized that a second gunship had seen enemy fire and rolled in on it. He hadn't seen us because of our lights being out. The copilot flipped on our running lights and I heard myself literally yelling "stop fire" over the radios. I was embarrassed at my lack of cool, but too busy to dwell on it. My UH-1E lead rolled left and I followed to depart the area. I looked out my right window and saw tracers bouncing along the ground in every direction. There was fire impacting over along the dark shape of a tree line and I was glad I wasn't down there as we left the area to regroup and try again. In a few minutes we were clear enough to the south to climb above the fog. We made our way over the area where we thought Fox was trying to lead us to his position. He was frantic as he tried again and again to show us his location. We had a hard time hearing him and I felt angry with him until I realized the noise in the background was bullets impacting all around him, causing him to shuffle and squirm for cover in the many bomb craters in the field. His ground unit was keeping close contact with us, urging us on to keep trying for their man. Fox came on the airways again, and his sobs of anguish sobered all those who listened in. He pleaded with anyone to get him and immediately responses came from everyone across Vietnam as they renewed efforts to pinpoint Fox's position. My UH-1E lead asked if we were ready for another try. We acknowledged and we both dropped down once again to follow him into the battlefield. This time I made sure everyone knew our bird was right behind the Huey so we wouldn't make the same mistake twice.

I was afraid following lead down there, because our closeness to the ground prevented us from turning very sharply for fear the rotors would hit something. We stayed in so close I had to be careful not to fly into his path as he weaved back and forth while trying to get Fox to acknowledge our existence. There was still impacting all around, and the thought of running into any spent flares still bothered me. Black shapes blurred past my window as I followed the lead helo and I realized we were into the trees along a rice paddy. Suddenly Fox screamed out that he saw us and started calling directions. But in his fear and confusion, he was calling out our location from him instead of the other way around, and our twisting and turning was taking us into the tree line where all the fire was taking place. Our lead broke it off once again and we retreated south to find a hole in the fog to climb through. After Fox had come up on the radio and it seemed as if we were going to find him, all hell broke loose. The airwaves were full of the flights checking in and the spotter trying to guide us closer to where he thought Fox might be. The grunts and the different planes waiting to cover us added to the confusion.

We climbed up near the spotter plane as he tried to calm Fox into helping us identify his location. More bullets impacted over the radio and Fox's pleas once again sobered everyone as all thoughts turned to saving the man. I wondered how many people were laying their lives on the line just to save one American ground-pounder. Amongst all the radio chatter and confusion, Fox once again came over the radios, questioning why we couldn't see his match he had just lit to show us where he huddled in some smelly bomb crater far below. I looked out my window and could see nothing but the action going on through the fog 2,000 feet below. There was no way we would be able to see a lonely match, and then we heard Fox squeal in terror as bullets impacted all around him.

He breathlessly pleaded again . . . "Won't someone please get me out of here . . . they're all around me and trying to kill me . . . " Our lead helo tried to calm him so we could make another run but Fox kept cutting in, pleading his case for life. Once again the airwaves were full of chatter as the beehive of activity over Fox searched for an answer.

My flight suit was soaked with sweat from our efforts and my collective developed a "light resistance to my pulling," telling me my copilot was keeping me from overtorquing the engine. I wondered what I would do without him and his support.

The radio confusion was overwhelming as we continued to orbit, waiting for a chance to get to Fox. Suddenly I felt a tremendous heat well up inside and I broke into the airwaves with absolute, total loss of control. I turned the airwaves blue with four-letter slang as I chided everyone on that channel to keep it quiet and let us do our job. No one could hear Fox and if we were going to get him, everyone had to "shut the f— up."

It was very quiet for the next few seconds and I called Fox to make contact. It took several tries until he came up, pleading once again for us to get him. I knew we had to get him to give directions or our case was lost. I told Fox to listen up and then I very carefully told him we were going to make one last run on him, and if he didn't give us the right directions we were going to have to leave him there. Then I told him if we did that he was going to die. I felt real bad saying something like that and I wondered how many pilots across Vietnam heard me being such an asshole. Fox came up again and said he would do everything he could to direct me, and his efforts to control his fear made me feel even more like a heel. I drifted to the south, getting ready to drop down and make my run. I checked with the crew and everyone was ready. I talked with them for a moment to clarify what we were going to do and they responded with an anxious enthusiasm that gave me courage to go on.

We started our drop, down to south of the pick-up area. When I asked for flares some far off voice told us they were on the way. My UH-1E lead was just behind to help suppress fire and we started weaving toward the pick up area. As the flares lit up little areas in the fog nothing looked the same as we swung low over the trees. Suddenly, Fox yelled out that he saw us and that we were just a little ways

USS Sanctuary *Hospital ship off coast of Cua Viet. (Quang Tri Prov) Note one CH-46 on pad with another coming in on short final (HMM-262). It was a "bad day." (Courtesy of Peter G. Sawczyn)*

away from him. I growled at Fox to call his position from us and he rapidly called directions, correcting himself as we slowed to turn, several times the wrong way. I was focusing on the tree line up ahead when I realized we almost hit a dike along the edge of a rice paddy as I rolled to Fox's directions. The moving flare light was causing shadows that were confusing to us. With our speed the trees looked like shadows, and the shadows looked like almost anything that could hide the enemy. Fox called another turn and I rolled too sharp, missing him again. I wanted to avoid the tree line where all the action was taking place, but I felt like Fox was trying to move us into that area. He radioed that we were within 100 yards as I turned back to my previous heading. The copilot kept reminding me of the RPMs as I rolled the UH-34 first one way, then another. Fox was getting very excited as I closed on his position. Suddenly my crewchief called that he saw Fox at our four o'clock and I rolled sharply right and yanked on the nose to slow down. The crewchief's call was excited now and we hoped for a miracle. My collective kept moving as the copilot helped me out.

Fox was yelling on the airwaves that we were there, and my crewchief yelled directions to a landing. We sat down a lot harder than I had planned and I looked all around for any kind of movement, friendly or not. My engine was whining at an RPM a lot higher than I was used to and I looked down at my gauges. I was doing everything wrong and I told the co-pilot I needed more help. There was no sound from him, just a gentle pull on the collective to keep the turns down. Suddenly the crewchief said, "We got him," and surprised, I yanked up hard on the collective to get out of the zone. After all the confusion and struggle I still had not seen Fox, even when he jumped onboard my bird. We quickly accelerated as I made a beeline to the south. There was more chatter on the radios as everyone wondered if we were OK and whether we had made the pick up. I no longer was mad, I just wanted to get out of there. We were looking for the friendly lights of Dong Ha when finally the ground troops cut in on the airwaves, frantically asking if we had found their man, Fox. I replied we had him and that we would drop him at Dong Ha at the medical battalion. We were out of the battle area, and now the relative peace and quiet made the previous efforts seem like a lifetime away. The engine finally was at a drone I could recognize and I slumped down in my seat, still feeling shame on my performance. Then, in the black of night, a call from the grunt came over the radio, and in words I still can hear, said, "We sure thank you for picking up our man, Highboy, you've got the biggest balls we've ever seen."

Sitting there in the dark, I was finally able to eke out the semblance of a smile, and I hoped and prayed the crew couldn't see the tears in my eyes.

A MOMENT TO REMEMBER
Gene Salter

At the end of September, 1967, HMM-363 had done their "time in the barrel" around Con Thien. We had run re-supply missions delivering water, ammo, and food supplies, as well as the replacements, to that huge bunker setting in the middle of a bulldozed strip that ran for 11 miles along the DMZ.

Re-supply and medevac up from Marble Mountain to Dong Ha. Pick up supplies and get clearance to "run the trace." The trace is that straight shot from Cam Lo to Con Thien. "The trace" is where the NVA had every artillery piece, mortar, and rifle pre-registered just to get a shot at me, coming or going. Sometimes coming and going.

Anyway, we had done our time in the barrel and were back in the Da Nang area full-time. We supplied and re-supplied all of the outposts that a truck couldn't get to (such as the lookout position atop Marble Mountain). We also did our medevacs and some inserts and extracts with the recon folks.

I guess it was the last part of October or the first part of November of 1967 when we were briefed on a recon mission "over west of Da Nang." Way west! Like in West Ashau. It was to be four UH-34s, each loaded with a recon team that carried only their weapons, ammo, and water. No packs. No rations. Just drop them in the designated landing zone and pick them up an hour later (unless they called for an emergency pickup).

Sounded simple enough to me until the briefing officer mentioned that the landing zone was an NVA concentration area.

It seemed that the NVA would come down different routes from North Vietnam through Cambodia, and muster in this one area "just west of Da Nang" for briefings, re-supply, and assignments.

"Someone" had decided that this NVA rendezvous should be broken prior to any devilment they might cook up. "They" had decided that the 105s of the artillery were to concentrate on that assembly area, then the Marine Air (fixed-wing type) were to make a few fast runs over the area, dropping whatever they could bring to the party, and then Gene Salter and the "Red Lions" of HMM-363 would immediately drop off the recon teams as the last of the fixed wings flew off into the morning mist.

Sounds like one hell of a party unless you're one of the Red Lions or part of the recon team. And you don't know how many NVA will hang around 'til the end of the party.

Everything started and we timed our take-off so that our reaching the landing zone would coincide with the last fixed-wing dropping his ordnance. We could see the smoke in the distance and arrived at the bombed out area just moments after the last explosion.

We landed next to some bomb craters so that the recon troops could exit the helicopters and immediately find cover prior to starting their sweep through the area. We received a few sniper rounds but saw no organized resistance. In fact, it appeared to be an abandoned site. Nevertheless, as soon as the last recon man had left the helicopter we relinquished our parking spot to whomsoever desired it.

We flew back to Marble Mountain, refueled, and waited for the call to extract the recon team from the NVA rendezvous area. On call, we returned to the area and began our approach to a landing zone at the opposite end from where we had inserted the recon troops.

As I touched down with the lead helicopter, I saw a canvas tarpaulin leap into the air. I thought we had landed in an ambush.

"Machine gun nest at three o'clock!" I yelled over the intercom.

My crewchief had seen the tarp and opened up with his machine gun. Then the recon troops grabbed the ground and opened up on the tarpaulin with their weapons.

Then we saw what the tarp had covered - a camouflaged food dump.

After the recon teams had destroyed the enemy food cache, we completed the recon extraction and returned the troops to their area. No troops were hit by the distant sniper and no helicopters were hit. The only calamity was the stained skivvies of one old Major.

REMEMBERING T.D.
John F. Norman, D.D.S "Lt. Hawg"

The fog was beginning to thicken, forcing us to hover ever closer to the side of the mountain. We must have looked like an angry palm tree in a hurricane as the rotors pulled the mist under us in great cloudy swirls and the scrubby shrubs and grasses rolled in concentric waves away from us. T.D. was at the controls and seemed to be thoroughly enjoying himself. As co-pilot my job was to worry (of which I was doing plenty) and follow along without actually touching the stick so I could grab it if he got hit. I expected this to happen at any second. T.D. had kicked out the crewchief, gunner, corpsman, both M-60 machine guns and any other loose gear back on the runway at Khe Sanh to lighten our load. That meant we were completely helpless if Charlie stepped out from behind a bush and opened up on us.

Somewhere above us a platoon of Marines had been trapped for a couple of days by the weather. The monsoon overcast had drifted down below the tops of the surrounding hills. No aircraft could reach them from above and the bad guys were waiting below. Of the squad that tried to break out the day before, two were dead and several more badly wounded. If someone didn't medevac those guys before nightfall, there would be at least two more unwelcome telegrams heading back to the world the next day. It's a strange and wonderful feeling for a 21-year-old second lieutenant to know what a difference he can make in the lives of some family back home by the action he takes in the next few minutes. Even though it wasn't my decision to make,

I was still proud and excited when T.D. agreed to try. In all the time I knew him, he never turned down a medevac no matter what the risk, and never gave up trying unless his plane was shot up so badly he couldn't continue. There is no doubt in my mind that a lot more names would be on that wall in D.C. if not for my friend T.D. Wilson.

The voice on the radio had given us the landmarks to look for. There would be a crashed UH-1E gun ship on our left and a crashed CH-46 on the right. Comforting thought! Concealed out there in the soup were NVA troops and a mortar team who already had the LZ pinpointed. There was no way to know if we had been sneaking in under the clouds at the base of the hill or if they guessed what we were up to, but I anticipated incoming rounds any time now. The roar of the engine and beating rotor blades had become deafening as we crept up the slope. You could tell we were getting close because the guy on the radio was screaming into the microphone to overcome the noise making it impossible to understand what he was trying to tell us. "Just calm down there, ole buddy, and keep your voice down," T.D. said. "We can hear you just fine." Every pilot in Vietnam wanted to sound perfectly calm on the radio no matter what was happening, and T.D. was a master at it. He had a gutteral drawl and sometimes tried to talk like John Wayne, but after awhile other guys were trying to talk like him.

I thought I could make out some dim figures moving about in the haze above and hoped they were our guys. There were all kinds of debris being swept up by our rotor wash and our blades took off the tops of some low bushes as we spotted the wrecks of the choppers which had come in before us. The grunts were trying to hold their helmets on and protect their eyes from the stinging sand. We were being directed further up to get as close as possible to the wounded. I kept waiting for the mortar rounds to come slamming in on top of us, but they never did. After what seemed an eternity we settled down on the uneven red earth.

Sitting in the left seat as I was, you didn't always get a good look at what was being loaded on your aircraft. Most of the time I would rather have not seen it anyway. This time, however, they brought one of the stretchers from a bunker up ahead of us and in their rush to get us out before we drew more fire in on them they stumbled and down they went. Among other injuries, the patient had had his left hand blown off and when they went down the stump was jammed into a muddy pothole. I couldn't see the other wounded but I didn't have to. Most medevacs have the sweet, salty smell of men who haven't bathed for weeks at a time. To know what the really bad ones smell like, you need only walk into a butcher shop. This bunch must have been pretty bad.

We had six men on board who should have been in the hospital 24 hours ago. Even without our crew, that was a lot of weight for this old bird at this altitude and humidity. The ground seemed to tremble and the aircraft screamed in protest as T.D. brought up the collective. Using every ounce of horsepower we could muster the wheels would barely clear the ground. As we settled back down, the radio was yelling that we were taking small arms fire. Quickly we had the grunts strip the wounded of helmets, packs, weapons and anything else they had thrown on the plane with them. The enemy couldn't see us so they were firing blindly at the noise. If they started in with rockets or mortars they could still ruin our afternoon.

Some men can drive an automobile, motorcycle, or an airplane and make it do things beyond the wildest dreams of the engineers that designed them. They seem to just become an integral part of the machine and can feel its limitations as well as they feel their own. T.D. was one of these men. He knew instinctively the strengths and weaknesses of his aircraft and could use them better than anyone I ever knew. He knew, as well all did, that it takes less power to turn a UH-34 to the right than to go straight or to the left. Somehow he seemed to milk just a little more power than everyone else. Today was no exception.

Once the wheels were off the ground he started a slow right-hand spin. I saw a few tracer rounds flashing by that told me Charlie was getting more accurate. There was a bank of red clay with sand bags and concertina wire on top that we had to get over just to get beyond all the surrounding obstacles. Each revolution we made gained us about 10 feet of altitude, but we would drop hard as soon as we started any forward motion and left that cushion of air beneath us. Finally, when we were high enough, he just dumped us over the side of the mountain into the fog in hopes that we would pick up flying speed before we hit the side. As we broke out of the overcast at about 100 feet my backside released the grip it had on the seat cushion. The "pucker factor" for this mission would have been an easy 100%.

By the time we got back to the runway at Khe Sahn we had used enough fuel to allow us to pick up the guns and crew. There was a nice long runway so we would have no trouble getting airborne now. Our guys were happy to see us and even happier to be getting out of Khe Sanh. Now all we had to do was climb up over the cloud cover without hitting any mountains, fly to Dong Ha, and then drop back down through the clouds without hitting any other aircraft. It's the same feeling you get driving on a dark night with no lights on and hoping no one else is dumb enough to be doing the same thing, at the same time, in the same place. There was little or no air control in that area. It all had to be done with dead-reckoning navigation; meaning, I reckon you better navigate right or you'll be dead!

The doctors and corpsmen were waiting for us on the landing pad when we arrived. Everyone of the wounded were still alive when we got there and that usually meant they would survive the war. Since we were the medevac aircraft for HMM-163 for that day, we would return to Quang Tri and await the next call. I hoped it wouldn't come before I could gulp down some C rations. I was starved.

This was only one story of many I could tell about the outright heroism of T.D. Wilson. While most of us were scared to death most of the time, he seemed to thrive on combat flying. The tougher it was, the better he liked it. He even took his old H-34 into North Vietnam one night to go after a downed Marine F-4 RIO that the "jolly greens" wouldn't touch. "Why don't you boys just go on home and to bed and I'll go get him," he said over the radio.

It's hard for me to think about the war without remembering T.D. The world was a better place with him in it. He was like the heroes in the movies who seemed to be able to do the impossible. The difference was that with him it was for real.

T.D. died of cancer about 10 years after we got home. He was a good man, and I'll never forget him.

REUNION

Ernest P. Sachs

Bill's letter arrived in a pile on my desk - handwritten in ballpoint on awkwardly-folded lined paper. He'd been a crewchief in the HMM-362 Ugly Angels, and through the haze of years I remember him as a skinny kid working all night to install a new engine before we went aboard the *Iwo Jima*. "Do you think it'd be OK to write to Lieutenant Carley's wife?" he asked. "I've been having a pretty hard time and I'd like to do something for her."

A flood of memories rushed over me, Mike Carley, the Connecticut Irishman from Brown University, had been killed near Nui Loc Son in the rainy gray of a February afternoon in 1967. Flying copilot for Jim Hippert, his H-34 had gone down in a muddy field beyond a VC position after suffering dozens of hits from small-arms fire; Hippert took a round in the leg. Mercifully, their wingman had managed to set down just to the left of the damaged bird. Despite the volume of fire, the crew managed to evacuate Hippert and the guns, but in the fog of war Mike, obviously dead, his body horrible blown apart, was left hanging in the straps.

Communication has strange powers, and stranger limitations. The crew of another downed 34 rode back in the wingman's belly with Hippert and the crewchief. Arriving at the Med Pad, one of the lieutenants asked about Carley.

Bill: "He's dead, sir. He's still in the plane."

Deak: "Sh—, we shouldn't have left him there."

And Bill spent much of the next 25 years feeling guilty that a dead Mike Carley had been left slumped in the seat of a dead Sikorsky.

I called the Brown Alumni Office, located Mike's widow, and learned he had a son who'd never known him. Connie's response was immediate, "Goodness, we have to do something for this man!" She and young Michael joined me and Deak in Cen-

Nine day rain at Marble Mountain, November 1968. (Courtesy of Michael Wolter)

tral California that fall to spend the weekend with Bill. We talked, remembered Mike's foghorn voice, and listened to a remarkable tape recording. Deak and his wife had corresponded to and from Ky Ha with three-inch reel-to-reel tapes, and he'd managed to find the one he sent home the day after Mike's death. The timber of his voice, a quarter-century younger, echoed through Bill's modest apartment as we retreated to our private memories of a lost friend.

Connie thanked Bill for sharing so personally his memories of Mike's last minutes; she'd never known how he died. Michael, 26 years old, met for the first time men who had flown, fought, and partied with his father. Bill finally came to believe that nobody blamed him for anything. And we returned to our homes with a greater understanding that the Marine Corps takes care of its own, even years after leaving the battlefield.

We later learned that a recon team, "duckbill", had recovered Mike's body about an hour after we evac'd the wounded crew.

ROCKET ATTACK
D. L. Stiegman, Col. USMC (Ret)

In the spring of 1967, HMM-363 was deployed to Dong Ha - the first squadron there on a full time basis to support Marine operations around the DMZ of the northern boundary of Vietnam.

There wasn't a hell of a lot up there when we arrived enmasse - a couple of hootches the group had used for Hq/Ops/Comm forward, a couple of hootches designated "Officers Country" and a large open area along side the very short runway which we used to disperse our Lucky Red Lion UH-34 helicopters. The troops of 363 set about erecting a compact tent city for their billeting.

Bunkers were nonexistent except for a half hearted attempt someone had started in the hootch area, but never completed. That first night, we all wondered why.

Shortly after midnight I came bolt upright in the rack, looked out through the side of the hootch and watched the shitter being blown away in a hail of red sparks and fire as the first round landed in our midst.

By the time the second or third shell detonated, we had all made our way to the "sort of" bunker, each man trying to get deepest in the shallow hole. For the next hour, we endured an artillery, mortar and rocket attack ... kind of a welcome aboard 363, from your neighbors, the VC.

It was quiet in the bunker (except for the crash-bang going on around us), each man with his own thoughts, praying the big one wouldn't land in our laps. Me and God became real close pals that night.

Our losses were minimal; no lives, a few shrapnel wounds, no aircraft. However, our resolve was intense. The next day Marines of HMM-363 filled 18,000 sand bags and constructed substantial bunkers all through the squadron area.

SNAKE IN THE COCKPIT!
Huey C. Walsh

During our tour in Vietnam we all had numerous events that live with us. One of mine was very unusual as it involved a snake in the cockpit! Duane Jensen and I were flying to Khe Sanh, to stand SAR North, about late 1966, when Duane jumped out of his seat and ran to the rear yelling, "Snake! Snake!" Of course I was about to jump out the window when he stuck his head back in the cockpit and pointed to a snake on his rudder pedals.

He got his K-Bar and struck at it and it fell to the floor and disappeared. When we landed in Khe Sanh, we all tried to find the sucker, to no avail. Thankfully we weren't dispatched on SAR North that day, as we would have been more nervous than ever.

On the flight back to Marble Mountain that evening we were very uncomfortable. Approaching Da Nang, the snake stuck his head out of the instrument panel near Duane. He immediately unstrapped and ran to the rear again. He leaned in the cockpit and struck at it with his knife and it disappeared into the instrument panel. I knew that it was headed my way so I yelled for Duane to get back in his seat.

A few minutes later, there he was on my side! And I unstrapped and ran out! I hit at it with my knife and he disappeared again into the instrument panel. For the remainder of the flight we were very nervous as we didn't know which side he would come out next!

When we landed at Marble we downed the aircraft for "Snake in Cockpit!" After many hours of effort by the crew, which included taking up the floorboard, the snake was found. Thank goodness it was a bamboo viper, a "10 stepper," or we would have been embarrassed had it been non-poisonous. The aircraft had been used the previous day to haul ARVNs and their junk around and the snake had been left, either intentionally or unintentionally, who knows?

SOMETIMES THE NAVY IS ALL RIGHT!
R. M. Nebel, "The Blue Max"

It must have been later summer or early fall of 1966 while I served, very proudly by the way, with HMM-363, the Red Lions. We had been operating as the helicopter side of the SLF working off the good old USS *Princeton,* a real, though converted, LPH as opposed to the newer but ill-conceived, single screw, round-bottomed, "helicopters don't need wind across the deck" waste of limited Defense Department money. So much for political comment, but you've got to admit, the old trusty four-screw, straight-deck, wood-plank conversions beat the living hell out of the Iwo class. This tale makes my point. Not only in ship's capabilities but in crew's attitude.

I'd led a flight of four UH-34s across the beach to Dong Ha in support of either the end of Hastings or, as it became called, Prairie. We worked runs to Khe Sanh and the Rockpile resupplying and medevacing or whatever was required. As you recall, not an unusual "schedule" for a day's flying. On toward evening, we stopped to top off at the pits at Dong Ha. A fairly heavy fog bank was rolling in from the South China Sea and it became questionable whether to fly out or RON Dong Ha. My second section leader opted to remain ashore while I, a Captain with some smarts, thought that clean sheets, a hot shower, and dinner in the ward room was much more attractive than a canvas cot, tepid water, and "Cs." With the best wishes to my section leader and his motley crew, I took my two-plane section airborne and climbed above the encroaching fog and darkness. Smart? You bet! H.S. smart!

Oh, yes. No sooner were we clear of the airfield when (as we learned in our basic weather course) the temperature dropped a degree or two, as it seems to do almost every evening as the sun sets, and just happened to coincide with the dew point! Do we remember what that causes? Well, of course - fog! Seeing the fog rolling in might have been a good indicator to the expected weather phenomenon had I not been quite so H.S. smart. So, here we are, above a fog bank that had just closed the field and extends for as far as the eye could see given 1,500 feet and approaching darkness. I guess the one good note was having just refueled. Options? Try a no-facilities instrument approach back to Dong Ha? In section or single plane? And leave my wingman without any guidance? Nah. Look for the ship! Sure. It's buried in the fog too and the top of the blanket is about 500 feet. Well, we could make an instrument approach to the ship but it was zip zip, zero zero, nada baby, and not recommended. OK. Next. We can head down the coast, or where the coast would be if we could see it, to Phu Bai. It ain't clean sheets and a late movie but it is relatively safe. And, if they get socked in, as it looked with the weather moving southward, we still had enough fuel to make Marble and they, at least, had GCA.

As we headed southwest along the imagined coast, clear of and parallel to the mountain tops to our right, all of a sudden a darkening area is noted to port. Turning seaward toward the spot, what should we behold but a jolly St. Nick and his eight tiny reindeer? No! But, we did spot a hole in the fog. Not some rinky-dinky little break in the clouds but a hole a good quarter mile wide, like someone had just made a doughnut hole over the sea. I guess it was only a mile or two off shore but it was far enough that I felt encouraged to notify the *Princeton* of what we had discovered. Had it been the *Iwo*, it wouldn't have been worth the call 'cause one, she wouldn't be that close to shore and two, she would respond to a request from one of her flock with disdain. But, luck had it that we were operating with the *Prince(ton)* and she was a willing and gracious hostess to two wandering birds wanting to come to roost.

They immediately notified me that the skipper had not only approved but encouraged a positive response to my request and they asked for steers to the opening. Between their radar and our steering on the reciprocal DF, it was about 15 or 20 minutes when we saw the bow of that mighty warrior breaking free of the foggy shroud and its deck gleamed in the final rays of the sun painting rainbow-colored diamonds from stem to stern. Our approaches were normal, landings as smooth as a baby's ass, and thanks were passed to all concerned as we shut'em down and headed for the showers. Sometimes, the Navy IS all right.

SPECIAL MEMORY
Martin L. Poleski

"Helicopter flying overhead, do you copy?" was the transmission we received when flying over a small base camp while enroute to Marble Mountain. It was twilight and we were returning from a resupply run to Hill 881, or was it Hill 882?

Our pilot replied to the affirmative. The ground radioman informed us that they had a "grunt" whose father was gravely ill and needed a ride to Da Nang for an emergency leave flight back home. Could we pick him up?

Our pilot replied to the affirmative. As we spiraled down I heard that familiar "pop . . . pop" and keyed the mike to tell the pilot we were taking fire; he was already adding power. Back up to altitude, our pilot radioed to ground for a report on the ground fire. There was a slight delay. The radioman replied that the night perimeter guard was just coming on duty and was test firing their weapons.

Our second descent was completed to the accompaniment of several more "pops." We picked up the grunt and as we took off the perimeter laid down a field of fire.

Back at Marble Mountain I informed our passenger that he could sack in our hootch for the night, then we would arrange transportation to Da Nang at first light. In the meantime he was welcome to join us for some chow and a few beers at the E.M. Club.

After a few beers I told the "grunt" that I thought for sure that we were taking fire when we came down to pick him up. He said we were. After our first descent was met with sniper fire and the pilot asked what was going on, the radioman turned to the "grunt" and said, "How bad do you want to go home?" The "grunt" said, "I want to go home now." The radioman answered the pilot's inquiry with the story about perimeter testing their weapons.

At least they did give us good cover fire on our departure.

QUE SON MOUNTAINS
F. William Valentine

I have many vivid memories of the war, and 24 years later they still remain among the high and low points of my life. The most enduring memory occurred on November. 16, 1970, when Lieutenant Colonel William Leftwich and 13 Marines were killed in an aircraft accident in the Que Son Mountains of Vietnam.

It wasn't an ordinary accident, to be sure. The accident occurred during an emergency extraction of the seven-man reconnaissance team "Rush Act." The team was stranded by the monsoon season at the 3,000 foot level of the rugged Que Son Mountains. Low clouds prevented numerous attempts to extract them and extended the team a week beyond the scheduled pick-up time. One of the team members was injured which hampered a descent down the mountain below the cloud layer. Their location was in an area controlled by the North Vietnamese army. Their only link was their radio, the CH-46 helicopters and their protecting Cobra gunships that had been trying to pick them up and the tactical aircraft that would provide close air support if they became surrounded. The team was out of food and water and was totally exhausted.

Finally we got a sudden break in the clouds. The extract CH-46 with Colonel Leftwich, the 1st Recon Battalion Commander on board, lowered a SPIE extraction device through the jungle canopy. The team hooked up and the aircraft started pulling the team out. We never really knew what happened at that point. The clouds rolled back in as the extraction was in progress. The monsoon rains began again in earnest. They might have taken fire. We only know that we found them all the next day. Fourteen shipmates, dead on the mountainside.

It is the most enduring memory for a number of reasons. It symbolizes the close relationship between the Marines in the air and those on the ground. Even the worst of times the helicopter flight crews never gave up the faith. They would push the envelope beyond what was expected and do whatever was necessary to help their colleagues on the ground.

More significantly, this tragedy reminds me of what our nation lost. I was with recon at the time and it was my squadron performing the extract. I lost 14 friends that day. But the Marine Corps may have lost its greatest leader. We now pay tribute to Bill Leftwich with the Leftwich Award, the annual leadership award in the USMC. A statue of Leftwich remains the only one at the Basic School. He was exceptional not just for his bravery or for the things he accomplished but for his leadership example, his quiet dignity, honesty, and selflessness. He symbolized what America lost, the best of what we had to offer.

"METRIC SNEAK" EXTRACT
D. M. Petteys, Captain

On February 23, 1967 at 2130, the flight was launched on an emergency extraction of a Special Forces Reconnaissance team, "Metric Sneak," which was surrounded and under fire at coordinates YD314086. This was our second attempt. The first attempt was aborted due to an unacceptable landing zone, heavy automatic weapons fire, and bingo fuel.

The TACAN position was 265°/32NM Ch 69. I was seventh in the flight of nine. As we lifted off Hwy. 267 at Hue, we had to come up through a broken layer of clouds at 600 feet altitude! And until we were within five miles of the A Shau Valley, the flight was VFR on top of an overcast. It was also nearly a full moon, which helped considerably.

There was a large coordination problem. With nine CH-46 and 2 UH-1Es (Oakgate 16-1, 16-2) the sky was full of rotating beacons. It was difficult to tell which one was your particular flight leader. The air waves were filled with many pleas for a blink of the rotating beams for identification. At about 25 miles out, just short of the last ridge line, the flight set up a right-hand orbit. The positions in the flight were finally ironed out when Major Sellars (Rosanne 1-1) instructed the Bonnie Sue to anchor at 5,500 while the other birds anchored at 5,000.

Meanwhile, at the LZ, the TACA (Hillsboro on 44.6) was coordinating four flights of fixed-wing fighters and Spooky II, who kept a constant string of flares illuminating the area. The FAC was Covey 73, although I could not differentiate his function from Hillsboro.

After the final flight of fixed-wing expended their napalm, the Bonnie Sue departed the right-hand orbit and headed in toward the LZ. We were to attempt an extraction. If we were unsuccessful the other birds, loaded with troops, were to make a reinforcement insertion of 80 or so Montagnards with advisors. Bonnie Sue 30-3 was leading, with 30-2 and 30-4 following. The flight leader asked Metric Sneak if he could hold out until morning. He said no, he was desperate and had to get out of there.

The zone was a single-aircraft zone. Orbiting over the zone became quite confusing. There were the Bonnie Sues, the TACAs and the Oakgate guns, all trying to orbit around the same point not to mention the Spooky flares that were dropping down through the orbital pattern. I don't know what I would have done without the copilot's (Lieutenant Roy D. Hammock) outstanding lookout doctrine, and the gunner and crewchief calling out aircraft that were near us.

The 30-3 was the first to go into the zone. Things had gotten so confused in the orbiting that I wasn't aware of it until he made the pickup and was on his climbout. The 30-3 told me that 17 troops had stampeded aboard his aircraft and that it was a "bad zone, pardner." Therefore, I came up on the air and told Metric Sneak I would take 12 troops. He answered that the troops were in a state of panic and he couldn't control their boarding of the helicopters. Therefore, I told Corporal Leonard, the crewchief, to tell me when we had 12. I planned to lift when 12 men were aboard.

It was my turn to roll in. I called, "30-2 rolling in" and commenced my letdown. The zone was around 2,000' elevation and consisted of a small clearing in the trees.

The zone was situated on the crest of a ridge line, near the western end of the ridge. I came into the zone on a northern heading. The advisor on the ground had a blessed strobe light and Spooky's flares were exactly overhead lighting the area superbly. Therefore, I had everything I needed for a successful approach ... the zone was pinpointed by the strobe and the terrain by Spooky's flares. The approach worked out just right. Later, the ground forces told me they could hear the enemy firing on us from the hillside below their position.

As I touched down in the zone, everybody started talking on the radios at once. As I couldn't hear the crewchief at all, I instructed the copilot to turn off my mixbox switches, which he did. We couldn't put all three landing gear on the deck, only the starboard rear. Corporal Leonard told me that we were striking trees with the rotorblades on the left. But from my side I could see that we were chewing up shrubbery on the right, so I saw no choice but to hold what we had. My aircraft had no #2 boost pump, and a hydraulic leak. But due to the emergency classification of the mission, I felt the risk was justified.

We had been hovering for a minute or so. I asked Corporal Leonard how many he had. He said none. So I instructed him to see what he could do to board the troops quickly. After another minute (it seemed like hours) he said we had eight aboard and one KIA. I knew we could take 12, so I waited longer. Then Corporal Leonard said, "There's no more coming, sir! Let's go!" So I lifted out of the zone. I could hear popping. I couldn't tell if it was our blades hitting trees or automatic weapons. At any rate, I instructed the crew to lay out .50 caliber fire as we lifted out of the Zone. I attribute the fact

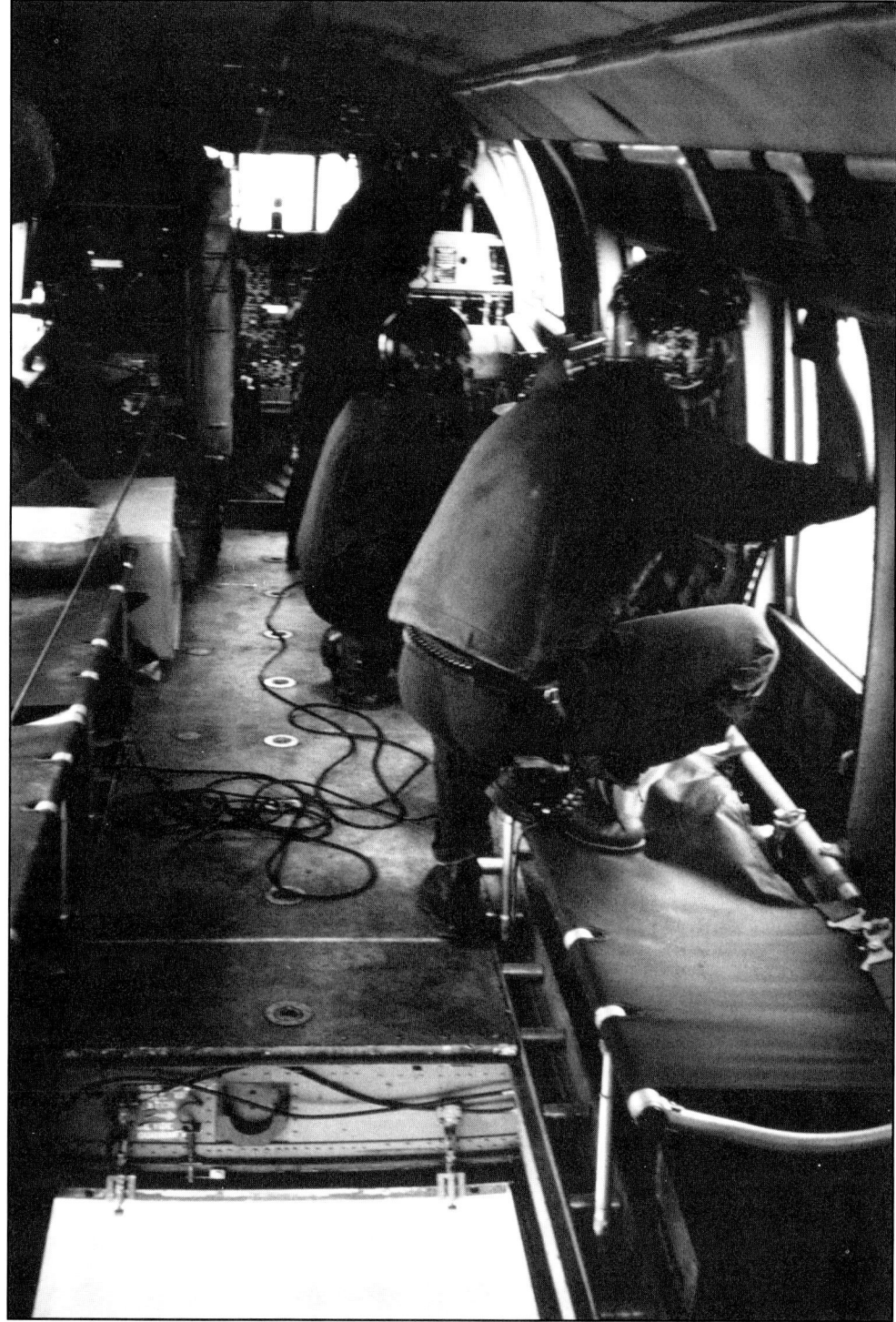

Medevac chopper crewchief and .50 cal. gunners looking for the Hot L.Z. and muzzle flashes. The "hell hole" is open for a hoist pick up, 1969. (Courtesy of J.J. Houser)

we took no hits to the outstanding suppressive fire delivered by my crew.

We were about 40 knots and the airplane had a terrific beat in the rotors. We had dropped to 90% NR (rotor blade rotation in RPM%). The copilot warned, "There's a UH-1E making a run!" I had to roll in 45° angle of bank and narrowly missed the stream of tracers that we would have passed through. This aggravated the flight problems I was having. I had to get more airspeed and NR, so we started descending into the A Shau Valley on a northwesterly heading. I began thinking about what we'd have to do if we would have had to set it down on the valley floor.

However, I did not want to give up the airplane without a fight. I found that at 70 knots, 92% NR and a little right rudder, we established a rate of climb of about 800 fpm. I circled and did some lazy eights trying to gain enough altitude to clear the ridge that separated us from Phu Bai. All the while, the airplane was shaking and vibrating and I had to fight the panic that kept trying to rise in my breast.

I needed an escort. We were headed for Phu Bai and I didn't want to turn back to all the orbiting and confusion to find my flight when I wasn't sure I'd even make it to Phu Bai. So, I called for any aircraft to escort me back to Phu Bai, but I received no answer. I also called my flight leader 30-3 to ask him his position, but again received no answer.

So I had to head out on my own. I decided to call Landshark Charlie and tell them I was inbound to Phu Bai, but they would not answer. I could hear them talking to Rosanne 1-1. I heard conversation mention Rosanne flight going to 42.8. I switched to that frequency, but could get no one to talk to me.

About this time, I notice that our TACAN was inoperative, both azimuth and DME. As we were VFR on top of an overcast, and down to 40 minutes of fuel, I became quite concerned. I immediately went to 38.2 and asked Rosanne Bravo what the weather was. They said the field was still clear. I called Hue tower and asked them for a short count for a VHF ADF. I checked the #1 needle switch to ARA-25 and went to the ADF position. But it didn't function.

I could see the ground in patches. I could see long strips of sand and long lakes. For a moment I thought we might be over the finger lakes region. At this point I became completely disoriented. I was about to start driving in triangles and squawk emergency IFF, when I looked at our 3 o'clock low and saw the runway lights of Phu Bai right at the edge of a break in the overcast. I called the tower to tell them I had the field in sight, but I had no side tone and the tower wasn't reading me. I could hear them calling me on the guard channel. So I switched to Guard myself and told them I was safe on the deck at Phu Bai. We rolled into the parking area and shut down at 2300.

The Smoothest Night Medevac
James L. Shanahan

The smoothest night medevac I ever flew, and one with a surprise ending, started about 0200 in mid-January 1967 at Marble Mountain. The phone went off with the familiar three-ring "rruup, ruuuup, rruuup," signalling a medevac mission. I got the briefing for an emergency (translates, probably fatal if not evacuated) US medevac with severe head wound from an outpost protecting a bridge over the river east of An Hoa. This was monsoon season and the weather was cold with drizzling rain and a ceiling varying from 100 to 150 feet, not exactly field grade night flying conditions. Consoling ourselves with that part of the briefing that said no enemy activity in the vicinity of the LZ, we fired up the UH-34, took off, checked in with the DASC, and orbited a bit SW of the Da Nang runway while I raised the grunt outpost on the radio to chat about conducting the mission. Their version of enemy activity was a lot different from what was in the briefing ... they were under attack by fire from small arms, machine guns and mortars; they couldn't illuminate the LZ due to incoming fire, and the wounded Marine was fading fast. Our walk-in-the-park mission had become anything but that. We were lucky that the grunt radio operator and whoever was in charge of the outpost were two real sharp folks.

In a few minutes we worked out conducting the mission thusly:

In view of the enemy fire we had to get in, get the medevac aboard, and get out ASAP. We would come in from the north and since they couldn't light up the whole LZ they would mark where the helo's right main landing gear should go. They would do this by piling sandbags around a lit Coleman lantern so that it could be seen only from the north. They would put a field jacket in front of the lantern so it couldn't be seen at all until we called for them to remove it. When we did, some brave soul would move the field jacket back and forth (making the light blink) until we called saying we had it in sight. They would have the wounded Marine on his stretcher as close to the LZ as possible so he could be put aboard the helo ASAP and we could depart, with minimum time on the ground.

We had the bearing and distance of the outpost from Da Nang Tacan so we started the eight to 10 mile run to it and began taking occasional fire as soon as we crossed the river south of Da Nang. Since the terrain was relatively flat all the

SSgt. Hall and Cpl. Goddard checking out engine on H-34. (Courtesy of Richard H. Hall)

way, we climbed into the cloud cover a bit, turned out all the exterior lights, and using the radar altimeter to keep out of trouble, continued until we were a mile or so north of the outpost. We dropped down clear of the cloud, called for the light, saw it blinking, called again, went straight in and put the right gear next to the Coleman lantern. We were no sooner on the deck than the stretcher bearers had the medevac at the door of the helo and our crew had him aboard in less time than it takes to read about it. We were off the ground and climbing out in less than five seconds.

We took the casualty who had head wounds to a medical facility called Bravo Med. About half way there there was a commotion in the belly of the helo. I called the crew chief to see what was going on and got a "wait one" reply. Shortly thereafter the corpsman came on intercom and explained that the medevac had leaped off the stretcher and started thanking everybody in sight for getting him off that outpost and on the way home. It seems he had been in RVN only four or five months and already had two Purple Hearts. Earlier in the evening he had gotten about a five-inch laceration in the scalp on top of his head from a mortar fragment. This was a fairly deep laceration and since the scalp bleeds badly from even minor cuts this laceration looked near fatal to the outpost corpsman who was having his first night in the field in RVN. Since three Purple Hearts meant going back to the States, the wounded Marine had no intention of correcting this mistaken diagnosis. So he lay still as death on the stretcher until he was in the air in the helo half way to Bravo Med and could no longer contain himself. We all had a good laugh about it as we neared Bravo Med and called them for landing. When we did land they had their normal lifesaving crew out to meet the helicopter ... doctors, nurses, corpsmen, a dolly for the stretcher and assorted life-sustaining equipment. These marvelous people had saved the lives of dozens, hundreds, even thousands of medevacs but I don't think they were prepared for this chap - he leaped out of the helo, his head drenched in blood from a minor wound, dancing around waving his arms in the air and embracing everybody who couldn't get away from him. Eventually they got him inside out of the rain.

I went over to see him later the next day but he wasn't there. They had fixed him up and transported him to Da Nang to catch a medevac flight out of RVN. I hope he was as lucky, and as happy, the rest of his life as he was the night we medevaced him off that outpost.

WAR STORY
Walter E. Pinkerton, Jr.

My contribution in this category is a verse I wrote after leaving the Marine Corps and reflective of a night medevac mission in which I participated. It goes with the melody from *Ghost Riders in The Sky* and I call it *Ghost Riders in My Eyes*.

It was a hot, dark and clammy night, some resting in their beds
Tonight it's night medevac so some kid won't lie dead.
The radio it crackled loud, I jumped up fast half stunned
The call, emergency medevac and I prayed we'd see the sun.

We launched quickly from our pads, the messages coming fast;
The "pos" pushed up by shackled code, we plotted as we speed
We got our briefing on the go at fifteen hundred feet,
With Hercules lighting up the skies, preventing their retreat;

The Cobras they were following began to make their runs,
Rockets launching with red glare to muffle Charlie's guns;
And they made a let-down pass, when a quiet voice, it said,
Change in classification boys, permanent routine now instead.

Our heartbeats then quieted as we returned to our beds,
Back listening to the tunes, like the Grateful Dead;
Blood, Sweat and Tears screams aloud, another message blurred,
This time asking a question, war, what good is it for?

Yippee Eye Yay, Yippee Eye Oh, Ghost Riders in My Eyes.

WAR STORIES/ REMEMBRANCES
Q. R. "Goose" Meland

Anyone who flew missions in that stink-hole called Vietnam can recall, sometimes more accurately and vividly than one might want, all too many memorable scenes and poignant events that describe "His Tour." To pick out one or two is not necessarily difficult. What is difficult is to choose a category: Humorous, Technically Difficult, Horrifying, Frustrating or, all of the above.

I suppose the two remembrances I do wish to share fall into the horrifying category. Both took place somewhere south of Ky Ha or Chu Lai, it doesn't really matter where since one place is pretty much the same as another. Our sister squadron summed it up by the motto painted on the tail of their H-34: "Who Gives a S—." The reason the events are important to me is that they finally gave me a reason to press on; to risk not coming home to my young wife and two little girls.

As I recall, both stories began while on medevac duty. Each call came in as an "emergency medevac," the most urgent. Upon arriving at the first pickup point we were informed that our cargo was to be a number of Vietnamese children who had been machine-gunned by the Viet Cong as they were returning from school. After loading the precious little ones in our UH-34 we sped back to the field hospital. I think one girl survived.

The second medevac mission was to pick up several Marines who were badly wounded during a fire fight. As one was being loaded into our H-34 I looked out my window at him. I will never forget the captain looking up at me through tear-stained eyes and mouthing, "Thank you, Thank you." I don't know whether he lived.

Counting the days remaining before I rotated home never stopped but I knew from that day forward why I was there. I was there to help the helpless and my fellow Marines.

A LONG, LONG YEAR
R.W. McAmis

This is no sh—! I arrived at HMM-262 in Phu Bai, November 1969, during the monsoons when cold rain and heavy grey clouds seemed to go on forever. Five days later my first mission was as co-pilot on afternoon chase medevac at Marble Mountain. The weather was so bad that our relief could not launch so we were assigned night medevac as well. I couldn't even start the aircraft without a checklist. I had never seen an armor-plated seat or worn body armor. We launched several times throughout the night. I was totally lost, sure I was about to buy the farm, and of no help to the pilot, Lieutenant C.K. Doi. But that is another story.

A few days later I was assigned as the SDO. At around 1800 Sergeant Major Duke approached me with a LCpl, I'll call "Jones," in tow, and informed me that threats had been made against this young Marine. I was to protect him until he could be shipped south in the morning.

Jones was very quiet and later in the evening asked to stay in the same room where I was working on admin tasks. He sat on the floor not talking. At 2200 hours I answered the field phone in the ready room next to where I was working. When I returned to the room he had a hammer in both hands and was using the claw to beat in the top of his head. Blood and pieces of scalp were splattering all over the place. The sound was that of ripe watermelon being thumped every time he hit his head.

I screamed, "Get a corpsman!" to the two enlisted men playing ping-pong in the ready room. When I yelled, Jones put down the hammer and I grabbed him by the upper arms. He was bleeding like a stuck pig, covering himself and me in blood. He looked at me with a vacant stare and said, "What did you do it to me for?" My first thought was, "Oh sh—! Nobody saw this but me and I'm brand new."

I managed to get him to lie down in the ready room. We were attempting to apply a dressing when he became more agitated, jumped up, ran out of the ready room and down the line to the hangar where the nightcrew was working on aircraft. He ran into the hangar screaming, "Help me, help me, they're trying to kill me." He looked like someone had almost succeeded, and close behind him was this lieutenant who no one knew, out of breath, and covered in blood from his elbows to his fingers.

I instructed the nightcrew to watch him, then I dashed back to the ready room and called the MPs. Gasping for breath, I called the O Club to inform the CO, who had just taken over command a couple of days before, then ran back to the hangar. The MPs and the ambulance arrived at the same time. Jones made a break for the door but the men on the nightcrew grabbed him and he was strapped to a stretcher and carried out screaming incoherently. Someone was trying to kill him, and it was him.

By midnight I had cleaned up all the blood and taken statements and written my own. I had just finished when a guard came in and said, "Sir, we've got a man out here in the jeep that just took a whole bottle of aspirin and washed it down with plexiglass cleaner." I realized it was going to be a long, long year and I had been there only a couple of weeks.

Welcome to Vietnam
Bill "The Chicken Man" Ring

I guess I could write about getting "shot down" or "shot up" or saving peoples lives or other acts of bravery, exhibition of flying skills, or shear adrenalin pumping in a young man's body. But I'm not, I'll save those for others to tell.

No, I am going to tell you how a young naive 21 year old green copilot fresh from the States got an official welcome to Vietnam, Phu Bai style. And it wasn't from Hanoi Hanna.

There I was minding my own business, having a beer or two at the Phu Bai O Club. Naturally I had hung my trusty .45 up on the hook, so I was unarmed and definitely unaware of what was about to take place.

Hoping to hear and learn from the "older" more experienced pilots congregating around, my ears were like giant radar stations, trying to hear and learn whatever I could but not wanting to be too obvious.

As I sat back enjoying a cold beer with several of my fellow squadron "new guys," I began to notice the club filling up with plenty of "seasoned" pilots.

Now I enjoy an occasional bet or a chance to throw the bones, but what I began to hear was amazing. There was definitely an argument brewing and big time wagering taking place, over what, I wasn't quite sure, but one of the "old guys" started to ask us "new guys" how much we weighed.

Apparently bets were being made as whether or not a team of two "old guys" could pick up several of us "new guys" (huddled tightly together) off the floor. Sounded amazing, but who was I to argue when asked by an "old guy" to participate.

As I huddled on the floor holding on to my fellow "new guy" buddy, holding on to his "new guy" buddy, etc., I didn't quite notice all the pitchers of beer in the hands of all the old guys surrounding us.

As the betting got more intense and cheering grew louder, I actually thought that these two guys were going to be able to pick us all up. I was even thinking about betting some of my own money on them.

But, to my surprise, as I was clinging for all I was worth and the countdown hit three, I've never seen so much beer coming down on me in my life - along with a cheer from the "old guys," "WELCOME TO VIETNAM."

"Does this mean I'm officially here?"

Monsoon
Gary Cook

How do you suppose we got this way
our childhood washed
by Monsoon rain that wept grey
adrenaline as the helicopter bellowed
engine strained like our sanity
caught between reason and panic.

Obsolete warriors running a gauntlet
of eyes almond in surprise
we fled through slate-grey mist
twenty feet above dry riverbed,
down an alley in the jungle
rank and humid as our fear.

Not enough oxygen or muscle
to lift us from that moss-green fold
where bodies thrown soddenly down
crept white and puckered as color
rushed their eyes. Whump of rotor blades
left only pastels. We grew old clawing to
escape the stench of graves.

It didn't matter; we ate well.
Our dreams edible. Nourishment
at seventy miles per hour
twenty feet above the trail.

"Look, Up In The Sky. It's A Bird, It's A Plane, It's.........
Roger Herman

It was 1965 and I was in flight training as a Marine helicopter pilot in Pensacola, Florida. When I began training in May of that year, I think that I might have heard the name Vietnam mentioned once or twice. I hadn't really paid much attention to it. Fifteen months later, when I had completed training, I was on my way there. The war had escalated significantly during this period. As a result, helicopter pilots were needed "in-country" as soon as possible. In order to get us over there in the shortest amount of time after we got our wings, some of the additional training that was consid-

J.J. DeHaan

ered optional was deleted from the schedule. In my case, I never got to go on a stateside training mission in which the crewchief and gunner fired their M-60 machine guns from the belly of the H-34 helicopter that I was being trained in. This would prove to be significant.

I joined HMM-361 in January 1967 as the squadron was returning to Vietnam from a short stay in Okinawa. We were based out of the Marine helicopter facility called Marble Mountain. It was located on the beach just east of Da Nang. We immediately began flying medevac, re-supply, and troop insert/extract missions. Our main operations were to the south and west of Da Nang. After a couple of weeks of flying, I had a pretty good feel for the job and also felt comfortable with knowing the area. I had been lucky so far, and had not been shot at yet. The same couldn't be said for a lot of my fellow squadron pilots. They were coming back from the same types of missions with lots of bullet holes in their aircraft, and some wild tales to tell. A lot of the guys were getting wounded also. A couple of more weeks went by and my good fortune continued. Not a shot was fired in anger at me. Days turned into weeks, I had now been in-country nearly two months and still had not been shot at. Everyone else was continuing to take fire on various flights. Now I was really starting to worry. This wasn't how it was supposed to be. I was being kidded in the squadron about my "virginity." I just knew in the back of my mind, that the odds had to catch up with me some time soon. Everyone was saying that I was way overdue, and that when I took that first enemy fire, the "bad guys" were really going to lay it on me to make up for lost time. Now I was to the point where I just wanted to get shot at the first time so I could get it over with.

It was an overcast day and we were returning to Marble Mountain after another mission that went pretty smoothly. Normally we would fly above 1,500 feet to stay out of small arms range. This day, however, due to the bad weather, we had to fly low. We were only several hundred feet above the ground as we navigated north, paralleling Highway 1. We were about 10 miles south of the Da Nang area when it happened. Instantly my world, as I had known it, ended forever. From out of nowhere the deafening sounds of automatic weapons fire erupted all around me. They were the loudest noises I had ever heard. I came several inches out of my seat in total fear. I'm sure that I must have been yelling as loud as I could, but that it was being drowned out by the machine gun fire. I looked up to see that the windshield was covered with blood. Blood and what appeared to be body parts were everywhere. The additional sounds of emergency radio calls filled the earphones in my helmet. It was total chaos. I must

Gerald W. Johnson, Vietnam, 1965-66.

have been hit pretty badly. I can remember thinking to myself, "So this is what it's like to die. Oddly enough, I'm not in pain." Several seconds later, but what seemed like an eternity, the situation began to settle down. I was still alive! So was the rest of the crew. After checking myself over, I realized that I wasn't even hit. In fact no one was wounded! How could this be? What had happened? How could we have survived this intact? The answers started coming slowly as things continued to calm down and I started putting it all together.

The extremely loud machine gun fire I had heard was our own. The crewchief and gunner had opened up simultaneously with their M-60's on enemy positions that they had spotted firing on us. They hadn't forewarned us of it because there just wasn't time. Having never heard our machine guns fire before, and expectantly waiting two months for what would be my first exposure to enemy fire, I had thought the loud automatic weapons noises were the VC guns shooting at us. Soon thereafter I would learn that you can usually smell the enemy fire first as it penetrates and burns the magnesium skin of the helicopter. You hardly ever hear it. There was no time to analyze the situation at that moment, however, and surmise that realistically I wouldn't be able to hear the VC weapons firing from such a distance. And of course, how could you hear anything over the loud noise of our reciprocating 1820 engines anyway? So much for that portion of the mystery. But what about all that blood....... and those body parts? Well, as luck would have it, at the very moment our M-60's started firing we had a direct, head-on, mid-air collision with a couple of seagulls! They were splattered all over the front of the windshield. Yes, there was lots of blood and internal body parts everywhere, but on closer examination much of it was in the form of crumpled wings and feathers as well.

I had learned a lot of lessons in a very short time on this one flight. We had taken several hits, but no serious damage had been done, except to my nerves. Our body count for the mission was two confirmed VC seagulls. Welcome to the war. The remainder of my tour consisted of many exciting adventures in the H-34. It was a durable helicopter that handled combat missions and bird strikes in stride. I've flown a lot of different aircraft in the years since, but none comes close to my affection for the Sikorsky UH-34D. Now that was a helicopter. There was never another mission quite like this one either. I haven't looked at a seagull quite the same since 1967. Come to think of it, I'm not too fond of sudden loud noises either. But every time I catch a glimpse of a '34........... well that's another story.

Tarbush Night Medevac
Rod Carlson

The flight schedule was posted a day in advance, so I'd known for 24 hours that I had drawn my first night medevac mission. During that time, the closest I had come to being able to relax was a long run on the sand up to China Beach. With everyone else flying day missions, the hootch was deserted, but I couldn't sleep. I tried writing a letter, but gave up when I felt like William Holden writing his final words to Grace Kelly in "The Bridges at Toko Ri."

At 1600, I put on a fresh flight suit and headed for an early dinner at the Marble Mountain Officers' Club before flagging down the HMM-361 driver for a ride around the perimeter road to the squadron area on the other side of the runway. As we passed the north end of the runway, a CH-46A flew twenty feet overhead. At the boundary fence, it banked to the left and headed inland toward Da Nang. Its rotor wash rocked the truck and smelled of JP-4. The 46 made a smooth graceful whooshing sound that I missed.

After finishing flight school in 1967, I had been assigned to one of training squadrons at Santa Ana, California, knowing that I would soon join an operational squadron in Vietnam as a 46 copilot. Once in country, I flew three hops with Marine Medium Helicopter Squadron-265. But then there were additional orders which abruptly changed everything.

With the UH-34Ds being phased out the copilot pipeline had been prematurely reduced to a trickle, and now there was a 34 copilot shortage. Along with 6 other new 46 copilots, I was traded to HMM-361, an experienced 34 squadron that continued to perform miracles with a dwindling number of ancient "dogs," as the D models were called by those who had mastered them.

Although it was flattering to be in demand, flying the ancient 34 without any advanced training was painful for everyone involved, especially for us shanghaied copilots. Even if we'd had the customary two hundred hours of Stateside snapping-in, the transition to flying the 34 in combat was at best an aeronautical boot camp.

With increased NVA activity in the Da Nang area, MAG-36 was losing pilots daily, but fortunately 361 was leading a charmed life. So far we had been spared the customary memorial services that plagued the other squadrons in the air group.

After a short bumpy ride, the truck clattered across the marston matting and stopped in front of HMM-361's flight line. The screen door slammed behind as I entered the darkness of the ready room. I took off my sunglasses to read the status board.

Bad news! I was still on the schedule. Captain Ron Sabin would be the Helicopter Aircraft Commander who would be the section leader of the two-bird night medevac mission.

Except for the last acey deucey game of the day, the ready room was abnormally quiet. Pilots scurried in and out to complete their after-action reports in order to catch the next ride back to officers' country, the 0-club and an evening of air-conditioned revelry that I would miss.

Picking up my helmet and flight bag, I headed for the maintenance shack and checked the aircraft log for YN-17. 1 shuffled through the pages documenting the most recent flights.

"ASE intermittent."

"TACAN marginal."

"#I boost fluctuating 200 psi." *(Note: With engineering-like precision, the maintenance chief had printed, "#1 boost okay. Gauge replaced "),*

"ASE still intermittent."

"Possible over boost. 53 inches at 2950 rpm, 3 seconds." *(Maintenance had countered. "Engine checked, seals and pressures within limits.")*

More bad news. The aircraft was flyable. I was running out of time.

I walked outside toward the revetments to preflight YN-17 while it was still daylight. Grabbing the hand grips, I climbed up and put my gear through the left cockpit window.

"How does it look?" I asked the crewchief who opened the clamshell doors.

"Okay, Sir" he muttered in a tone that implied, "It'll get us there and back if some green lieutenant doesn't kill us first."

Using my flashlight to scrutinize the recesses of the engine compartment, I examined every hose, fitting, line, safety wire and coupling. When l couldn't find a "downing" discrepancy, the crew chief closed the clam shell doors, and I climbed topside to study the rotor head, the dampers, the transmission, the swash plates, boost actuators and the shaft that powers the tail rotor.

He buttoned up the top while I eyeballed the fuselage back to the tail rotor. Instead of being smooth and straight, the skin was as rough as a washboard. The surface looked like the back of a turtle. Everywhere there were 2 inch square patches covering a lifetime of bullet holes. I stood in front of the bulbous nose that gave it the shape of a pregnant guppy. The small pilot windows peered down at me like beady, disapproving eyes. With several hundred gallons of high octane avgas, you might worry about dying, but not about being maimed for life.

"Looks good to me," I said. The crewchief grunted something unintelligible.

Walking toward YN-17 with the assigned Navy corpsman, our gunner carried two metal boxes of 7.62 mm rounds for the 34's side-mounted M60 machine guns. The corpsman carried a large first aid pack and a Thompson .45 submachine gun. Over his bullet bouncer. he wore an old-fashioned magazine carrier with six extra clips and a large K-Bar fighting knife.

By now, the sun had slid behind the mountains west of Da Nang and it was getting dark fast. There was nothing left to do but go back to the ready room and wait.

Sharing a Southeast Asia style "sea" hut with the CO's office, the ready room was outfitted with sagging stuffed chairs brought along when the squadron rotated from the States in '64; benches built by the Navy Seabees who had built everything on the flight line; large topographic map with a compass card and string for determining the radial and distance of a ground position from the Da Nang Vortac; another map covered with red dots indicating where somebody had been shot at; pigeon holes for pilots' mail; a small desk with a hand-crank field telephone which connected the squadron duty officer up the line to the Group and the Direct Air Support Center, and the status board which had already been updated to include tomorrow's mission assignments.

I reported to Sabin who was semi-asleep on one of the threadbare easy chairs. Having told him that the aircraft was in good shape, he asked if the crew had all its equipment and was ready to go.

I answered, "Affirmative, Sir."

When the red lights were turned on to preserve our night vision, everything in the room tumed into the nauseating color of clotted blood. And with both temperature and humidity at one hundred, body heat and the sudden absence of a cooling breeze off the South China Sea the ready room was a sauna. Skin oozed greasy sweat and collected in beads and streaks on foreheads and arms and trapped an occasional mosquito.

At 2000, the DASC called to inform the duty officer that until 0800 tomorrow, the squadron officially had the emergency medevac standby and the responsibility of evacuating wounded Marines who were expected to die before daylight unless they were evacuated. Due to the extreme danger of flying into hot zones in total darkness, all less critical cases were given the status of priority or routine and made to wait until daylight.

I had already thrown on my shoulder holster and bullet bouncer and was standing before I was fully conscious of the fact that the telephone had awakened everyone in the ready room.

"Roger," the duty officer said, "220 degree radial Da Nang Vortac at 18 nautical miles." I followed Sabin to the door with the two pilots that would fly our medevac chase.

"Hostage gunships are on station with two hours of fuel. Meet them on button yellow," the duty officer added.

Sabin was a shadow sprinting ahead of me down the line of revetments toward YN-17's running lights that the crew chief had turned on to guide us. By the time I strapped in and put my clear visor down, Sabin had completed the check list. "Clear," he ordered. The Pratt and Whitney 1820 under our feet cranked reluctantly, belched, and then roared. A constant blue-white flame from the exhaust stacks extended past my side window like a huge blow torch.

"Rotor brake," Sabin said over the static of the radios. I released the lever and the turning rotator blades rocked the helicopter gently as they accelerated to 2000 rpm.

"Ground, this is Tarbush medevac with two for taxi," Sabin said. "Tarbush medevac, taxi approved, contact tower at your discretion. Have a good flight, sir." Sabin clicked the mike in acknowledgement, and I switched the selector to the tower frequency.

At the taxi way, he called again, "Tarbush medevac, ready for take off."

"Roger, Tarbush medevac, approved for take off." Sabin twisted on more rpm's, and I checked the magnetos. The instant the selector was back to "both," he added full power and the helicopter rose 15 feet off the ground, tilted forward, accelerated and then started climbing to 1500 feet.

"Two aboard," medevac chase announced.

"Welcome aboard, two. We're blacking out and switching to yellow." Sabin flipped the light switches overhead and, except for the bright flame from the exhaust stacks, everything vanished in total darkness. I felt as helpless as though I were in a state of free fall. Everything I needed to fly was gone including: airspeed, altimeter, rotor rpm's, manifold pressure, and an external reference. But the sensory inputs I needed for survival, Sabin didn't seem to need or miss in the slightest degree.

"If we get hit, I'll decide if we abort. If, one of us gets hit, we will abort immediately and go direct to the Navy hospital. You know where G4 is don't you?" I clicked the mike in the affirmative.

Below us, lights blinked like the small towns and farms we flew on night hops at Pensacola. But tonight, each light was the flash of a bullet being fired at us. "Lotta folks up tonight," the crewchief chuckled. "It's a big sky," Sabin replied with an air of confidence not shared by me.

"Hostage one, Tarbush medevac. We're five minutes out, what have we got?"

"Zone's hot and taking fire from southeast to southwest. No fifties so far. We're hosing the tree line, and we've got a flight of F-4's standing by. Should be a piece of cake. We're in contact with the grunts on fox mike oscar." My hand searched for the mix box and flipped the switches to hear simultaneous conversations on the UHF and the FM radios and dialed in the right ground frequency for "oscar."

"You're on them now, Hostage, nice shooting. Tarbush medevac, this is Rich Widow Six Actual, we've got to get our guy out of here fast, you copy?"

"Roger, Rich Widow Six. How will you mark the zone?" There was a long pause.

"How 'bout a strobe light. We'll put a strobe light in the center of the zone."

"Hostage, this is Tarbush 45 seconds to commence approach. Give me another 45 seconds and hit the tree line, but be sure to save something to cover our egress."

"Roger, Tarbush, we've saved a few two-seven-five's for the occasion," the Huey gun ship section leader acknowledged. We were at fifteen hundred feet directly over the zone where the Marines had circled their wagons for the night. The standard approach procedure was to spiral down directly over the zone in order to present the most devious target for the shortest time.

During the day, this falling approach was dangerous and required a high skill level.

At night, I was sure that it would be impossible. There were no visual references, no horizon. no lights, no instruments. Surprisingly, the impossibility of performing the impending feat had transformed my fear into curiosity.

"Okay. commencing the approach. No shooting. I'm not sure where the grunts are. And let's get loaded and take off in twenty seconds. No sense hanging around until the mortars find us." Sabin said.

He pressed the transmitter key on the stick, "Rich Widow. Tarbush coming down, turn on your beacon." He bottomed the collective, screwed back on the grip throttle, dropped the nose, and banked steeply to the right. I could feel the helicopter spiraling down like a duck with a shot wing. I looked out Sabin's window at the strobe light which occasionally came into view as he varied the angle of bank. Without any horizon or instruments, the blinking light was the only clue as to the direction of "down." After five complete revolutions, he stopped banking. The strobe light was now bright and dead ahead.

Streams of tracers etched the tree line as the Hueys attacked in a tight circle to cover our approach with continuous fire support. I could feel Sabin raising the nose to slow our forward movement and twisting on full power to stop our descent as smoothly as a new elevator. There was no artificial horizon, engine and rotor rpm, or manifold pressure gauges to keep the destructive forces within extremely narrow critical limits. *(Excessive manifold pressure and the engine blows, too few rpm's and the blades cone, too many and they whirl off into space.)*

The brightness of the strobe told me that we were low and close to the landing zone. Sabin gentled the aircraft downward hoping to land in a narrow space between the Marines. Then he moved ahead, and then to the right. "You're following him all over the zone. He's running to get away. The idiot put the damn strobe on his helmet," the crew chief yelled through our ear phones.

Sabin landed with his side of the 34 toward the shooting so that the exhaust stacks on my side would not be a target. "Sorry about the strobe," Rich Widow Six Actual said over the fox mike.

"No sweat, I'm just glad I didn't land on him."

The crewchief interrupted, "All aboard, Skipper."

Sabin twisted on every drop of energy the 1820 had saved up. rose swiftly through a hover and gained translational lift. With a series of brilliant flashes, the night was brighter than high noon. Just above the zone, the Hueys had fired their rockets like huge flashbulbs in rapid succession as we sneaked away quietly and began climbing toward Da Nang.

I realized that I hadn't taken a breath since we started the approach, exhaled, and sank back in my seat. "How's he doing, Doc?" Sabin asked, "I've got my hand inside his chest, but he'll make it."

It took only minutes to get to G4, the Navy hospital just off the west side of Marble Mountain's runway. As we neared, flood fights were turned on around the landing pad. A welcoming committee of doctors, nurses and stretcher bearers was already at the hatch door before the landing gear struts were fully compressed. Somebody else's hand was now inside the Marine as the corpsman yelled above the engine noise to share his diagnosis with those who would finish the job.

The IV bottle was held cautiously below the rotor blades while the group trundled their new patient toward the sandbagged entrance of the hospital. As they moved away, the wounded Marine shot Sabin an enthusiastic smile. "That's what it's all about," Sabin said as he waved back at the Marine.

That night we flew another 4.1 hours and made 11 more landings. On our last run, with just a hint of horizon, Sabin turned the controls over to me. Except for a couple gentle nudges on the collective, I flew the whole mission myself and it didn't stink.

After our last drop off at G4, we landed at Marble Mountain, stopped briefly at the fuel pits, then pulled into the taxi way, and shut down. Although we would be on call for another two hours, somehow we knew that the night was over.

When I swung down from the cockpit with my gear, Sabin was waiting for me. "Not bad," he smiled. "We got almost five hours of night time."

Walking ahead of us toward the line shack with the gunner and crewchief, the corpsman turned back and said, "Thanks for the ride, gentlemen, catch you next time."

There would be many "next times." And anyone who was there will tell you that, all things considered, it was time well spent.

ANGRY EYES AT A SHAU
George Twardzik

The following experience occurred in the A Shau Valley around March 25, 1966, which demonstrates the down and dirty gut spirit of Marine helicopter pilots and aircrew, especially the can-do, will-do character of my outfit, Marine (Medium) Helicopter Squadron -163.

At the time, I was a PFC flying as a gunner in the UH-34Ds, based at Phu Bai. Around March 23rd, we received a frantic call from an Army special forces unit of about 220 men who were in a very critical situation at their outpost in the Ashau Valley. The special forces were under siege from an enemy force estimated to be 3000 men. An Army helo sped in to assist them and was promptly shot down. It was determined that the outpost could not be saved because the risks were so great, and that all military units were to stand down and stay away from the Ashau outpost.

We could hear them begging over the radio for medevacs, ammo and water for three days. Finally, our squadron skipper had had it. He strode rapidly from the radio shack to one of the H-34s when he stopped in his tracks and turned to the waiting pilots and aircrews. He said he was "going for a ride," and if anyone else wanted to go he was not going to stop them. Everyone raced for their birds and powered up for the flight to Ashau.

The skipper's helo and three others in the first wave, were immediately shot down. Since this happened late in the day, the other helos returned to Phu Bai. Next morning, HMM-163 "Angry Eyes" helos returned to Ashau. Americans were dispersed all over the zone, singly and in groups. We sighted three or four running near the old runway, so we spiraled on in to pick them up. We could hear a .50 cal. machine gun popping away at us on our approach. It missed us, as we scooped up the soldiers and spiraled up and away. The .50 was shooting at us all the while but once again, we were very lucky and flew out of trouble.

We had a lump in our throats when we spotted some of our aircrews in the zone. We went on in and hoisted them aboard into the cabin. An M-60 machine gun came up on the hoist first. I grabbed it and set it up out the aft escape hatch where a rescued gunner put it into action.

I had on an Army flack jacket since we were short of the USMC regulation ones, and was secured in the cabin helo by a gunner's belt hooked to a "D" ring on the deck. Our luck ran out, when the enemy .50 cal. machine gun found our helo and I was hit squarely in the chest. I lost it. I was told later by the crewchief that the force of the round drove me completely out of the cabin doorway and into space. When the safety belt reached the end of its travel, it snapped me back right into the cabin!

Other rounds hit the helo; one of them penetrated the radio compartment and ignited a 5 gal-

47

lon can of engine oil we had stored there. It burst into flames and filled the helo with acrid smoke. The pilot idled the engine and autorotated down to a safe landing below. As we touched down we managed to pitch the burning oil can overboard and put out the remains of the fire.

After we pulled the seat cushions out of our butts, the pilot engaged the rotors and took off, dragging the main gear through the tree tops as we headed for Phu Bai. I must have smoked a whole pack of cigarettes on the twenty minute return flight! After we landed and secured, I got out of the 34 to view the battle damage. The aircraft was sieved with bullet holes. My flak jacket had a burn mark in the center of the chest area. Thank God for small miracles!

Air Force A-1 "Skyraiders" dropped their loads and strafed the surrounding enemy while supporting us. Even when they ran out of ammunition, they continued to make runs with their landing gear down, to take the heat off of the trapped GIs and aircrews. One of the "Spads" as they were called, was hit and had to make an emergency landing on the old strip. One of his buddies immediately landed behind him; and stuffed the downed pilot in his lap, and took off safely. I heard later on that the rescue pilot got the Congressional Medal of Honor for his brave deed.

As it turned out that day, we saved all the HMM-163 aircrews who went down the evening before. I learned that Sgt Puckett, an aircrewman who was forced down, grabbed his M-60 machine gun, leaped out of his helo and dived into a bunker while under enemy fire. He then set up his M-60 and joined the other "grunts' who had a mortar and another machine gun blazing away. Puckett was the only one to get out of that bunker alive, as it turned out. Fighting was furious.

As it turned out, we also saved around 190 of the 220 men who were trapped in the zone, which looked like an aircraft parking lot by the time we were finished the mission. By the time it was all over and we took inventory, HMM-163 only had five flyable choppers left on the line!

The special forces presented HMM-163 with a special award for their extraordinary efforts to rescue the beleaguered troopers from their precarious zone at A Shau. I personally received an award of a special patch which was sewn onto my flight jacket and remains there to this day.

This action at A Shau was the high point of my Marine Corps career. It exemplifies what I feel that it was all about. Semper Fi!

Marine aircrew take a rare break in the action in the hootch area at PROVMAG-39 at the Quang Tri helicopter facility. Sept. 1968.

An Hoa Valley and River from Medevac chopper on mission with Huey gunship on flank. (Courtesy of J.J. Houser)

Glossary Of Terms

AH-1G, AH-IJ: The initial Marine AH-1s a were Army G models modified only with Navy compatible radios, Marine green paint and a rotor brake for shipboard operations. Later Marine models such as the AH-1J were specifically designed for Marine aviation requirements and were twin-engined.

ALO: Air Liaison Officer.

AMO: Aviation Maintenance Officer.

AO: Area of operations, similar in definition to TAOR discussed below.

ARVN: Army of the Republic of Viet Nam.

BDE: The abbreviation for a US Army brigade.

BLT: Battalion Landing Team, a US Marine infantry battalion specifically task organized and equipped to conduct amphibious or helicopter-borne landings from the sea.

CH-37C: Powered with twin reciprocating engines, this heavy helicopter operated on a limited basis in the early years of the war, mainly for the retrieval of downed aircraft.

CH-46A, CH-46D: A twin gas turbine powered medium helicopter that replaced the UH-34D for troop and cargo lift, medevac, and other assigned missions.

CH-53A, CH-53D: A twin gas turbine powered heavy helicopter that replaced the CH-37C for the retrieval of downed aircraft, as well as the movement of heavy and large items of equipment such as trucks and artillery. The CH-53D was the improved version with more powerful engines.

CHARLIE RIDGE: A prominent ridge of mountainous terrain approximately 20 miles southwest of Da Nang that afforded the Viet Cong, also known as "Charlie," a route from Laos into the Da Nang area. It was the site of many Marine operations aimed at disrupting Charlie's movement of men and supplies.

CHINOOK: The nickname of the US Army's CH-47 heavy helicopter.

CIDG: Civil Indigenous Defense Group, a paramilitary, militia-type unit made up of local Vietnamese who participated in the defense of their own village or hamlet.

CO: The Commanding Officer, or Commander of a specific unit.

CP: Command Post, the location from which the CO commanded his unit. This could range from an extensive permanent building complex to a hole in the ground.

D/CS: Deputy Chief of Staff. The staff officer responsible for a specific function such as D/CS AIR (Aviation) at Headquarters, US Marine Corps, responsible for all aviation matters.

DASC: Direct Air Support Center.

DET: Abbreviation for detachment, usually a parent organization's smaller detached unit capable of self-sustained operations. For example, DET, HMH-463 would indicate a small number (4-6) of CH-53 helicopters operating independently of the parent unit, HMH-463.

DMZ: The Demilitarized Zone separating North and South Viet Nam that was established by the United Nations at the time French IndoChina was partitioned into the two countries. It was generally ignored by both sides during the war.

EAGLE FLIGHT: (also PACIFIER, KINGFISHER, SPARROWHAWK) A package of aircraft, on either ground or airborne alert, designated to respond to emergency situations or targets of opportunity by either inserting ground units or attacking by fire or both. The group usually consisted of a command helicopter, troop lift helicopters and attack helicopters. In some instances, fixed wing attack aircraft were also added to the package.

FAC: Forward Air Controller.

FNG: F*@cking New Guy.

GUARD Channel: The universal radio channel monitored by all aircraft on which emergency transmissions and requests for assistance are made.

H&MS: Headquarters and Maintenance Squadron, a unit of a Marine Aircraft Group that performs both administrative functions and intermediate level aircraft maintenance. In some cases, an H&MS would operate small numbers of specialized aircraft, such as the CH-37C from 1965-1967.

HARDHAT: An acronym for the aircrew member's protective helmet.

Hill 55: (and others such as Hill 845 etc) A means of identifying hill formations on the metric maps used in Viet Nam. The numbers indicate the height above sea level of the highest point of that particular formation.

HMH: Marine Heavy Helicopter Squadron, the first H means helicopter; the M means Marine; the second H means heavy.

HML: Marine Light Helicopter Squadron, the first H means helicopter; the M means Marine; the L means light.

HMM: Marine Medium Helicopter Squadron, the first H means helicopter; the M means Marine; the second M means medium. (NOTE) The three numbers following these letters usually identified the original parent Marine Aircraft Group and the sequence in the squadron was first commissioned. HMM-161 was the first squadron commissioned in MAG-16. HML-367 was the seventh squadron commissioned in MAG-36. There were exceptions. HMH-463 was not the third squadron commissioned in MAG-46. MAG-46 did not exist in the active force structure. it was then and still is a Marine Reserve Aircraft Group. As a wartime expedient both HMH-462 and HMH-463 were commissioned in other aircraft groups and, when operational, were transferred to MAG-36 and MAG-16 respectively.

I CORPS: Viet Nam was divided into 4 geographical areas known as Corps in order to delineate responsibility for the military operations therein. From north to south they were I Corps, II Corps, III Corps and IV Corps. Early in the war, The US Marines were designated responsible for I Corps, which extended from the DMZ in the north and included the provinces Quang Tri, Thua Thien, Quang Nam, Quang Tin and Quang Ngai Province in the south.

IFR: Instrument Flight Rules, a condition during which a pilot is flying in the clouds on instruments, and without reference to either the natural horizon or the ground.

KIA: Killed In Action.

LPD: Landing Platform Dock, a Navy amphibious ship, capable of supporting and operating a small number of helicopters for an extended period of time. Usually, 4 to 6 on board with 2 being operated simultaneously.

LPH: Landing Platform Helicopter. a Navy amphibious ship, capable of supporting and operating a squadron of helicopters for an extended period of time, and capable of transporting and off-loading a battalion of Marines at the same time.

LZ: Landing Zone, an unimproved site where helicopters landed in the performance of their assigned mission.

M-60: The standard light machine gun of US Marines in Viet Nam, in both ground and aviation units. In helicopter squadrons, M-60s were mounted in fixed forward firing positions on UH-lE gunships, and on flexible pintle mounts in the UH-1E side doors as well. They were also employed in the door or windows of transport helicopters (UH-34, CH-46, CH-53).

MAB: Marine Amphibious Brigade, a temporary headquarters superimposed over such amphibious units as a Regimental Landing Team and a Provisional Marine Aircraft Group. The MAB was identical in mission and structure to the Marine Expeditionary Brigade (MEB.)

MABS: Marine Airbase Squadron, the housekeeping unit of a MAG.

MACV: Military Assistance Command Viet Nam, the senior US headquarters charged with overall responsibility for conduct of the war.

MAG: Marine Aircraft Group, the unit immediately senior to aircraft squadrons, A helicopter MAG typically would have the following squadrons: 3 HMMs, 1 HML, 1 HMH, 1 VMO, 1 H&MS, and 1 MABS attached.

MEB: Marine Expeditionary Brigade, identical to the MAB discussed above.

MEDEVAC: Medical evacuation, the term generally used to identify the mission of Marine helicopters involved in rescuing wounded, injured, and sick personnel.

MMAF: Marble Mountain Air Facility, the home of MAG-16 from August, 1965 until May, 1971, located east of the Da Nang Air Base, on the beach, between China Beach and the Marble Mountains.

MSO: A US Navy wood-hulled mine sweeper.

MUV: Marine Unit Vietnam, the temporary identification of Marine helicopter squadrons operating from Da Nang Air Base in 1965 prior to the arrival in country of MAG-16.

NSA: Naval Support Activity, a US Navy organization responsible for various support functions in the Da Nang area, such as port facilities, fuel storage areas, and hospitals, i.e., the NSA Hospital.

NVA: North Vietnamese Army.

NW: Northwest

0-1, 0-1C: A light single engine observation aircraft used for forward air control, artillery spotting and general reconnaissance, from 1962 to 1969.

OV-1OA: A twin turboprop, twin boom observation aircraft that replaced the 0-1s in 1968. It had significantly greater performance and carried a larger payload.

POW: Prisoner of War.

PROVMAG: Provisional Marine Aircraft Group, a temporary group organized for a limited period of time to meet a specific tactical or operational need.

QEC: Quick Engine Change, an expedient procedure developed to quickly change the engine of a downed helicopter in the field.

RLT: Regimental Landing Team, a US Marine infantry regiment specifically task organized and equipped to conduct amphibious or helicopter-borne landings from the sea.

RPG: Rocket Propelled Grenade, an enemy grenade fired from a device utilizing a small rocket propellant charge, greatly increasing its normal range.

RVN: Republic of Viet Nam.

SAR: Search and Rescue, the mission assigned to either dedicated aviation units or other available aviation units related to locating and extracting downed aircrews and other personnel. The term CSAR identified Combat Search and Rescue units of the Air Force and Navy who were specifically trained and equipped to operate in heavily defended North Vietnamese airspace to conduct SAR missions.

SAR NORTH: The SAR mission north of the DMZ assigned in the early years of the war to Marine HMM squadrons, prior to the assignment of the specific CSAR units discussed above.

SHU-FLY: The name coined for the initial introduction of USMC helicopters into Viet Nam in 1962.

SLF: Special Landing Force, the designation of the USMC BLT and HMM squadron assigned to the Seventh Fleet Amphibious Ready Group. The SLF regularly conducted amphibious operations across Vietnamese beaches into areas of suspected VC and NVA activity.

SW: Southwest.

TAOR: Tactical Area of Responsibility, the geographical area assigned to a military unit having responsibility for all operations therein, i.e., the First Marine Division's TAOR was the city of Da Nang and surrounding areas defined by specific features such as rivers, roads etc.

TF: Task Force, a unit temporarily organized to carry out a specific short term mission.

UH-1, UH-IB, UH-1E: The "Huey" was the standard US Army troop carrying helicopter in Viet Nam. The Marine version, the UH-1E was able to operate from shipboard. It performed command and control, liaison, observation, gunship, and medevac missions.

UH-34D: The standard USMC medium helicopter at the beginning of the war. It served in Viet Nam from April, 1962 until August, 1969.

UA: Unauthorized Absence, both a disciplinary term, and an acronym for anything or person not in the place it is expected to be.

UTT: Utility Tactical Transport, the name of the Army's UH-1B gunship unit early in the war when the helicopter gunship was still in an experimental state.

VC: Viet Cong, or Charlie, the original enemy in South Viet Nam.

VMO: Marine Observation Squadron, the V means fixed wing; the M means Marine; the 0 means observation.

VNAF: Vietnamese Air Force.

WIA: Wounded in Action.

XO: Executive Officer, the second in command of a unit.

ZERO/ZERO: Zero ceiling, zero forward visibility, as in IFR flight conditions.

HMM-161's CH-46s enroute to a major troop emplacement along the Ben Hai River in the DMZ, 17 Sept 1968.

USMC/Vietnam Helicopter Pilots & Aircrew Biographies

A CH-53 from HMH-463 slings in a load of supplies to OP-1 at the Con Thien outpost along the eastern DMZ. (Photo credit: HB Staff)

LARRY G. ADAMS was born on April 20, 1940 in Kellogg, ID. He graduated from high school in 1958 in Kamiah, ID; received a B.S. in aeronautics from Embry-Riddle Aero University; taught school in Milwaukee, OR; joined the USMC as a MARCAD, October 1961 and received his wings and commission on Nov. 8, 1963.

He departed Santa Ana with HMM-361 in May 1965 to RVN; October 1965 to HMM-161 (the Grand Pers Switch-Er-Roo); June 1966 to New River, HMM-162 and HMH-461, then to RVN 1969, VMO-6 (0-1 Bird-Dog), HMH-463 and 361; to HMX-1, 1970-74 then to Embry-Riddle Aero University, Daytona Beach; 1975 to Okinawa, HMH-463; 1976 to Command and staff; 1977 to CINCPAC Airborne Command Post, Hawaii until 1979, then transferred to U.S. 3rd Fleet as Fleet Marine Officer.

Adams retired as a Lieutenant Colonel in August 1982. He flew 5,001 flight hours and 880 missions. He received 44 Air Medals, Joint Service Commendation, two Navy Commendation Medals and the Cross of Gallantry. He retired to Spokane, WA, sold real estate, was the Fire Commissioner and in a graduate program; airport manager, Pullman-Moscow Regional Airport, 1986-1990; and manager of Walla Walla Regional Airport 1990-present.

He has two sons, both former Marines, and two grandchildren. Adams is single and prefers it, but is constantly looking for someone to change his mind. He is on the board of directors of the Walla Walla Chamber of Commerce, Washington Airport Managers Association Board of Directors, American Association of Airport Executives, AOPA, American Legion and Rotary International.

RICHARD JOHN ADAMS Lieutenant Colonel, USMC (Ret) was born on Nov. 18, 1931 in Endicott, NY, He graduated from Union-Endicott High School and received a BA from Trinity College, Hartford, CT in 1954. He enlisted in the Marine Corps and went to boot camp at Parris Island and attended aviation schools at NAS Jacksonville, FL and Millington, TN prior to receiving orders to flight school in February 1955. He was commissioned a Second Lieutenant in August 1955 through the AOC Program. He received his wings at Chase Field, Beeville, TX in September 1956.

A tour with VMF (AW)-513 at Atsugi, Japan was followed by a tour at Cherry Point, NC. Retraining in Transport Aircraft was accomplished with VMR-353 followed by tours with VMR-252; Headquarters Squadron, FMFLANT and 1st MAW at Iwakuni, Japan. After attending Amphibious Warfare School, Quantico, VA, he was assigned to VMGR-352, El Toro and DET "A" 352, Futema, Okinawa and Da Nang. 1967 brought retraining in the CH-46 and assignment as S-3 and later XO of HMM-161 at Quang Tri, RVN.

A HQMC tour was followed by Armed Forces Staff College; CO H&MS-16; CO HMM-161 and S-3 MAG-36. Next came a tour as adjutant MCAS El Toro and finally retirement as AC/S G-1 3rd MAW on Dec. 31, 1981.

His personal decorations include: Meritorious Service Medal, two Distinguished Flying Crosses, Bronze Star with Combat V, two Single Mission Air Medals, 57 Strike/Flight Air Medals, Purple Heart and Vietnamese Cross of Gallantry and the Boeing Vertol Rescue Award.

He served a total of 32 months in RVN and in 28 years flew 23 different aircraft with more than 6400 accident free hours.

He is married to Barbara and has two daughters, Charlene and Laura and three grandchildren. He is a member of TROA and the VFW.

EDWARD J. ANDERSON was born on Oct. 11 1936 in White Plains, NY. He was raised in Mt. Kisco, NY and graduated from Trinity College, Hartford, CT in June 1959. He was commissioned a USMC Second Lieutenant in December 1959 and designated Naval Aviator in May 1961 (played basketball for Pensacola Goshawks).

He was with HMM-163 during the next two years at MCAF Santa Ana and in RVN (Soc Trang and Da Nang). He was assigned to the Development Center in Quantico for three years as Assistant Helicopter Systems Project Officer (promoted to Captain 1964); he played one year of basketball for the base team and spent four months at MCAF New River training in UH-1E before assignment to VMO-2 at MCAF Marble Mountain, RVN. He split a tour as Provost Marshal when he was promoted to Major. His final tour as OIC was at Camp Pendleton Airfield 1968-70. He received 31 Air Medals plus the usual RVN decorations.

He spent two years as a special agent with the FBI in Seattle and Portland and entered the private sector in 1973 in corporate security management. He worked for 17 years in various capacities (Investigator to Director of Security) with several firms and obtained an MPA degree in criminal justice administration/organizational development in 1978. He had a career change into outplacement services in 1990 and presently is vice-president for California Employment Resource Group in Santa Ana, CA. He has been married three times and has three children.

MARC J. ANDERSON was born on Feb. 22, 1943 at The Dalles, Oregon. He graduated from Willamette University with a B.A. degree in history in 1965, attended OCS, Quantico, CO. B 1st Platoon and was commissioned a Second Lieutenant USMCR. He was designated a Naval Aviator in 1967, Pensacola, FL.

He joined the Ridge Runners of HMM-163 in September 1967, Quang Tri, RVN.

Anderson earned the standard "two rows," including Air Medal with 16 Strike/Flight Awards, and PUC for support of the 26th Marines at the Siege of Khe Sanh. He returned to CONUS in November 1968 and joined HMM-264 at New River, NC where he completed his active military duty.

He joined the Oregon State Police in 1971 and currently serves as a narcotics detective assigned to a DEA task force. He is currently in his 25th year of marriage with two children (ages 22 and 15).

RICHARD L. ANDERSON, "RICK," was born on June 28, 1943 in Wyandotte, MI and graduated from Michigan State University in 1966. He was commissioned a Second Lieutenant in June 1967 with the 44th OCS Quantico. He was designated a Naval Aviator on Aug. 23 1968 at Pensacola and served with HMM-162 New River from September 1968 to December 1968 and qualified in the CH-46 Sea Knight.

He joined HMM-262 "Chatterbox" in Quang Tri, Vietnam in January 1969. With GOD and a little help from his hooch "The Morgue" he defied the war demons and flew more than 700 combat hours.

Working with and in support of other Marines, he received the Distinguished Flying Cross and two Single Mission Air Medals. He was also awarded 37 Strike/Flight Air Medals, the NDS, VCM, VCG and MUC.

Anderson returned to Santa Ana in January 1970 and served as a CH-46 flight instructor with HMM-163 while managing the base officers club. California will never be the same; the 46, dirt bikes, skiing, camping, wine, women and song. Ken, George, Clyde, Gene Stoney, Mac, Ron and many others were part of the fast-paced post-Vietnam years that offer memories of adventure, excitement and "good times." Captain Anderson USMCR completed active duty in December 1971.

From 1972 to 1990 various corporate positions in sales, marketing and management in the medical equipment business led to business ownership. His company, TheraMAX Medical, Inc., was formed in 1990 and specializes in medical equipment for Human Balance and Mobility.

Now after 23 great years with his Marine-era bride, Mary, their life continues to accelerate. Their two sons, Alexander and Spencer, are NROTC Midshipmen. Alex, a senior in engineering at Auburn University, is working toward a Navy flight slot. Spencer, a freshman in engineering at North Carolina State University, aspires to join the UDT/Seals. Mary and Rick live in Winter Springs, FL.

RICK HILL ANDERTON, "DOUGHBOY," was born on April 25, 1943 in San Pedro, CA. He joined the USMC in October 1965 in Quantico. He was stationed at Pensacola and New River in the States.

While serving in-country, he was with HMM-163 from 1967-68 in H-34s and was involved in Tet, and the Khe Sanh Siege 1967-68.

He received 35 Air Medals and the standard Vietnam medals. He achieved the rank of Captain at the time of discharge in February 1970. He is an airline pilot and now lives in Marietta, GA and is married with four children.

QUINTON G. ANGLIN "GUY" graduated from Auburndale High School in June 1962 and from the University of South Florida, Tampa, FL in April 1966 with a BA in botany. He joined the USMC on Feb. 14, 1967 and completed the 46th OCS in December 1967. He was commissioned a Second Lieutenant; completed Naval Flight School in Pensacola May 1969; was stationed at MCAS New River, NC from June 1969 to January 1970 (Squadron Pilot for HMM-365 and HMM-264); Marble Mountain, RVN Feb. 11, 1970 to August 1970 with HMM-161. He transferred to HMM-262, September to December 1970 FAC Duty, 1st BN 5th MAR. He was stationed at MCAS Cherry Point, NC in February 1971 and in November 1972, he was released from active duty.

In December 1974 he joined VTU FL3 in Tallahassee. In October 1976 he joined HMM-767, New Orleans, LA (in 1977 became HML-767) September 1985 and in September 1985 he joined MTU FL3 with the rank of Lieutenant Colonel. He retired on July 1, 1994.

After release from active duty (Nov. 15, 1972) he was employed by the Florida Dept. of Agriculture as a plant protection specialist until September 1990. From September 1990 to February 1992, he was a jurisdiction botanist for the Florida Dept. of Environmental Regulation. In February 1992 he was a botanist/ecologist for the National Forests in Florida and is presently employed by the USDA Forest Service.

He now lives on a 52 acre farm near Monticello, Jefferson County, FL with his wife Cindy, and daughters, Kelly 19, Sallie 17 and Carrie 12. His hobbies include: hunting, fishing, hiking and raising and training fox hounds.

JOHN C. ARICK, Brigadier General, USMC (Ret) was born in Washington, D.C. on Jan. 19, 1940. He was commissioned in 1962 following graduation from the Naval Academy. He served his first two years as an artillery officer and was designated a Naval Aviator in 1965.

He flew the UH-1E for the next two years, one of which was spent at Ky Ha with the "Klondikers" of VMO-6. After a tour as a student at the Naval Postgraduate School, he returned to Southeast Asia where he flew the AH-1G, this time with HML-367 "Scarface" at Marble Mountain.

His combat decorations include the Distinguished Flying Cross (four awards), the Single Mission Air Medal (two awards), the Strike/Flight Air Medal (83 awards) and the Purple Heart.

Following his return to the U.S. and a tour at Quantico, he returned to attack helicopters as the Operations Officer with HMA-269 at New River. During the next 18 years, Arick served in various staff billets and he commanded HMA-169 at Camp Pendleton and MAG-29 at New River. He retired from the Marine Corps in 1992 after 30 years of commissioned service and more than 4,300 mishap free pilot hours, 1,540 in combat and 570 as a Naval Test Pilot.

Immediately following retirement, Arick became the dean of Academics at the Marine Military Academy in Harlingen, TX where he and his wife, Terry, reside on campus.

DEMETRIOS (JIM) ASSURAS, "GREEK," was born on March 27, 1942 in McGill, NV. He graduated from the University of Nevada, Reno in June 1964 with a BS in engineering science. He was commissioned on May 28 1964 through the PLC (Aviation) Program. He was designated a Naval Aviator, Pensacola, FL on Nov. 10, 1965 and served with HMM-264, MCAS New River, NC, December 1965 to May 1966.

He joined HMM-263, MCAS Futema, Okinawa in July 1966 and served with HMM-263 as H2P, HAC and Section Leader in UH-34s exclusively. He was stationed at Marble Mountain, RVN, July 1966 to February 1967; MCAS Futema, Okinawa, February 1967 to April 1967; on board the USS *Okinawa* (LPH-3) conducting joint operations in RVN with BLT 1/3, April 1967 to May 1967; and Ky Ha, RVN, May 1967 to July 1967.

He was awarded the Distinguished Flying Cross, Single Mission Air Medal, 24 Air Medals, Presidential Unit Citation and Sikorsky Winged "S" (two).

He was Aerodynamics Instructor at NABTC, Pensacola, FL, September 1967, RELAD Nov. 1 1968, as Captain, USMCR. With the CIA, November 1969 to January 1986. He had several foreign and domestic assignments including a tour as Air Operations Officer, Long Tieng, Laos and working with several former Marine friends flying helicopters for Air America. He transferred to the Department of Energy in January 1986 and is currently employed as a Security Manager for the Bonneville Power Administration in Portland, OR. Assuras is married with one daughter.

JAMES O. ATKINSON JR., "JIMBO" OR "BOURBON," Major USMC (Ret) was born on July 27, 1943 of southern ancestry in Wyandotte, MI, where his father was serving with the USN during WWII. He graduated from high school in 1961 in Tampa, FL, moved to St. Petersburg, FL and attended St. Petersburg Junior College from 1961 to 1963. He enlisted in the Marine Corps in November 1963 and went through boot camp at Parris Island and as a PFC, 1st 1TR at Camp Lejeune, NC. He reported to NAS Pensacola, FL. as a MARCAD for flight school and received his Naval Aviator Wings and commission to Second Lieutenant in February 1966.

He served with HMM-162 and HMM-264 at New River, NC, 1966-67 qualifying as a HAC in the UH-34D and CH-46A respectively. He completed one and one half Caribbean Cruises; flew with HMM-361 (UH-34D) in Vietnam 1967-68; was promoted to Captain; and received awards that included the DFC and 24 Air Medals.

Atkinson completed tactical jet transition to the A-4 Skyhawk with VMT-103 at MCAS Yuma, AZ, 1968; was a jet flight instructor in the Naval Air Training Command with VT-7 (T-2A) NAS Meridian, MS, 1969-70; completed Amphibious Warfare School and Special Weapons Course at MCB Quantico, VA, 1971; and flew with VMA-311 (A-4E) at Bien Hoa Air Base, Vietnam, 1972. His awards include five Air Medals and a Navy Commendation Medal with Combat "V."

He flew with VMAT-102 (TA-4F, A-4E, F, M) MCAS Yuma, AZ, as an instructor 1973-75; promoted to Major; received BA from University of South Florida, Tampa, FL, 1975-77; and served with HQMC, 1st and 2nd MAWS, 1977, until retiring in December 1983. His last duty was with H&MS-32 (OA-4M) as operations officer. He was designated FAC (A) and TAC (A) and had a 20 year career with 13 PCS transfers, 23 separate commands and 3,500 flight hours.

Atkinson is married with three sons and three grandchildren. He has been employed with the U.S. Postal Service since 1984. He is an active member of the USMC/Vietnam Helicopter Pilots & Aircrew Reunion.

WILLIAM D. BABCOCK, "BABBER," was born on May 22, 1944 in Indianapolis, IN, the fourth of eight children. He graduated from Cathedral High School in 1962 and received a BS in biology from Marian College in 1966. His USMCR duty started with the 42nd OCS at Quantico, VA in October 1966. He was commissioned as a Second Lieutenant on Nov. 1, 1966. He played basketball for the Pensacola Goshawks and received his Navy Wings on July 30, 1968 from Squadron Eight NAAS Ellyson Field.

After CH-46 training in New River, NC, he joined HMM-262 at Quang Tri, Vietnam in March 1969.

Then First Lieutenant Babcock earned 46 Strike/Flight Air Medals and a Single Mission Air Medal with more than 700 combat hours. He joined SOES at Quantico, VA where he was a pilot instructor in T-28s till his tour was over in October 1971.

He married Carol in 1970 and they have two sons, Michael and Steven. Babcock is a Human Resources manager for a Fortune 100 company.

ROBERT L. BABOS, "BOB," Major USMCR (Ret) was born on Aug. 4, 1935 in Elizabeth, NJ. He graduated from Linden High School and holds a BS (with honors) from Utah State University, 1957. He was commissioned Jan. 5, 1959 as a Second Lieutenant from Quantico. He was designated Naval Aviator on Dec. 21, 1960 from Pensacola and served with HMM-361 at Marine Corps Air Facility, Tustin, CA 1960-64.

He was assistant flight ops. for 5th MEB during

the Cuban missile crisis in 1962. He was in Da Nang, Vietnam in 1963-64 flying UH-34Ds as assistant flight ops.

Babos was a flight instructor (check pilot) Pensacola, FL in 1964-66 and with HMM-764 in Los Alamitos, CA in 1966-75. He flew for NASA Apollo project (para wing), VIP Pilot for Gov. Reagan, and was a flood disaster pilot in 1969.

He ended his career as C.O. of Public Affairs Unit 12-99 in Los Angeles. He was awarded the Air Medal with star, Navy Unit Commendation, Rep. of Vietnam Campaign Medal, Rep. of Vietnam Meritorious Unit Citation, Rep. of Vietnam Meritorious Unit Citation, Marine Corps Expeditionary Medal and others. At present he is a commercial pilot for Continental Airlines and is a California R.E. broker. He has three daughters and six grandchildren.

GARY L. BAILEY "BEETLE" born on July 20, 1945 in Orange, TX, enlisted on Sept. 30, 1963 and went to boot camp at MCRD, San Diego, CA. He reported to Memphis for Jet Engine and Helicopter School and joined HMM-164 at MCAF Santa Ana, CA in October 1964.

He arrived in RVN at Marble Mountain with HMM-164 (1st CH-46As in Vietnam) in March 1966. He returned to MCAF Santa Ana in April 1967.

He was discharged from HMH-462 on Sept. 29, 1967 achieving the rank of Corporal. He received Combat Air Crew Wings, Bronze Star, six Air Medals, Letter of Commendation from U.S. Army Special Forces and Good Conduct Medal.

He worked for Bell Helicopter O.M.C. Amarillo, TX for five years as a flight line mechanic. He has been working as a packing machine adjuster for Maxwell House Coffee in Houston, TX for 16 years. He is still single.

GLEN BAILEY "CHIEF" was born Aug. 4, 1949 in St. Paul, NE, and joined the USMC in March, 1968. March 1972 discharged as an E-4, being assigned as a CH-46 gunner and electrician in HMM-161. His service stations included New River, NC, Vietnam and NAS So. Weymouth, MA. His most memorable experience was the little dead girl he held in his arms in some rice paddy and her brother who caught a sniper round. He couldn't save him on the way back. Also, picking up Garringer's helo along with Squirrelly's helmet.

His awards include the CACW, eight Air Medals, UCG, US&C, USAF Commendation Medal. He became a registered nurse after discharge and joined the USAF serving from August, 1976-November, 1980 as an 0-3. From August, 1987-present he serves with the USAR, Nurse Corps as an 0-4. He was in the Persian Gulf December, 1990-March, 1991. He returned to school in 1992 receiving his mortician's license. He and his wife reside in Central City, NE.

NATALE J. BALLATO, "GUINEA GARDNER," Lieutenant Colonel USMCR (Ret) was born on Aug. 28, 1943 in Westerly, RI. He graduated from high school in 1961 in Stonington, CT and holds a BA from the U. of Dayton, 1966. He was commissioned June 2, 1967, as a Second Lieutenant from Quantico and designated a Naval Aviator on Aug. 16, 1968 in Pensacola. He served with HML-267 at Camp Pendleton from September 1968 to January 1969. He qualified in the UH-1E.

He joined HML-167 in Vietnam in January 1969 and flew more than 900 combat hours.

Ballato was awarded the Single Mission Air Medal and holds a total of 37 Air Medals plus the CAR, NUC, MUC, Vietnam Service and Campaign Ribbons, the Vietnamese Cross of Gallantry, CT Medal of Merit and the Meritorious Service Medal. He completed active duty as a Captain with VMO-1 in November 1971.

He flew with reserve squadrons HMM-767 and 774. He was promoted to Major and served as operations officer and was assistant program director for the Saudi Arabian Civil Defense CH-46 OMT contract. He was promoted to Lieutenant Colonel in August 1983. He also served with the CT National Guard as a UH-1E pilot. He had a total of 26 plus years of service and more than 3,500 flights.

As a civilian he has served as a deputy sheriff in Jefferson Parish, LA, held positions in the field of airport management at various locations and is now employed by the State of Connecticut, Gambling Regulations Unit, Foxwoods Casino. He has one son.

JOHN D. BARBER, "JD," was born on Dec. 31, 1948 in the Dorchester section of Boston, MA. He arrived at Parris Island on May 1, 1967 and attended metalsmith school in Memphis. Upon completion, he was assigned to HMM-261 at New River.

While with HMM-261, he was deployed on two Caribbean cruises. During Carib 1-70 he cross trained as a CH-46 mechanic and crewchief and was awarded the Boeing Vertol Rescue Award and the Navy Commendation Medal for rescue operations involving a passenger ferry and a DC-9 Airliner.

Upon returning from this cruise, with nine months left on his enlistment, he was sent to Vietnam where he was assigned to HMM-263 at Marble Mountain.

As a crewchief and gunner, he was awarded Combat Aircrew Wings and 21 Air Medals. He was discharged on April 21, 1971 with the rank of Sergeant. He is now a Lieutenant with the Quincy Fire Department in Quincy, MA. He is married and has a son in college. He is Secretary/Treasurer of the USMC/Vietnam Helicopter Pilots & Aircrew Reunion.

ALAN HALSEY BARBOUR, "BIG AL," was born on June 19, 1941 and raised on a potato farm in Sagaponack, NY. He graduated from high school in East Hampton, NY in 1958. He earned BS degrees from Syracuse University in biology (genetics) and New York State College of Environmental Science and Forestry in forest biology (1963). He joined Platoon Leaders Class in 1959 and was commissioned Second Lieutenant in the USMCR in June 1963. He completed the Basic School, Quantico, VA in December 1963, began flight training at NAS Pensacola in January 1964 and was designated Naval Aviator in May 1965. He was assigned to MCAS New River, NC with VMO-1 and qualified in O-1 and UH-1E.

He had orders to FMF PAC (III Marine Amphibious Brigade) with VMO-2 (Deadlock) at Marble Mountain (April 1966), flying UH-1E slicks and gunships. He moved to Dong Ha in July 1966 and remained with VMO-2 (Forward), alternating between Dong Ha and Khe Sanh until May 1967. He participated in Operation Jay, Hastings, Prairie and Prairie II and amphibious landing with Operation Prairie. His primary missions were as gunship escort, force recon team insertion and extraction, tactical air control airborne, outpost re-supply and medevac.

Barbour completed active duty in April 1968 as a tactics and flight instructor with VMO-5 at Camp Pendleton.

He completed training as a flight officer with United Air Lines in October 1968 and has flown the B-727, B-737, DC-10, B-757 and B-767. Currently (1994) he is a Captain on a B-767 at Dulles International Airport in Washington, D.C. He is married to the former Susan G. Wusthof of Palo Alto, CA and they have four children. He spends time off making furniture for his many grandchildren at Poorhouse Hollow Woodworks (his retirement studio) in Marshall, VA.

AUSTIN BREWSTER BATES, "A.B." Captain USMC/Colonel USAFR (Ret) was born on July 11, 1937 in Akron, OH and grew up in Phoenix, AZ. He enlisted in the Marine Corps in 1959 and was awarded his wings at Pensacola in 1961. He was stationed in Quantico, Pensacola and Santa Ana.

He had initial orders to HMM-363 and later was assigned to HMM-163 and sent to Vietnam in the early advisory force. His memorable experiences were flying escort for President Diem, flying the Presidential family and in particular landing in Communist territory under white flags with President Diem's brother and sister-in-law "Madam Nhu" for peace negotiations. He also worked extensively with combat photographers and journalists, Horst Faust of *W.W.II Shanghai*, Richard Tregaskis *Guadalcanal Diary* and *Vietnam Diary*, and Dickie Chappel of *National Geographic*.

He completed active duty in 1964 and flew with reserve squadron HMM-764. He joined the Arizona Air National Guard and served as Phoenix KC-135 Squadron Commander and Director of Operations Arizona National Guard. Awards include: six Air Medals and six Sikorsky "S" Rescue Awards.

As a civilian he was employed by Bonanza Airlines in 1964 and three mergers later is flying Captain on DC-10s for Northwest Airlines.

SAMUEL K. BEAMON, "SAM," was born on July 27, 1947 in Waterbury, CT. He entered the Marines on Aug. 26, 1965 at Parris Island. After Memphis, he was assigned to H&MS-26 in New River. He was transferred to HMM-262 in August 1966 where he qualified as crewchief.

In November 1966 the squadron was deployed to Vietnam, flying from Cherry Point to Chu Lai by C-130 and onto Ky Ha. In February 1967 he was transferred to HMM-164 at MMAF. While operating from the USS *Princeton* (SLF), they flew missions to Hills 881 and 861. In January 1968 while at Phu Bai, he flew during Tet and the Siege on Khe Sanh.

He was discharged in June 1969. He was awarded 16 Air Medals, Combat Aircrew Wings, NUC, PUC and CAR with more than 600 hours of flying time.

Beamon joined the Waterbury Police Department in July 1970 and is now a Lieutenant and in charge of the Juvenile Division. He has a son Sam and a daughter, Susan.

WILLIS HERRERA BEARDALL, "BILL," was born on Oct. 9, 1943 in Colon, Rep. of Panama. He graduated from Balboa High School, Panama Canal Zone in 1961 and earned a BA from the University of Arkansas in 1966. He was commissioned a Second Lieutenant in December 1966 at Quantico and TBS in June 1967. He served with TACC, 3rd MAW, El Toro, 1967-68 and was designated a Naval Aviator in December 1969. He served with HMH-461 New River during the CH-53 transition and with HMH-463, Marble Mountain 1970-71. He had more than 600 combat hours.

Beardall completed active duty as a Captain at VT-1 Flight 18 Pensacola. He was awarded the DFC, three Single Mission Air Medals, 27 Air Medals, CAR, NUC and all other "I've been there too" Medals. Reserve duty was with HMM-774 in Norfolk, VA.

He graduated from NCSU, Raleigh, NC in 1977 with a master's degree in landscape architecture. He is currently employed as landscape/site property manager, York Properties, Raleigh, NC. He helped found N.C. Vietnam Veterans, Inc. and is dedicated to promoting a positive image of Vietnam Vets. He was instrumental in design, fundraising and dedication of the Memorial to NC Vietnam Veterans Capital Grounds, Raleigh, NC. He is married to Judith Smith of Winston-Salem, NC, daughter of Dodson Smith, WWII Marine.

ISAAC J. BEERY, "JACK," was born on July 11, 1949 in Harrisonburg, VA. He joined the service on July 10, 1967 and was stationed in Parris Island, Camp Geiger, NAS Memphis, MCAF Santa Ana, RVN, 1969 and Quantico, VA.

He was a crewchief and gunner in the CH-46A with HMM-164 in Vietnam.

Sgt. Beery was discharged on July 4, 1971. His awards include Combat Aircrew Wings, 10 Air Medals, Vietnam Service Medals, Campaign Medal and Vietnam Cross of Gallantry w/Palm.

He is married to Suzanne and they have daughters, Amanda and Emily. Today he is a heavy truck mechanic.

PATRICK R. BELITSKUS was born on March 2, 1947 at Twenty Nine Palms, CA. He enlisted in April 1965 at Bridgeville, PA. He was inducted into the USMC on June 10, 1965 and was stationed at Camp Lejeune and El Toro in the states. He was stationed with MASS-341 as an engineer equipment mechanic.

While in-country he was stationed at Phu Bai and Ky Ha and was a gunner on CH-46s. His memorable experiences are Shining Brass and Hickory Nut.

He was discharged on April 1, 1969 with the rank of Corporal. He was awarded the Vietnam Service Medal with two stars, Combat Crew Wings with three stars and the Air Medal with three stars.

Today he is a union carpenter with Local No. 152 in Martinez, CA.

HARVEY L. BELL, "769," Colonel ARKARNG (Ret) was born on Aug. 21, 1942 in Water Valley, MS. He graduated from the University of S. Arkansas in 1964. He joined the USMC on July 2, 1962 and was inducted on June 2, 1964. He was designated a Naval Aviator on Dec. 3, 1965 and was stationed in Pensacola and New River.

He went to the Republic of Vietnam from June 1966 to July 1967 with HMM-163 and was in HMM-365 at New River and in the Caribbean.

He completed active duty as a Captain and was transferred to ARKARNG and flew numerous Army helos while completing law school, 1969-1971. He graduated from law school (LLM) at SMU in 1972. He transferred to the Army National Guard Infantry, served as IG, then JAG at Brigade and retired as Colonel, state judge advocate in June 1994. His awards include 14 Air Medals, DFC and others.

He held the position of Arkansas State Securities Commissioner and State Savings and Loan Supervisor before "White Water." As a civilian, he has engaged in private law practice. He is author of the Vietnam Helo soap opera called "1369." He wrote the motto of the organization, "The Best Medal is a Live Man's Smile." To all, DarkMoon 9:19. He has two children.

RICHARD JACK BELL, "DING DONG," was born on June 20, 1944 in Los Angeles, CA. While attending San Diego State University, he succumbed to the John Wayne fly-boy Marine recruiting poster. After graduation and a carefree summer in 1967, he began training with the 46th OCC at Quantico, VA. Next came Pensacola and the Naval Flight School. Learning with the T-34, real flying started with the T-28, its sounds, the open skies above Alabama, and aerial combat training and carrier quals. By choice, helicopter bound, the really fun flying started with the TH-57, legendary UH-34 (Ka-poc-a-ta-poc-a-ta) and culminated in his goal, assignment to CH-46s.

In November 1969, he had the honor of becoming a "Purple Fox" when he joined HMM-364 in Vietnam. During his tour he progressed to the level of flight leader.

Particular memories include losing an engine while on a short final, with a max load of grunts returning to their hill and the only resulting inconvenience to the grunts was a short walk; listening to the mix of Elvis singing Christmas songs and incoming small arms fire on a Christmas Eve spent in the field with a wounded medevac ship; the "Batmobile" and its acquisition; Sheila's homemade chocolate-chip cookies that arrived in several thousand small pieces...but tasted great; and, of course, the friends, made from the beginning. Some are now gone, but not forgotten and others, still here, still friends.

He earned 41 Strike/Flight Awards, various combat medals and is most proud that other Marines survived because of his direct medevac interventions.

He is now serving as a Police Sergeant with the city of Costa Mesa, CA. He lives with his one and only wife, and their last, unmarried, daughter.

STEPHEN C. BENCKENSTEIN, "BECK," was born on Dec. 27, 1941 in Orange, TX. He enlisted in April 1964, and was commissioned in June 1966. He earned a BBA from Lamar University. He was awarded his wings in October 1967.

From February 1968 to March 1969 he was in RVN stationed with HMM-265 and MAG-36, Search/Rescue MCAS Beaufort and Reserve Squadron HMM-768 New Orleans. He was a HAC and qualified on the CH-46A, UH-1B, UH-34, and UH-1H while serving in-country.

He left active duty in November 1970 and resigned with the rank of Captain in 1976 at NAS New Orleans. His medals and awards include: 30 Air Medals, Combat Action, National Defense, Vietnam Service, Vietnam Campaign and Vietnam Cross of Gallantry.

From 1971 to 1979 he sold turbine aircraft; from 1980 to 1992 he was in banking/finance and from 1992 to present he is self-employed in financial services.

RICHARD R. BENDER, "RICH," Captain USMCR, was born on Aug. 11, 1940 in New York, NY. He graduated from Walt Whitman High School in Huntington, NY in 1958. He earned a degree in aircraft operation technology from the State University of New York at Farmingdale in 1961. He started preflight as a MARCAD in August of 1961; carrier qual aboard the USS *Antietam* in a T2J September of 1962; and was designated a Naval Aviator in February of 1963.

55

He joined HMM-365 at Santa Ana in April of 1963 and qualified in the UH-34, which he flew in Vietnam from October 1964 until August 1965.

He holds a total of 12 Air Medals, Vietnam Service and Campaign Ribbons plus VNAF Wings. He will never forget the truly great comradeship of all those fine men of HMM-365.

At the present time, he is a B-767 International Captain for American Airlines and president and chairman for Shelingham Tennis & Recreation Ranch, Inc. He and his family reside in Roanoke, TX.

BILL BENNINGTON was born on Dec. 9, 1946 in Cleveland, OH. He joined the USMC in November 1968. He flew CH-46s aboard the USS *New Orleans* with HMM-164.

He was discharged in November 1973 with the rank of Captain. He received the Vietnam Service Medal. He is now a Lieutenant in the Police Bureau and resides in Portland, OR.

ROBERT L. BENTON, "BOB B," was born on Aug. 8, 1939 in Coleman, TX. He enlisted on Nov. 16, 1956. He completed boot camp at MCRD, San Diego.

His first in-country tour was with HMM-362, the second was as a member of HMM-161. His third and final in-country tour was as the Maintenance Control Chief of Detachment Alpha, HMH-463 (Beeler's Bombers). Highlights of his Marine career include christening the first UH-34 helicopter for HMR(L)362, serving as a member of the West Coast Presidential Helicopter Detachment, serving with the Embassy Marines and participating on the Board of Inspection and Survey (BIS) trials for the CH-53A Sea Stallion helicopter.

He holds the Combat Aircrew Wings, 19 Air Medals, the Vietnamese Cross of Gallantry, three Winged "S" Awards and the 1500 Hour Service Award.

He retired in March 1971, and lives in Mesa, AZ with his wife, Mabel. They have two children and two grandchildren.

FRANK D. BERMUDEZ, born in Phoenix, AZ, entered the USMC Aug. 14, 1959, being assigned as a Crewchief UH-34D, and a Plane Captain C-117D, and Section Leader CH-53D. His service stations included MCAS Tustin, CA, MCAS New River, NC, MCAF Futema, JA, Da Nang, Marble Mountain, Alameda CA, MCAS Kanoehe Bay, HI, MCAS Iwakuni, JA, Ky Ha, Hue-Phu Bai. His most memorable experience was joining HMM-161 in 1966 for his second in-country tour. He was assigned to the 1st Marine Aircraft Wing in Da Nang, then MAG-16 at Marble Mountain. After a grueling 24 hour trip, he was sent back to Okinawa where HMM-161 was based.

His awards include 15 Air Medals, NCM, NAM, PUC, MUC, Combat Aircrew Wings, and many others. He was honorably discharged on February 30, 1990 as a Master Gunnery Sergeant. He and his wife Karen reside in Oviedo, FL. He is a vehicle parts manager in Orlando, FL.

RICHARD L. BIANCHINO, Lieutenant Colonel USMC (Ret) was born on March 23, 1941 in Albany, NY and graduated from Albany High School, received a BS from Cortland (NY) State, and an MA from Pepperdine in Malibu, CA. He was commissioned on June 5, 1963 and served two overseas tours as an infantry officer. He was designated a Naval Aviator in September 1968 in Pensacola. He attended TBS, AWS and C & S schools in Quantico.

He flew CH-46s in Vietnam with HMM-364 until he was shot down in April 1969. Later he served in HMMT-302 as S-1, HMM-165 as S-3, HMM-163 as XO, and CO of HMM-268. He was XO and CO H&MS-36, served as a flight instructor in HT-8 and was a track coach of the U.S. Olympic Team in Montreal in 1976. Other assignments included: USMC Director of Athletics, HQMC; Manpower Analyst, HQMC; Executive Assistant to Chief of Staff, CINCPAC, associate professor, Naval Science and Executive Officer of the NROTC Unit, University of Washington, Seattle.

He retired in 1987. His awards include the Silver Star, Purple Heart, Defense Merit Service, Merit Service, Air Medal, Navy Commendation, CAR.

He resides with wife, Brenda, and daughters, Nicole and Leah, in Anaheim Hills, CA and is employed as director of marketing, Interstate Specialty Marketing, Inc. Tustin, CA.

LEONARD BIEBERBACH "LEN," joined the Platoon Leaders Class, USMCR on Feb. 9, 1959, while a student at West Chester State College. During his college years he served as LCpl. with VMA-441, NAS, Willow Grove and attended PLC training, MCS, Quantico. He was promoted to Corporal in 1961, commissioned upon graduation from college in June 1962 and attended the Basic School, MCS, Quantico.

Assigned as a commander, 3rd Plt. G. Co., 3rd Bn., 4th Marines, Kaneohe, HI, he went on to serve with G Co. as weapons platoon commander and XO until he decided to become a Marine Corps Aviator. On May 7, 1965, carrier qualified in T-28 aboard the USS *Lexington* and designated naval aviator T-9413, August 1965. He applied for and received a regular commission in the USMC.

Assigned to HMM-264 as a "34 Driver," transferred to VMO-1 for training in the UH-1E gunship and in June 1967 was in RVN. Was a UH-1E pilot in VMO-2, he flew 480 combat missions and also served as the personal pilot to General Robert Cushman; Major General Norman J. Anderson, CG; and was pilot for heads of state and distinguished governmental officials including General William Westmoreland and General Omar Bradley.

Returned to the States and transitioned to multi-engine aircraft and flew the C-117. In August 1969, he requested a reserve commission and joined HMM-772 at NAS Willow Grove to again fly the UH-34. The squadron transitioned to the CH-53 and he was appointed XO, then Squadron CO. Following his nine year tour, he served as member of CINCLANT Joint Staff, NAS Norfolk, and Marine Liaison Officer Naval Readiness Command before accepting the position as CO of Mobilization Training Unit NJ-2 and PA-2.

Awarded the Bronze Star w/Combat V, Air Medal, 24 Air Medals, Navy Commendation Medal w/Combat V, Marine Corps Combat Action Ribbon, Vietnam Cross of Gallantry w/Silver Star, the Organized Marine Corps Reserve Medal and the Armed Forces Reserve Medal.

Married to the former Diane Behler and has two children, Christopher and Mrs. Lee Diane Tutchton, and grandson Jason Christopher Tutchton.

FOREST R. BJORNAAS Captain USMC (Ret) was born on Nov. 6, 1927 in Duluth, MN. He graduated from high school in 1945 in Proctor, MN. He joined the U.S. Navy and worked in the air traffic control field attaining the rank of Staff Sergeant. He entered flight school as a NAVCAD in September 1951. He completed flight school, was designated a Naval Aviator and was commissioned a Second Lieutenant USMCR.

He flew fixed-wing jets until July 1957 and then transitioned to helicopters and served with HMH-461 and HMH-462. His Vietnam service was with Sub-unit-1 of MAG-16 flying the CH-37. His total time flying the CH-37 was in excess of 2,000 hours.

He retired as a Captain on Sept. 1, 1971 with 26 years service and 6,100 hours flight time. His awards included W.W.II, Korean and Vietnam Service Ribbons.

He has been married for 38 years to a Pensacola lady, the former Mary Lou Bey, and they have five children, and seven grandchildren. Since retirement he has worked as an insurance broker specializing in the employee benefits area. He will retire eventually to Prescott, AZ.

JAMES R. BLANICH, Lieutenant Colonel (Ret) was born on Oct. 14, 1945 in Aitkin, MN. He joined the USMC through the PLC Program in 1967 after graduating from the University of Minnesota. He completed Naval Flight Training in November 1968. He was then transitioned to CH-53s in New River and deployed to Vietnam with HMH-361 in 1969.

He was stationed at Phu Bai and Marble Mountain. He then was transferred to HMH-463 in 1970. He flew 500 combat missions in-country and participated in the evacuations of Cambodia and Saigon in 1975 while assigned to HMH-462. He attended AWS and the Naval War College, and was executive and commanding officer of HMH-462 from 1982 to 1985. His tours of duty included a total of six years in California and nine years in Hawaii. He served in HMH-361, 363, 462, 463 and HMMT-301.

He completed a 24 year career as a Lieutenant Colonel in 1991 without a tour at HQMC. (That is a success in itself!)

He relocated to Minnesota after retirement and settled in as a wildlife artist specializing in wood carvings. He has two grown daughters and a stepson.

CHARLES A. BLOCK Colonel USMCR (Ret) enlisted in the USN on June 18, 1951. His education consisted of Junior College, Aircraft Maintenance Officer Course (Honor Graduate); Naval War College (With Honors); BS Auburn University (with highest honors) and MS University of Southern California. He was commissioned and designated a Naval Aviator in August 1954 from the NAVCAD program. He was squadron pilot (AD-4) VMA-324, Miami, FL in August 1954. He was a flight instructor (SNJ and T-34) Pensacola, FL in September 1955.

His first MAW assignment was in September 1957 with VMA-533 in A4Ds in June 1959. He was project officer for Short Airfield Tactical Support (SATS) at Bogue Field, NC, and served with HMM-163 in October 1963. From February 1965 to October 1965 he served in Vietnam as a maintenance officer and was part of the infamous "Day in the life of YP-13" flight featured in *Life Magazine*. In January 1966 he was Fleet Logistics Officer, Naval Air Systems Command, Washington D.C. and from October 1969 to October 1970 he was CO HMHT-301 (CH-53A). From December 1969 to January 1970 he was CO HMH-361, Vietnam (CH-53D). He developed "Barrel Bombing" and the slogan that "Napalm is Nature's Way." His call sign was "Dimmer 6." He was XO MARTD, Alameda, CA in January 1970, USMC liaison to the Army at Fort Rucker, AL in June 1974 and deputy director, Weapons Systems Directorate, Department of the Army in July 1976.

He retired as a Colonel on May 31, 1979.

He presently serves as test coordinator for OH-58D(I), Kiowa Warrior and RAH-66 Comanche Helicopters at Army Aviation Technical Test Center, Fort Rucker, AL. He is married to Bert Block and they have two grown daughters.

GEORGE F. BOEMERMAN, born Oct. 15, 1932 in Brooklyn, NY and joined the USMC on Dec. 7, 1952 being assigned MCAS Miami, Parris Island, Kaneohe, Cherry Point, etc. His most memorable experience was Nov. 10, 1964, a flood evacuation south of Da Nang, Vietnam with lots of wind and lightning where he rescued lots of indigenous civilians and shared a birthday cake with them.

His awards include the DFC, lots of Air Medals, Bronze Star, Vietnam Cross of Gallantry. He retired in December, 1976 with the rank of Lieutenant Colonel.

He is retired, residing with his wife Shirley in San Marcos, TX. They have a son, Steve, in Virginia, and a daughter, Sue, in Chicago.

MICHAEL K. BOICE, "PIGPEN," was born on Oct. 26, 1946 in Oakland, CA. He grew up in N.W. Washington and enlisted in December 1965 on a 120 day delay. He went to boot camp at MCRD, San Diego and then went on to Memphis for aviation training (recips, then helos). He was assigned to HMMT-302 MCAS Santa Ana, CA.

He was transferred to Vietnam in December 1967 and was assigned to HMM-364, after a week he was reassigned to HMM-165 and started crewing in late March. He was shot down on June 6, 1968 and ended up in Yokosuka, Japan (two months). He transferred to Okinawa (one month) as troop handler and then went back to H&MS-36 at Phu Bai to finish his tour. On returning to CONUS, he was assigned to H&MS at MCAS Santa Ana on the flight line, then later to Q.A.

On April 17, 1970 he was discharged with the rank of Sergeant. He received the following awards: Aircrew Wings, seven Air Medals, Navy Commendation with Combat "V" and the Purple Heart.

He married a California girl, Mary, and upon discharge they moved back to N.W. Washington. There he was hired in an aluminum reduction plant where he has been employed for 23 years. He is married and has two daughters, Kristine and Kathleen, and two granddaughters.

GARY W. BOLLER, "BIG G," Captain USMCR (Ret) was born on July 24, 1944 in Flushing, NY. He graduated from high school in 1962 and attended the Academy of Aeronautics in New York from 1962-1964. He left to attend Naval Flight School in Pensacola, FL, and was designated a Naval Aviator and commissioned as a Second Lieutenant in October 1965. He served with HMM-264, New River, NC, flying the UH-34 from 1965-66.

He joined HMM-263 in Vietnam in 1966 flying the UH-34. He served in HT-8 Ellyson Field Pensacola, FL as a flight instructor from 1968-69. He returned to Vietnam in 1969-70 and served with HML-367 flying AH-1G "Cobras." He flew more than 500 combat hours. Served with the New Jersey and NY National Guard as a pilot flying UH-1E and AH-1G.

He completed active duty as a Captain with MASS-1 Cherry Point, NC in 1970-71. He was awarded the Navy Commendation with Combat "V," DFC, Air Medal W/35 Bronze Stars, Vietnam Cross of Gallantry with Palm, Vietnam Service Medal with two stars and Vietnam Campaign Medal.

As a civilian, he has worked as a real estate professional from 1972 to the present time. He is married, with five sons, one daughter and six grandchildren.

JAMES L. BOLTON, see page 108.

NORMAN H. BOMKAMP, "NORM," Lieutenant Colonel USMC (Ret) was born on March 26, 1938 in Carlinville, IL. He graduated from Carlinville High School in 1956; Eastern Illinois University, 1960 (BS education); and Pepperdine University, 1976 (MA human resources management). His assignments included platoon commander, Company XO; squadron maintenance officer, operations officer, XO; Commanding Officer, Marine Air Reserve Training Detachment, NAS Whidbey Island, Oak Harbor, WA.

Served with HMM-165 in Vietnam from May 1967 to April 1968 flying CH-46s. His right flight boot sole stopped an AK-47 slug while on a "routine" medevac, providing a memorable experience, a "cheap" Purple Heart, and a great conversation souvenir.

"Early in 1968, HMM-165 was operating out of Phu Bai, a few miles south of Hue. My wingman, Dennis Dunnigan and I were launched on a mission with multiple Marine recon team inserts and extracts. As I recall, we had almost completed the frag and Dennis was making the final recon team pick-up. I orbited high, since I already had a recon team aboard my aircraft. The only problem-this team was in direct contact with the VC! The extract was successful, but as my wingman climbed out of the hot zone, the recon team leader counted noses only to discover he was missing a Marine-someone was still on the ground in the hot zone: This info was immediately passed on to the crewchief who relayed the dreadful news to his two pilots.

'SPACE 1-1, 1-2.' 'Go,' I replied. 'There's still one Marine in the LZ.' 'Roger that, what's your fuel state and how many PAX do you have aboard?' The radio transmissions went something like that. I soon realized my aircraft was the lightest in spite of having a recon team aboard. There was no choice as to who would make the pick-up. "What a sandwich this is!" I thought to myself. As I started the descent, all kinds of wild thoughts were going through my mind. My mouth was cotton dry, similar to feelings I had had while running a leg on my high school's mile-relay team, when I had also preferred being somewhere else. The old "pucker factor" had really set in. I said a quick prayer . . . Good Marine flight training told me that a high-speed, low-level approach to the zone from the opposite direction of the VC's last position was the right decision. Just prior to the zone, I did a flat autorotational flare to kill the airspeed quickly, keeping the crest of the landing zone between my aircraft and the VC, followed by a rapid low-level turn on the spot. My crewchief had the ramp going down as we touched the two main mounts to the ground and the solitary Marine sprinted aboard the helo! In seconds we were again airborne to safety as the ramp was being raised. "Thank You Lord!" I've often thought about how "abandoned" that Marine must have felt when he saw his fellow team members being lifted to safety while he remained alone with "Charlie" in the zone and how thankful he must have been that his team leader made a quick "nose count," albeit maybe a little late.

Although I recall this as one of the most frightening events in my Vietnam tour-of-duty, no one received any awards and no one expected any. It was just a routine example of Marines supporting and taking care of Marines."

Since retirement he has been self-employed as co-owner of Defensive Driving School, Inc. in Seattle WA. He and his wife, Judy, raised purebred Limousin beef cattle on Whidbey Island. They have four married children and three grandchildren.

JAMES M. BOONE, (JIM) "DAN" was born on Nov. 13, 1931 in Gadsden, AL. He joined the service on July 27, 1950. He was stationed in Korea with the 1st MAR DIV, FLAG DET USS *Estes* AGC-12, (1950-

51), NAG Korea 1955-1957, MWSG-37 MCAS El Toro 1957, 1st MAR DIV (1958) MCSC Barstow (1958-1959), HMH-462 MCAF Tustin (HR2S) 1960-1962, HMM-362 (1962-1963), HMM-365 (1963-1965) HMMT-301, transition from 34 to 53 (1966-1967) MAG-15 Futema, Okinawa, HMH-463 Marble Mountain (1967-1969) HMMT-302 (1970-1973).

He was crewchief, section lead, QA, HMH-462, and HMM-362, MCAS Tustin, 1960-1962 while serving in-country. In 1963 he served with HMM-365 (Shu-fly), Da Nang, 1964-1965 (USS *Princeton*) HMMT-301 (H-34) MHTG-30, crewed the first HMMT-301 CH-53A from Sikorsky to Tustin. From 1965 to 1967 he served with MAG-15 Futema, Okinawa in 1967 and HMH-463 Marble Mountain 1968-1969.

He was discharged May 31, 1973 and retired with the rank of Master Gunnery Sergeant. His awards and medals included: Bronze Star with combat "V" (A Shau Valley medevac Christmas 1964), 13 Air Medals, CAR, (two) PUC, NUC, (two) MUC, AFE, Vietnam Service with six stars, Vietnam Campaign Medal, Vietnam Cross G, Vietnam Palm.

He spent 10 years with NAESU DET CH-53 A/D Airframes REP MCAS Tustin, CA (1978-1987). He has been with the Chadwick-Helmuth Company, Inc. since 1987 as an applications engineer, in sales, training and customer support. Chadwick is a builder and distributor of both helicopter and fixed wing balancing rotors and propellers. He is married with three sons, one daughter and four grandchildren.

BRUCE PAUL BOUSQUET, "FLEA," was born on Sept. 27, 1948 in Worcester, MA. He enlisted in the Marine Corps on Sept. 27, 1965 and went though boot camp at Parris Island. He reported to Memphis for aviation training in helicopters. He was next assigned to HMM-261 at MCAF New River.

After a Caribbean cruise in December of 1967 he was transferred to Vietnam. From March 1968 to April 1969 he served with HMM-265 in-country as a crewchief.

During his tour, he was awarded the Marine Corps Combat Aircrew Wings, 40 Air Medals, the Vietnamese Cross of Gallantry and the Vietnam Civil Action Citation. He was discharged as a Corporal at New River on Sept. 2, 1969.

In May of 1980, he joined the Army National Guard and was commissioned a Second Lieutenant through the OCS program at Ft. Benning. He served as a Platoon Leader, Tact. Intel. Officer, Executive Officer and Company Commander. In December of 1986 he transferred to the U.S. Army and promoted to Major on Jan. 28, 1993. He was assigned as vessel operations chief in Boston, MA. He has also completed the Army Command and General Staff College.

He has been a police officer since January 1974. His present position is that of Operations Captain with the Worcester Police Dept. He is married, with three daughters and one granddaughter.

THOMAS A. BOWDITCH, "SHADOW," was born the son of a USMC Lieutenant on July 16, 1943 in Baltimore, MD. Entering the Corps through the PLC program at Bates College on May 6, 1965, he completed flight training in 1966 and reported to HMM-164 at Marble Mountain in January 1967.

During the following year-plus he flew as co-pilot and HAC in the then formidable CH-46, participated in the battles for Hills 881 and 861 and later the siege of Khe Sanh and was wounded twice, once flying the '46 and once as a FAC on the Ben Hai River north of Con Thien. Bowditch watched the Da Nang ammo dump go up, saw the Ann Margaret concert (and the picture), helped crash a '46 with Bill Phillips into the Phu Bai perimeter mine field, and survived R & R in both Bangkok and Manila. Without question a banner year.

He was stationed in Pensacola, Yuma, El Toro, Quantico, Iwakuni, Willow Grove, Newport, and Norfolk.

Against all predictions and advice Bowditch stayed in the Corps, seeing SE Asia again briefly in 1972 while Aide to the CG, 3rd Marine Division, transitioning to the A-4, suffering a tour in Italy and later commanding both a squadron and a group.

He rose to the improbable rank of Colonel, which pleased his old man, survivor of the Pacific Islands campaigns of a generation earlier. Bowditch retired on July 31, 1991 following a hardship tour in Hawaii in 1991, after spending his final Marine Corps Christmas in Saudi Arabia. His awards included 19 Air Medals, two Purple Hearts, MSM, LOM, VCG and CAR.

He and his lovely wife Shari now live in Charlottesville, VA where they went back to school at the university. They have two children, one boy and one girl. Bowditch is prepared to continue service to his country as Secretary of State or Ambassador to Monaco.

JACK A. BRANDON Major USMC, (Ret) was born in Alberta, Canada in 1930. He graduated from the University of Texas, (El Paso) in 1954. He was commissioned a Second Lieutenant in June 1954 and was designated a Naval Aviator in August 1956.

He served with the VMF (N)-513, Atsugi, Japan in 1956-1957; VMF-232, MCAS Kaneohe, HI in 1959-1961; Marine Liaison, German Naval Air Arm-Schleswig, Germany in 1961; VMF-251, MCAS Beaufort, SC in 1962-1964; ALO, Da Nang, Vietnam; 3rd Marines in 1965; VMF (AW)-235, Da Nang, Vietnam, 1966; VMO-6, Quang Tri, Vietnam, 1969-1970; VMGR-352, El Toro, CA in 1970-1974 (part of that time on detached duty with C-130s out of Futema, Okinawa).

He retired in 1974 with 20 years of service. His awards included Silver Star, Bronze Star, two DFCs, Purple Hearts, 100 plus Air Medals, plus various other awards.

As a civilian, he has been chief pilot operating Saberliner and Gulfstream II Aircraft worldwide: 1975-1980 out of Kano, Nigeria (Saberliner and Gulfstream II); 1980-1986 for Genstar Corp., San Francisco (Gulfstream II); 1987-present for Delaware North Corp., Buffalo, NY (Gulfstream II).

WALTER SCOTT BRANK III, "GRIT," born Aug. 7, 1947, in Raleigh, NC, entering the USMC on June 22, 1966 being assigned as a crewchief CH-46A and CH-46D. His service locations include New River, NC, Phu Bai 1967-68, Marble Mountain 1969-70, USS Tr*ipoli*, and the USS *Valley Forge*. His awards include the Silver Star for conspicuous gallantry and intrepidity in action while serving with Marine Medium Helicopter Squadron 165, Marine Aircraft Group Thirty-Six in connection with operations against the Viet Cong on May 5, 1968; 47 Air Medals; Bronze Star with "V;" two Navy Commendation Medals with "V;" Navy Unit Citation; Combat Action Ribbon; Combat Aircrew Wings with three stars; and the Vietnam Campaign Ribbon with five stars. He was honorably discharged on Aug. 3, 1970 at the rank of Sergeant. He has a 90% service connected disability.

He is divorced with no children and one large Rottweiler. He resides in Jonesboro, TN.

STEVEN D. BRAVARD, O.D. was born on Jan. 20, 1944 in Hinsdale, IL. He graduated from high school in 1962 in La Grange, IL. He was awarded a BS degree in 1966 from the University of Illinois and was commissioned a Second Lieutenant on Dec. 16, 1966 at Quantico. He was designated Naval Aviator on June 7, 1968 at Pensacola and served with HMM-561 at Santa Ana from June 1968 to December 1968. He joined HMM-262 in Quang Tri in March 1969 and flew 400 combat hours.

His most memorable "war story" involves a recon extract in "Indian Country" (Laos) as an FNG during his third week in-country. "We were hit by an RPG round immediately after picking up the recon team. The crash impact ejected me through the windscreen of our H-46 while still strapped in my armored seat. When I regained consciousness sometime later and reached for my survival radio, it was missing. Then, much to my surprise, I discovered that all my pockets were unzipped and my charts, survival gear, etc. were missing. The general consensus was that "Charlie" took me for dead, went through my pockets and then "dee dee mow'ed" as the gunships came in looking for me." Awarded the Single Mission Air Medal plus the Purple Heart for a recon extract on April 24, 1969. Also awarded a total of 30 Air Medals plus the CAR, MUC, Vietnam Service and Campaign Ribbons and Vietnam Cross of Gallantry. Completed active duty as Captain with HMMT-402 at New River in May 1971.

He graduated Magna Cum Laude and awarded a Doctor of Optometry (O.D.) degree from Southern College of Optometry in May 1976. He practiced optometry in Lafayette, LA until August 1991. He is presently living and practicing in Little Rock, AR. He has two sons, Steven Jr. and Christian who each attend the University of Arkansas at Little Rock.

H. PAYNE BREAZEALE III was born in 1944 in Durham, NC and joined the service in October 1962. He was stationed in various locations in the States as a pilot flying CH-46s and AH-1Js.

He was discharged in February 1976 with the rank of Captain. His awards included Air Medals, Purple Heart, DFC etc.

Today he lives in Sarasota, FL and is president of Paine Brazil, Inc.

BENJAMIN BRENNEMAN III was born on Feb. 27, 1944 in Augusta County, VA. After studying forestry for two years at Virginia Tech., he enlisted in the Marine Corps and entered flight school at Pensacola in 1965. He chose the chopper pipeline, received his wings at Ellyson Field in 1966, and was later named the Aviation Cadet of the Year 1966-beating all Navy and fixed wing students.

After making the transition to UH-1Es at VMO-1 in New River, NC., he joined VMO-2 at Marble Mountain in 1967. He flew about 800 missions in northern South Vietnam. Upon returning to VMO-1 at New River in March of 1968 he was deployed to the Caribbean on the aircraft carrier, USS *Guadalcanal.*

He left active duty in 1969. He received 38 Air Medals and two Distinguished Flying Crosses.

After leaving active duty he returned to Virginia Tech, to earn a bachelor's degree in aerospace engineering in 1972 and a master's degree in engineering mechanics in 1973. After briefly working for Sikorsky Aircraft, he has worked in the nuclear power industry, chiefly for Babcock & Wilcox in Lynchburg, VA, and as an expert in flow-induced random vibrations.

ROBERT BRUST "BOB" was born on May 10, 1949 in Kalispell, MT. He joined the service in July 1967 and attended boot camp at MCRD San Diego. He received his infantry training at Camp Pendleton and his aircraft training at the Naval Air Station in Memphis.

He served with HMH-361 under Colonel Andrus, as a crewchief; flew across the U.S. with 20 CH-53s; loaded aboard the LPH *New Orleans* in July of 1969 and landed in Da Nang Harbor, August 1969. He was stationed at Phu Bai and Marble Mountain after HMH-361 left the country. He served as a UH-1E crewchief/gunner with HML-167 until February 1971.

He was discharged in February 1972 at MCAS New River, NC with the rank of Sergeant. His awards were Combat Aircrew Wings w/three stars, 32 Air Medals with Bronze Star, Good Conduct, Vietnam Service, Rep. of Vietnam Campaign Medal, Vietnam Cross of Gallantry and Nat. Defense Service Medal. His Vietnam service was from August 1969 to February 1971. He is currently married with one son living in Fishtail, MT and has owned a country nightclub for the last 20 years.

EUGENE M. BRYANT "GENE", was born on Oct. 5, 1940 in Appalachia, VA. He graduated from Pennsburg High School in 1958 and attended Greenbriar Military Academy, Lewisburg, WV and Trenton Junior College, Trenton, NJ. He graduated in June 1961. He enlisted in the Marine Corps Aviation Cadet Training Program in April 1961 and reported to Preflight, Cadet Regiment, NAS Pensacola November 1961. In April 1963 he was designated as a Naval Aviator at HT-8, Ellyson Field, Pensacola. In May 1963 he joined HMM-162 at New River.

He was deployed to Haiti for hurricane relief work. He was deployed aboard USS *Boxer* during the Cuban Missile Crisis. He rotated with HMM-162 to Vietnam as an advisor with Operation Shu-Fly at Da Nang in May 1964. The Gulf of Tonkin changed the squadron's mission to that of combatant. He served with Readiness Forces aboard the USS *Princeton* in Vietnamese waters. He departed RVN in June 1965 and reported to New River for six months and then went to Ellyson Field as a UH-34 instructor. He was released from active duty in August 1968.

He flew as a pilot for Eastern Airlines on 727s, 1011s and 757s. He died on Aug. 4, 1984 in Newnan, GA. He is survived by his wife, one daughter and grandson, and one son.

WILLIAM DANIEL BRYANT, "ACTUAL," was born in Louisville, KY on June 25, 1946. He graduated from high school in 1964, Western Kentucky University in 1968 and was commissioned a Second Lieutenant in the USMC through the Platoon Leaders Class program. After attending the Basic School he was assigned to aviation training at the Army Flight School, Class 69-36. Upon completion of Army Flight School he was assigned to MCAS New River, HMMT-402 and designated a Naval Aviator in January 1970.

He was assigned to HMM-365 until transferred to VMO-6 in Okinawa. He transferred to HML-167 in Vietnam in 1970 and flew more than 500 combat hours in the UH-1E. He remained with HML-167 when the squadron was redeployed to MCAS New River in 1971. He was assigned to HMM-261 Composite for Med Cruise 1-72 then back to HML-167 and MAG-29 as Staff Legal Officer for the remainder of his active duty service. He was discharged as a Captain in 1973.

As a civilian he joined the Seattle, WA Police Department in 1975. He has worked in the Patrol Division, Special Operations Division, Narcotics Section and the Training Division. He is currently assigned as a Captain and Narcotics Section Commander.

JIM BULLOCK was born on March 4, 1948 in Malvern, AR. He joined the USMC on Dec. 15, 1966. He attended basic training at MCRD San Diego from December 1966 to February 1967; Avionics Training at NATTC Memphis from February 1967 to December 1967 and served with HML-267 at Camp Pendleton, CA from December 1967 to October 1968.

He served with HMM-165 at Marble Mountain, Vietnam from October 1968 to November 1969 and with HMM-161 Santa Ana, CA from November 1969 to December 1970.

He was discharged on December 9, 1970 with the rank of Corporal. His awards included Vietnam Service, Vietnam Campaign, Combat Aircrew Wings, five Air Medals and Vietnamese Cross of Gallantry (unit citation).

Today he is married to Decarla and they have two daughters: Deanna and Tina. They reside in McAlester, OK. He is an Oklahoma Highway Patrolman.

BRUCE WILLIAMS-BURDEN was born on April 3, 1948 in Lima, OH. He enlisted in the Navy on Nov. 28, 1966 and went through boot camp and Hospital Corps School at Great Lakes, IL and FMF Combat Medical Training at Camp Pendleton in 1967. He was assigned to dispensary duty at MCALF (LTA) in Tustin, CA during 1968 and to 1st MAW in Da Nang in April 1969.

He volunteered and was assigned to MAG-16 Medical to fly as a medevac corpsman until February 1970.

He was discharged as a Petty Officer Third Class in August 1970 at Camp Pendleton. During his tour he was awarded the Navy Aircrew Wings, the Marine Corps Combat Aircrew Wings, seven Air Medals, Navy Commendation Medal for Heroism, PUC, NUC, MUC, CAR, Vietnamese Cross of Gallantry and a Rescue Citation from the Boeing Corporation.

In 1974 he graduated from Northeastern University Physician Assistant Program and has worked for the U.S. Government in North Carolina, Washington State and Sudan. He worked for Alyeska on the Trans-Alaskan Pipeline and in private practice he has worked in California and Washington. He currently works in neurosurgery at the VA Hospital in Seattle. He also served as a volunteer Medical Team Leader in California after the Northridge earthquake.

He has three sons, Tristan, Sean and Josh and one daughter Gillian and lives in Lake Forest Park, WA.

JEFFREY W. BURSTEIN was born on June 27, 1946 in Philadelphia, PA. He entered the Marine Corps on Dec. 2, 1964, then went to boot March 1, 1965 Parris Island. After boot and ITR went to Memphis, then stationed at Cherry Point, NC in VMT-1, a starched wing outfit with TF 9-J Trainer fighter aircraft.

He received orders for Vietnam in the summer of 1966 and was sent to Marble Mountain-a Rotor base! Good ole Marine Corps!! From starched wing to rotor wing!! Makes sense, doesn't it?!! Not too illustrious there. Flew a bunch as a port side gunner on UH-34s; pulled his share of medevacs and at the base, worked as a supply guy.

Still a Corporal, came home and back to a starch wing, VMGR-352 had a few problems with the rocks that had to be painted and wound up as a PMI at the range. Was released from the Corps on Feb. 28, 1969 as a Sergeant. OOORAH!!! Oh yeah, he got a Good Conduct Medal. Go Figure!!!

He now resides in Spring House, PA and is a carpenter/contractor.

THOMAS BUSCEMI JR. Major USMC (Ret), born Oct. 13, 1944 in Manhattan, NY and graduated from the U.S. Merchant Marine Academy in 1962, went to OCS in October 1966, USN Flight School in January, 1967, and arrived in Quang Tri on March 24, 1968 as a UH-34 pilot with HMM-163 and HMM-363.

After a tour as a SAR pilot at MCAS Kaneohe, he transitioned to A-6s in 1972. His USMC career was completed with two tours at the USMC Tactical Systems Support Activity at Camp Pendleton.

His most memorable event occurred at a retirement party 25 years after Vietnam when thanked for being a medevac pilot by "grunts" in attendance. He helped some mother's son to make it home. He continues to fly H-34s as a weekend CFI. He and his wife Nancy lost their son, Paul Thomas who died in September, 1993. Their daughter, Tammy is engaged to a Navy helicopter pilot.

JAMES T. BUTLER, ESQUIRE was born on Dec. 9, 1942 in Panama City, FL. A prominent Tampa attorney for more than 20 years, James T. Butler's life reflects many celebrated acts of achievement, responsibility, commitment and generosity. As a student he participated as a first string football player on the only Plant High School team that was undefeated and untied in the school's history as well as serving as president of the student council. After graduating from the University of Florida, he was commissioned as a Second Lieutenant in July 1965 in the United States Marine Corps with OCS, Quantico. He attended flight school at Pensacola and New River, NC. He was promoted to Captain. He served in Vietnam as a helicopter pilot flying UH-34s and CH-46s from October 1967 to November 1968. During that year of service he flew 988 missions.

He was discharged on Dec. 15, 1970 with the rank of Captain. He was awarded various medals including one Silver Star, two Distinguished Flying Crosses, two Single Mission Air Medals, Navy Commendation Medal, 49 Strike Flight Air Medals and various battle ribbons.

Butler earned his law degree from the University of Florida in 1972 and worked in the areas of defense and personal injury as a trial lawyer. He formed his own law firm in 1981. Butler has been a very active and dedicated volunteer with the National Kidney Foundation for the past seven years, helping the Foundation raise more than $350,000 with its Tampa Bay area golf tournaments and the nationally recognized Pro-Am Surfing Contest held annually at Cocoa Beach, FL. He is married and has one son.

REGINALD FOSTER BYRNE, "REG," was born on Aug. 4, 1939 in New York City, NY. He was inducted into the Marine Corps on Sept. 18, 1960. He served at various locations in the States: Parris Island, Camp Lejeune, Pensacola, Santa Ana, El Toro and Kaneohe.

He served in-country as an aircraft commander/flight leader on CH-46 A/D for HMM-265. He was stationed at Marble Mountain, on the USS *Tripoli*, at Phu Bai, Da Nang and Iwakuni, Japan.

He was discharged on Feb. 18, 1970 and attained the rank of Captain. His medals and awards included 26 Air Medals, PUC, NUC, MUC, Vietnamese Service and Campaign Ribbons and the Vietnamese Cross of Gallantry.

Today he resides in San Diego, CA and practices as a real estate broker. He also writes as an avocation.

WILLIAM J. BYRNE, "BILL," was born on June 20, 1941 in Syracuse, NY. He attended Syracuse University and attained a bachelor's degree in landscape architecture in 1965. He joined the USMC in January 1966.

He was a pilot flying CH-46s and UH-1Es and an intelligence officer while in-country.

He was discharged June 1, 1970 and attained the rank of Captain. His awards and medals included: National Defense Service Medal, Vietnam Service Medal, Presidential Unit Citation, Vietnam Campaign Medal, Air Medal with Bronze and Gold Star, 38 Air Medals (760 Combat Missions) and Certificate of Commendation

He and his wife, Ralene "Babe" Atkins-Byrne, and their three daughters, Theresa Ann (Teri) 24, Shannon 21 and Kelly Marie 17, reside in Marietta, GA. He is Cobb County Commission Chairman and a planning consultant and project manager for Byrne Design Group. He is a member of these organizations: Girls Inc. of Cobb County (1989-1992) board of directors; National Sudden Infant Death Syndrome Foundation (SIDS) (1989-1992) board of directors; Cobb County Land Use Plan Advisory Committee, committee member; Government and Business: "A Needed Partnership"-Cobb County Chamber of Commerce, committee member; International Arabian Horse Association, Federation of Cobb Homeowners Association, Greater Acworth Business & Professional Association, Kennesaw Business Association and Smyrna-Cumberland Rotary Member.

JERRY CAMPBELL, Lieutenant Colonel, ANG (Ret), born September 27, 1939 in Livonia, MI. He was commissioned a Second Lieutenant on March 15, 1964 at Quantico, VA designated a Naval Aviator on August 15, 1965 at Pensacola, FL. He joined HMM 165 at Santa Ana as the Embarkation Officer. He qualified in the CH-46 and went to Vietnam in July, 1966, flying 800+ combat hours. His ship's control cables were shot while leaving a zone and he had to fly for 30 minutes before landing with only three out of 12 cables left. His awards include 28 Air Medals, MUC, MUC, Vietnam Service and Campaign Ribbons, Vietnamese Cross of Gallantry.

After a Med cruise, he completed active duty on August 15, 1968 as a Captain at Camp Lejune, joined VMO-4 at Selfridge ANG Base, MI, joined 110 TASS ANG in Battle Creek, MI flying O-2s and A-37s. He retired in 1989.

TOM CANNON, "ACE," was born on May 13, 1949 in San Francisco, CA and graduated from Manzano High School, Albuquerque, NM in June 1967. He enlisted in the Marine Corps that same month and went to boot camp at San Diego. After completing aviation training at Memphis, he was assigned to VMFA-251 at MCAS Beaufort, SC before joining MAG-36 at Phu Bai, RVN in December 1968.

He served with HMM-265 and H&MS-36 while in-country and flew CH-53s, CH-46s and UH-1s as a door gunner with HMM-265, HMH-462 and HML-367. He was transferred to HMH-462 in October 1969 and deployed with MAG-36 to its new home at Futema, Okinawa that same month. He continued to serve with HMH-462 at Futema for the remainder of his Marine Corps service.

"In May of 1969, I was detailed along with several other Marines to escort a four truck convoy going to Wing Headquarters at Da Nang to pick up supplies. We arrived at Da Nang without incident after a hot, dusty trip and picked up the supplies the next morning.

"Among the supplies being loaded were two six bys full of beer destined for MAG-36 Special Services back at Phu Bai. We positioned these two trucks in the center of the convoy for maximum security and started back.

We were delayed upon reaching Hai Van Pass, north of Da Nang as an action was going on in the pass. When it was felt OK to proceed, we started the climb up the pass in the company of several other Marine vehicles.

At a sharp bend in the pass, our little convoy was hit by a determined, but poorly executed ambush. We immediately deployed from the vehicles, and a after a sharp, but brief fire fight, the enemy broke contact.

Some of the trucks had suffered minor damage, (the cargo of beer was examined first by several anxious individuals), but no one was wounded and all the beer was safe and undamaged.

We made our way back to the MAG-36 compound at Phu Bai without further incident with our precious cargo intact!"

He was returned to Camp Pendleton in June 4, 1971 and was discharged with the rank of Sergeant, having completed two and a half years continuous service in the Orient. He was awarded the Marine Corps Aircrew Combat Wings, Air Medal and Combat Action Ribbon during his tour in Vietnam.

He enlisted in the Army in November 1972 and was commissioned through OCS in June 1976. After an eventful Army career that took him back to Asia, as well as to Europe, he retired from the Army as a Major on Aug. 31, 1993, after 25 years of service. He holds a BA in Asian studies and an MS in international relations. He has been married for 22 years to his sweetheart from his MAG-36 days, who is from Taipei, Taiwan. They have one daughter. He continues to reside in the Far East, and works as the director of Plans and Programs for the U.S. Army at Camp Zama, Japan, southwest of Tokyo.

ANTHONY A. CANZONERI, "TONY," was born on May 28, 1949 in Nesquehoning, PA. He graduated from Marian High School and enlisted in the Marine Corps in July 1967. He graduated from boot camp at Parris Island on Sept. 12, 1967. He reported to Memphis for Aviation Operations School and graduated on Feb. 2, 1968.

He reported for duty with H&HS-2, Wing G-3, 2nd MAW Cherry Point and transferred to Vietnam joining HML-167, MAG-16 in December 1969. While serving in the Operations Section S-3, he volunteered for flight duty as a door gunner on the UH-1E.

He received Combat Aircrew Wings on May 18, 1970. At the completion of his tour, he was awarded a Single Mission Air Medal and a total of 14 Air Medals. He was reassigned to the S-3 section H&MS-14, 2nd MAW, Cherry Point upon returning to the United States and was discharged on July 2, 1971 with the rank of Sergeant.

After discharge he attended Lehigh County Community College and Penn State University receiving degrees in business administration and accounting. He is currently employed as an accountant with the Commonwealth of Pennsylvania, Harrisburg, PA. He is married with two sons and resides in Lewisberry, PA.

ROBERT R. CARBARY JR., "TIM," was born on Sept. 27, 1945 in Baton Rouge, LA. He joined the Marine Corps as an Aviation Cadet after attending LSU from 1963-65. He was commissioned a Second Lieutenant and designated a Naval Aviator on Aug. 7, 1967. He was assigned to HMM-264 as a UH-34 helicopter pilot from August 1967 to November 1967.

He was transferred Dec. 4, 1967 to Vietnam and served with HMM-363 until Dec. 24, 1968. He accumulated more than 800 combat hours.

He completed active duty on July 1, 1970 flying the CH-46D with HMM-264. He achieved the rank of Captain and was awarded the Single Mission Air Medal as well as the Navy Commendation, 30 Air Medals, NUC, MUC, Vietnam Service/Campaign Ribbons and the Vietnamese Cross of Gallantry. Captain Carbary continued to fly with reserve squadron HMM-767 until 1977.

He graduated from LSU in 1971 and attended Loyola University Law School. Since December 1972 he has served as a trooper/pilot with the Louisiana State Police. Command Pilot Carbary has received numerous awards and recognition during his law enforcement career and in 1988 was promoted to his present assignment as the Aviation Commander of the Louisiana State Police.

Carbary and his wife, Tootsie, have been married for more than 25 years and reside in Baton Rouge, LA with two daughters.

JOHN CARBONE, "GUINEA," was born on Jan. 11, 1948 in Brooklyn, NY. He graduated from Sheepshead Bay High School in 1965. He enlisted in the Marine Corps on Sept. 10, 1965 and went through boot camp at Parris Island, Platoon No. 279. He reported to NAS Memphis, TN for aviation training, graduated as a jet helicopter mechanic and was assigned to HMM-261 at MCAF New River, NC. After a Caribbean cruise aboard the USS *Guadalcanal LPH-7* in 1967, he was transferred to HMM-161 also at New River. On April 20, 1968 he was a flight crew member when HMM-161 flew 24 Sea Knight CH-46Ds in a record breaking cross-country flight from MCAF New River to MCAS El Toro, CA.

The crew of HMM-161 then boarded the USS *Princeton LPH-5* which transported them to join newly formed Prov MAG-39 at Quang Tri, Vietnam. From May 1968 to June 1969 he served with HMM-161 and HMM-262 at Quang Tri as a helicopter gunner, check crew, reactionary platoon and perimeter guard duty.

During his in-country tour, he was awarded the Marine Corps Combat Aircrew Wings with three stars, six Air Medals, Vietnamese Cross of Gallantry, Civil Action, NUM, MUC, GC, ND Vietnam Service, Campaign Medals and the N.Y.S. Conspicuous Service Cross. He was honorably discharged as a Sergeant at MCAS, El Toro, CA on June 4, 1969.

He has been a police officer since June 1971 and has earned two Exceptional and three Excellent police duty awards while assigned to neighborhood foot patrol, Juvenile Bureau Motor Patrol and is presently senior motor patrol training officer with the Malverne Police Dept. He has attended the New York Institute of Technology for three years and is a life member in the VFW, DAV and he is also a member in the 1st MAW ASSOC and USMC/Vietnam Helo and Aircrew Association. Carbone has been married for 11 years.

ROBERT A. CARLSON "BOB" Major USMC (Ret) was born on Sept. 4, 1937 in Chicago, IL and grew up in Millbury, MA. He attended Rensselaer Polytechnic Institute under the NROTC Program, and upon graduation in 1960 was commissioned as a Second Lieutenant, USMC. After designation as a Naval Aviator in 1962, he was assigned to HMH-461 at New River.

His first Vietnam tour was in 1966-67, with H&MS-16, SU-1 (CH-37 Detachment). He returned to Vietnam in 1970 to fly OV-10s with VMO-2, borrowing his TAC(A) call sign "Hostage Junkman" from his previous Vietnam helicopter tour.

He holds the Distinguished Flying Cross and 37 Air Medals. He accumulated a total of 4,000 flight hours before retirement in 1980.

He is currently a marketing manager with Walter Kidde Aerospace, a manufacturer of fire detection and suppression products for military and commercial aircraft. He lives in Wilson, NC with his wife, Carole. They have two sons.

EMMETT L. CARSON JR. was born on Sept. 5, 1942 in New Iberia, LA. He graduated from Welsh High School in 1960, then from Northwestern State University of Louisiana with a bachelor of science degree in 1965. He joined the Marine Corps in 1965 and completed OCS in 1966 at Quantico, VA. He was designated a Naval Aviator in 1967 at NAS Pensacola, FL.

He joined HMM-364 flying CH-46s at MCAS Santa Ana, CA and was sent as part of an advance party to train with HMM-165 flying UH-34s in combat for a few weeks. He rejoined and helped train HMM-364 personnel when it arrived in-country as a cadre from the U.S. He was wounded during the 1968 Tet Offensive at Khe Sanh. He rejoined HMM-364 in Phu Bai after hospital recuperation in Japan.

"While we were standing by in the aircraft for an imminent 12 helicopter troop insertion, a fellow pilot, Joe Duckett, succumbed to the demands of a mild intestinal virus and hastily occupied a convenient three holer" outhouse along the side of the runway. As luck would have it, the call to "Launch all aircraft," came with Joe's flight suit at half mast. While rushing to suit up, Joe watched in horror as his web belt and pistol, which had been removed to unzip the flight suit and placed next to him on the seat, quickly disappeared into the adjacent black hole. All that we saw from our bird was the door to the outhouse burst open and a figure holding a brown snaky-looking object, at arms length, cover the distance to the helicopter at an ungainly gait and disappear into the cockpit.

"During the debrief after the mission, Joe said that he momentarily contemplated facing a court-martial for abandoning his weapon instead of actually having to reach in and retrieve it. "As legends go, Joe's nickname stuck... 'Honeybucket Duckett.'"

Carson was awarded 26 Air Medals, Purple Heart, Presidential Unit Citation and Vietnam Service and Campaign Ribbons. His Vietnam tour ended November 1968 and he was assigned to HMM-162 at MCAS New River, NC. He went on a Caribbean cruise with HMM-162 aboard the USS *Guam* in 1969. He left active service in April 1970 as a Captain.

He is presently employed by Chevron out of Lafayette, LA flying helicopters to offshore rigs to the Gulf of Mexico. He is still married to his college sweetheart and has one son, Ryan.

GEORGE G. CASEY was born on July 13, 1948 in St. Louis, MO. He was inducted into the USMC on Oct. 10, 1970. He was stationed at Quantico, VA, Pensacola, FL, Santa Ana, CA and New River, NC.

While in-country he was HAC on CH-53s and was stationed at Futema, Okinawa. His memorable experiences were the evacuation of Phnom Penh and Saigon.

He was discharged on June 1, 1976 with the rank of Captain. His awards included the Air Medal, Vietnam Service Ribbon and Shirley Highway Ribbon.

Today he resides in Chesterfield, MO with his "USMC Issue" wife and his four "NCOs" (kids). He is a broker with Smith-Barney-Shearson.

BERNARD D. CERRA, "BERNIE," was born on Aug. 22, 1946 and lived in Carbondale, PA. Upon graduation from the University of Scranton in June 1968, he received a commission in the USMC reserves and thereafter reported to flight training in Pensacola, FL. He was designated a Naval Aviator in September 1969 and then served with HMH-461 at the New River MCAF.

Qualified in the CH-53, he reported to HMH-463, MAG-16, Marble Mountain, Vietnam in April 1970. He flew more than 800 combat hours.

He was awarded the Distinguished Flying Cross, along with two Single Mission Air Medals. He holds a total of 41 Air Medals, Vietnam Service and Campaign Medals, NUC, MUC and RVN MCU Cross of Gallantry. He completed active duty with the rank of Captain in 1973, and then entered on duty as a special agent, FBI.

He is assigned to Washington, DC and lives in Virginia. Cerra is married, with two sons and one daughter.

HENRY J. CIPOLLA, "CHIP," Major, USMC (Ret) was born on May 25, 1938 in Providence, RI. He enlisted in January 1956 and served as an aircraft electrician until February 1964 when he was selected for 5th Warrant Officer Screening Class. His last enlisted assignment was HMX-1 as a member of the Presidential Executive Flight Detachment.

After OCS, he had duty with MAG-36 until September 1965 when he was deployed to Vietnam with HMM-363 as an AAMO. He served two additional tours in Vietnam as AMO of HMM-165 and HMH-463.

He served numerous stateside and overseas duty stations until retirement on July 1, 1978. His awards included Meritorious Service Medal, two Bronze Stars w/combat V, three Air Medals, three Navy Commendation Medals w/combat V, two Good Conduct Medals, National Defense Medal, Armed Forces Expedition Medal, Vietnam Service Medal w/14 Battle Stars, Vietnam Cross of Gallantry w/Palm, Vietnam Unit Medal, three Navy Unit Citations, Combat Action Ribbon, Presidential Unit Citation, Combat Aircrew Wings, Presidential Service Badge, Rifle and Pistol Expert Badges and Winged Sikorsky "S."

As a civilian, he was vice-president with MFC First National Bank, Escanaba, MI until May 1992. He and his wife, Mary Kay, currently reside in Escanaba, MI.

CHARLES T. CLARK, II, "PIGGY," CWO-3, USMCR (Ret) enlisted in the Marine Corps on Sept. 19, 1965 and went though boot camp at Parris Island. He then reported to Memphis for Jet Helicopter Maintenance Training. His next assignment was with VMO-5 at MCAF Camp Pendleton for UH-1E Helicopter Aircrew Training.

In May 1967 he was transferred to Phu Bai, Vietnam and served as a crewchief with VMO-3 and HML-367 until June 1968. He participated in 1968 in the Tet Offensive during the battle of Hue City and Khe Sanh. He returned to CONUS in July 1968 and served with VMO-1 at MCAF New River until September 1968. He returned to Vietnam for his second tour with VMO-6 at Quang Tri. During both tours in Vietnam he served as a crewchief.

During his two tours he was awarded the Marine Corps Combat Aircrew Wings and two Single Mission Air Medals. He holds a total of 39 Air Medals plus the Navy Commendation Medal with a Combat "V," Combat Action Ribbon, PUC, NUC, MUC, Civil Action Ribbon, Vietnamese Cross of Gallantry and Naval Reserve Medal. He completed active duty as a Sergeant in September 1969.

Clark returned to college and earned an AA, BS and MA degrees. He joined the Marine Air Reserves in July 1973 and was commissioned a Warrant Officer in October 1979 and attended the Warrant Officers Basic Course at Quantico in July 1980. He served as assistant aircraft maintenance officer with VMFA-321, NAF Andrews AFB, MD until his retirement on Aug. 1, 1989.

As a civilian he holds a position as a systems manager with AT&T in Bridgewater, NJ. He has 21 years service with AT&T. He is married and has two sons.

MIKE D. CLARKE was born on March 18, 1947 in Fresno, CA. He graduated from high school in June 1966. He reported to MCRD San Diego on June 6, 1966. He was assigned as a plane captain, mechanic at VMF-(AW)451 at Beaufort, SC.

He was transitioned to VMF(AW)-235 Da Nang and to H&MS-13 Chu Lai. His third tour was with HMM-265 at Phu Bai and aboard the USS *Iwo Jima*. He was also stationed at Iwakuni, Japan.

He was awarded Combat Air Crew Wings, 12 Air Medals and was discharged on Sept. 12, 1969 as a Corporal.

He joined the Marine Corps Reserves, Combat Engineers, 4th LAMB at Fresno in 1977 and transferred in 1979 to the 144th F.I.W. Air National Guard at Fresno. He was a crewchief on T-33s, F-106s, F-4s and currently F-16s as a full-time air technician, still on the flight line.

"A public thank you to our wives for listening for so many years about which one of us back in the CH-46 days at HMM-265 was the better door gunner versus crewchief. What was important was coming home, and our friendships over the miles and through the years."

He is enjoying golfing with his wife, Kathy, whom he met in high school, and they have one daughter, Missie. His best friend, Corporal John E. Hunt (Crazy Horse) from Antlers, OK, and he try to get together every year or so when Hunt's wife lets him off the reservation. To all USMC/Vietnam Helicopter Pilots and Aircrew Reunion members. *Semper Fi*.

RAYMOND MIKE CLAUSEN, JR, "PEACH BUSH PVT/BLOOD, SWEAT & TEARS," born October 14, 1947 in New Orleans, LA and joined the USMC on May 27, 1966 receiving training at the MCRD San Diego, CA then, Aviation Mechanic and Basic Helo Training at NAS Memphis, TN. In April, 1967 he was stationed at New River, Jacksonville, NC in HMM-365. He went to Vietnam in 1967 and served with H&MS-36 at Phu Bai. On January 31, 1970 he was serving with HMM-263, MAG-16, 1st Marine Aircraft Wing, when he received the Medal of Honor, the only USMC aircrewman to do so. It was awarded for conspicuous gallantry and intrepidity at the risk of his life above and beyond the call of duty while rescuing wounded Marines. (Details in Chronology) His other awards include the Combat Air Crew Wings, Purple Heart, 94 Air Medals with 4 Gold Stars, Combat Action Ribbon, Presidential Unit Citation, Navy Unit Commendations with Bronze Star, Good Conduct Medal, National Defense Service Medal, Vietnam Service Medal with Silver and Bronze Star, Rifle Sharpshooter Badge, Vietnam Gallantry Cross, Campaign Medal, and Civil Action Unit Citation.

RICHARD W. CLIFTON "DICK" was born on Dec. 25, 1941 in Brooklyn, NY. He graduated from Sweetwater Union High School, National City, CA in 1959. He was commissioned Second Lieutenant in 1963 upon receiving a BS degree from Cal. Poly, San Luis Obispo. Following completion of flight training at Pensacola, he joined HMM-363 at MCAF, Santa Ana, November 1964.

He served as embarkation officer for MAG-36 aboard the LPH USS *Princeton* en route to Ky Ha, RVN, August 1965. He flew 700 UH-34D combat hours. He was promoted to Captain and returned to HMMT-301 as flight instructor. He transitioned to the OV-10 and joined VMO-2 at Marble Mountain, RVN, December 1968. He served as an air liaison officer with the 4th Marines at Vandegrift Combat Base near the DMZ.

He was awarded 33 Air Medals plus CAR, PUC, NUC, MUC, Vietnamese Service and Campaign Ribbons. He returned to civilian life in 1970.

He presently is a medical technologist for the North Colorado Medical Center Laboratory in Greeley. He has been married to Sharon for 32 years and has two daughters, one granddaughter and a 1965 Beechcraft A23.

ERIC J. COADY, "JOLLY GREEN GIANT," Lieutenant Colonel USMC (Ret) was born on April 19, 1939 in Philadelphia, PA. He was graduated from high school in 1957 in Harrington, DE. He holds a BS from Auburn University, 1973 ("Bootstrap"). He enlisted in the Marine Corps Reserve April 19, 1956, completed boot camp at Parris Island in 1958 and entered flight training at Pensacola in May 1959. He reported to HMR(L)-363 at MCAF Santa Ana in November 1960.

He served three tours in RVN with HMR (L)-163 in 1962-63, HMM-361 and HMM-363 in 1967

62

and as ACOS G-2 in 1970-71. Other tours were as a flight instructor at Pensacola and staff assignments at FMFLANT and FMFPAC. His non-flying duties were primarily in aircraft maintenance and his last two operational assignments were as CO of H&MS-24 and H&MS-17, respectively.

His final active duty assignment was as an aircrew training officer at HQMC and retired on April 1, 1981 with the rank of Lieutenant Colonel. He was qualified as a HAC and section leader in the HUS-1 (UH-34D) during his first tour in RVN and as a flight leader during his second tour. During his third tour he was qualified as HAC in the in the UH-1E and copilot in the C-117. He had almost 5000 hours of flight time divided between rotary and fixed wing and has been awarded three DFCs, the MSM, one Single Mission and 44 Strike/Flight Air Medals, two NCMs with "V," the PUC and NUC.

After just over 10 years as an engineer with government contractors he is now a high school math and industrial technology teacher. He and his wife now reside Pensacola, FL. His wife, Sandy, has cheerfully supported his unsettled lifestyle since October 1960.

ROGER K. COOK was born on Aug. 27, 1938 in Beaumont, TX. He was graduated from Natrona County High School, Casper, WY, in 1956. He holds a BS/BA from Central State University, Oklahoma. He was commissioned in 1962 and entered basic school. He was designated a Naval Aviator in June 1964. He served with HMM-362, MCAS Santa Ana, CA.

He was deployed to Ky Ha, RVN with HMM-362 in August 1965 and flew more than 500 missions.

He was released from active duty in June 1968. His awards and medals were Air Medals and Navy Commendation w/combat "V" (two). He rotated to CONUS in 1966 and was an instructor in HT-8 at Pensacola.

He flew with Braniff in South America from 1969-1983 in the DC-8 and was a B-727 Captain. During a furlough with Braniff, he flew UH-1s and UH-34s with Air America, Udorn, Thailand, 1971-1973. After the demise of Braniff, he started over again with Continental in 1985 'til present. He flew with Air Micronesia and is currently flying South and Central America with a total of 22,000 hours.

WARREN G. CRETNEY Colonel USMC (Ret) was born on April 4, 1933 in Amarillo, TX. He enlisted in June 1952 and was commissioned in June 1953. Basic School and Naval Flight Training followed. He received his BA degree from San Diego State College in 1968 and attended MC Command and Staff College in 1972-73.

During his career, he served with VMFA-115, HMR-162, HMX-1, VMO-1, 2, 6 and had 1,100 combat hours. He commanded H&MS-39 and HML-367 (Cobras) in RVN. He served as Marine Corps liaison at Fort Rucker, AL. He then spent three years in Italy as a NATO planning officer. He returned to CONUS to command MAG-36 and this was followed by two Med cruises as CO 32 MAU.

He completed his career on April 1, 1981 at FMFLANT as air officer and chief of staff, 4th MAG. He holds the Legion of Merit w/V, five Distinguished Flying Crosses, three Air Medals with 53 Strike-Flight Awards for actions in Vietnam and the Defense Meritorious Service Medal and two Meritorious Service Medals. Today he resides in Panama City, FL and is teaching sailing/seamanship in an alternative school.

BRUCE M. CROW, "BUZZARD," was born on Oct. 18, 1946 in W. Reading, PA. He joined the service in August 1965 and attended boot camp at Parris Island. He was stationed at P.I., Camp Geiger, Millington NAS, Santa Ana and New River.

He was a gunner/crewchief while serving in-country for HMM-361, (UH-34s). He remembers pulling two wounded Marines out of a rice paddy when their squad was pinned down. It was rare to see the entire picture from the air, knowing exactly the dangers before dropping in for the medevac.

He remembers circling over a force recon team and listening to them whisper on the radio because the enemy were that close. Going in for the extract, he hung by his gunner's belt, while the gunner held his feet and legs, so he could reach them. He remembers night medevacs. A corpsmen instructing him to hold a WIA Marine on his side so he wouldn't drown in his own blood. The Marine died in his hands on the flight back. Memorable experiences also include Sept. 3, 1967 when an ammo dump was hit by NVA artillery. HMM-361 took 47 WIA.

He was discharged on June 11, 1969 with the rank of Corporal at New River. He received 15 Air Medals and a CAR.

He has been married for 23 years and has three children, two daughters and one son. He resides in Richmond, VA and has been in the building material sales and management area for 23 years.

THOMAS V. CROW, "BIRD," was born on Dec. 29, 1947 in Altus, OK. He joined the USMC on May 21, 1966. While in the States he was stationed at MCRD San Diego, Lakehurst NAS, New River, NC and Willow Grove NAS.

While in-country he was a door gunner and a "stinger" and flew the CH-46D. He was stationed in Marble Mountain, RVN.

He was discharged in May 1970 and achieved the rank of Sergeant. He received these medals: Jump Wings, Combat Aircrew Medal with three stars, Single Strike/Flight and the Distinguished Flying Cross.

Today he is married with two children and is a school facilities manager. He resides in Hatboro, PA.

JAMES A. CUTRI, "COOCH," was born on Oct. 21, 1947 in Long Beach Naval Hospital, CA. He was living in Syracuse, NY when he enlisted. After boot camp at PI in Platoon No. 207, and Camp Geiger, he went through Aviation Technical Schools at Memphis. Then on to El Toro and finally to LTA to form up with HMM-164. He was the one who dragged his guitar around and sang up a storm for his "bros." "House of the Rising Sun" was a favorite.

James "Cooch" Cutri was one of the original crew of HMM-164 (CH-46s). They shipped out in February 1966 on the USS *Princeton* bound for Marble Mountain. He was assigned to hydraulic shop and flew as port side gunner.

He received the Air Medal and the Navy Commendation Medal. He was discharged on Jan. 20, 1969 and achieved the rank of Corporal.

He still functions as a mechanical technician in the water treatment field, with certification in both mechanical technologies and water treatment. As for the music he sings country and ballads in supper clubs in Sonoma County, CA. He is married with three kids and lives near Santa Rosa, CA.

JAMES J. DALTON, II, "MAD DOG," was born on July 2, 1941 in Watertown, NY. He enlisted in the U.S. Army on June 9, 1959 and then transferred to PLC(A) USMCR Feb. 15, 1962. While in the States he was stationed at New River, NAS Atlanta and FMFLANT Norfolk.

While in-country he was a co-pilot/HAC on CH-46As and also flew UH-34s and UH-1Es.

"Vietnam on April 25, 1967, I was flying co-pilot for Capt Jack House (KIA June 30, 1967). Our crew consisted of Staff Sergeant "Red" Logan and Lance Corporal Dan Dulude. We were the second aircraft to attempt a landing near a pinned down recon team. We waved off and made a tight turn landing on Spiney Ridge, just up the hill from team. The team was unable to move to our aircraft because all the team members were wounded. We locked the brakes and backed the helo down the hill towards the team. Once close, crewchief Dulude jumped from the rear of our helo and began helping the recon team get aboard. We extracted all team members and never heard from them again until Fred Baker, one of their members, called to thank me in early 1989."

He was discharged on July 1, 1991 with the rank of Lieutenant Colonel USMCR. His awards include 19 Air Medals, PUC, NUC and an Army Good Conduct Medal.

He is now a banker/lawyer, financial advisor and resides in Riverdale, GA. He is married to Virginia Burnett Dalton and they have a son, Michael Joseph Dalton (Oct. 30, 1964), a daughter, Anne Marie (Aug. 2, 1966), and a son, Brendan Christopher (March 10, 1971). He has one granddaughter, Erin Caitlin Dalton, born on April 6, 1994.

DOUGLAS CHARLES DARRAN, "DOUG," was born on Oct. 26, 1941 in Easton, PA. He graduated from EHS in 1959 and had extended R&R at Penn State from 1959-64. The went through PI and ITR at Lejeune in 1964 then MARCAD and was commissioned a Second Lieutenant, Naval Aviator in March 1966 and joined HMM-162 at New River.

He served in HMM-363 in Ky Ha; aboard the LPHs *Iwo Jima* and *Princeton*, Dong Ha; Marble Mountain from September 1966 to October 1967; HMM-264 New River; and aboard LPH *Guadalcanal* in 1968; the Caribbean cruise ("Sandals Doth Not A Hippy Make"; Grunts Can't Drink Flaming Hookers") New River.

He was discharged in March 1969. He received 32 Air Medals and the Sikorsky H-34 1,000 hour pin. He was a Captain in the Reserve HMM-772 Willow Grove and Lakehurst; finished Penn State with a B.Sc. (business administration) in 1970 and went to Australia for R&R, laborer. He was an administrator for James Cook University from 1971-73; Key West-R & R, hospital business officer manager 1973-74; and at Penn State a Ph.D. candidate (ABD) 1974-77. He was assistant professor of business and administration and an administrator in 1978-89 for the University of South Carolina and received an MBA from the University of South Carolina in 1989. He was economic/financial litigation support officer for a CPA firm from 1989 to 1992.

He is currently administrator for Hospitality Resources Inc., an employee leasing and accounting services firm with more than 3,000 employees in 62 restaurants in 11 states. His primary duty is Workers' Compensation. He is also adjunct lecturer in economics at University of South Carolina. His secondary duty is R & R.

NED L. DAUTRIEL, "NEDDY," was born on Feb. 29, 1948 in Lake Charles, LA. He enlisted in the Marine Corps on June 12, 1966 and went though boot camp at San Diego. He reported to Memphis for aviation training in helicopters and was next assigned to HMM-365 at MCAF New River.

After a Caribbean cruise in July of 1968, he was transferred to Vietnam. From December 1968 to January 1970 he served with the Purple Foxes of HMM-364 in-country as a crewchief. His nickname was Neddy and his aircraft was Candy. He has been known to cool a few beers with CO2 fire extinguishers and warm up C rations in CH-46 engine compartments (note: punch hole in can before heating).

During his tour he was awarded the Marine Corps Combat Aircrew Wings, 33 Air Medals with Gold Star, Navy Commendation Ribbon, Vietnamese Cross of Gallantry and the Vietnam Civil Action Citation. He was discharged as a Corporal at New River on June 12, 1970.

He is presently self-employed as a plumbing contractor in partnership with Leon Dautriel (1958-61 USMC). He has flown a few missions with Medal of Honor recipient, Mike Clausen, who remains a close friend.

JON J. DAVIS, "REDS," was born on Jan. 18, 1948. He enlisted on Aug. 15, 1966 and went from Parris Island to Memphis for his flight training.

After a brief stint at New River, he was transferred to HMM-265 at Marble Mountain, arriving just before Tet in 1968. He served as a door gunner and crewchief until May 10, 1968 when his CH-46 was shot down attempting to insert troops into Ngoc Tavak, an artillery base on the Laotian border. Severely wounded, he was medically retired and had 16 operations over a five year period, before returning to the workforce as a salesman.

With the help of the VA and the DAV he returned to school in 1990 and got his BS in political science from Towson State in Baltimore. He is married, with four sons and one granddaughter.

CHARLES J. DAW "CHARLIE" was born on April 17, 1936 in Milwaukee, WI. He joined the Marines on Sept. 8, 1953. He served as a communications "grunt" until 1960, when he got tired of walking and retrained into aviation maintenance.

He flew in-country with HMM-162 in 1964 and 1965, with HMM-165 in 1966 and 1967 and served as maintenance chief of HMM-263 at Marble Mountain in 1970 and 1971.

Daw was awarded a Single Mission Air Medal and two Navy Commendation Medals with Combat "V," among others. He retired as maintenance chief of HMM-163 at MCAS Tustin in March of 1973.

In April of 1973 he elected construction as a second career and is presently a licensed general contractor in the states of California, Hawaii, Nevada and, Arizona. He is married, and has three children and seven grandchildren. What a guy!!

DANIEL J. DEBLANC Captain USMC was born on Sept. 13, 1944 in New Orleans, LA. He graduated with a BS degree from Southeastern Louisiana University. He was commissioned as Second Lieutenant on June 2, 1967 at Quantico and was designated a Naval Aviator on Aug. 16, 1968 at Pensacola. He served with HML-267 at Camp Pendleton from September 1968 through January 1969.

He joined VMO-2 at Marble Mountain, Vietnam in January 1969 (Hostage Romeo). He flew the UH-1E and AH-1J Cobra gunships. He had the dubious distinction of having the first USMC Cobra shot down in Quan Duc Valley in September 1969. He spent the last two months in-country as helo director in Da Nang DASC.

He was awarded the Silver Star, Navy Commendation Medal with Combat "V," 25 Air Medals, NUC, MUC, Vietnam Service and Campaign Ribbons and the Vietnamese Cross of Gallantry. He departed country January 1970 to HML-269 New River. He entered civilian life on Nov. 15, 1971.

He married Sharon on Nov. 20, 1971 and flew H-34s and UH-1Es for Air America, Udorn, Thailand and Saigon, Vietnam 1972-73. He was flight commander AH-1J Cobra, Isfahan, Iran in 1974-75. He is presently owner of Southside Cafe Bar & Grill in Slidell, LA.

JAMES J. DEHAAN, "JIM," was born in Mauston, WI on Jan. 17, 1946. He enlisted in the Marines on Dec. 16, 1965 and reported to San Diego for boot camp on Jan. 4, 1966. He attended recip and helicopter mechanic schools at NAS Memphis and reported to HMMT-302, a CH-46 squadron at Santa Anna, CA. He was also one of the original crewchiefs to join HMM-364, a new '46 squadron at Santa Ana.

After volunteering for Vietnam, he ended up at MCAF Futema, Okinawa in a UH-34 V.I.P. unit. He arrived in Vietnam in November of 1968 and was assigned to HMM-165 until they left Vietnam in mid-1969, at which time he joined HMM-364 for the remainder of his tour.

Cpl. DeHaan left Vietnam in September 1969 and was discharged upon returning to the States. While serving as a crewchief, he earned 57 Air Medals, including two Single Mission Air Medals.

Today he is employed by Sprint/United Telephone as a construction supervisor and lives in Rensselaer, IN with his wife and son. He also is a certified flight instructor and instructs part-time.

CLYDE K. DOI, "C.K.," was born on Sept. 22, 1944 in Honolulu, HI. He joined the service in March 1967 and was stationed at MCAS Santa Ana.

While serving in-country he was a pilot flying CH-46s. He was stationed in-country at Marble Mountain, Quang Tri and Hue Phu Bai.

He was discharged in December 1971 with the rank of Captain. He received the Purple Heart and Air Medal.

Today he is married with two girls and resides in Aiea, HI. He serves as an air traffic controller.

EDWARD B. DOLNEY was born on Sept. 2, 1948 in Milwaukee, WI. He enlisted in the USMC on July 26, 1966 and completed boot camp at MCRD San Diego, CA.

He arrived in Vietnam in April 1967, the first of four tours. He served with the 1st MAR DIV till December 1968 and transferred to HMM-265. He flew as a mechanic/aerial gunner until returning to MCAS Santa Ana, CA for transitional MOS training in July

1969. In December 1969 he returned to Vietnam to serve with HMM-262 until his discharge.

He was honorably discharged in July 1970 as a Lance Corporal. His awards include Marine Combat Aircrew Wings, 35 Air Medals, w/Single Mission Star, VCM, VSM w/11 Campaign Stars, NUC, PUC w/star, Vietnamese Cross of Gallantry and Vietnamese Civic Action Medal.

In 1972 he again returned to Vietnam, this time as a civilian advisor to perform contract maintenance on U.S. Army CH-47 Chinooks. In January 1973 allied forces withdrew from Vietnam and he was reassigned as an advisor to the S. Vietnamese Air Force, training personnel in aircraft maintenance and assisting in building new lift squadrons. In September 1974 he returned home and completed certification as an FAA airframe and powerplant mechanic in May 1975. From 1976 to 1979 Dolney worked as a civilian helicopter mechanic and enlisted in HML-767 MARTD New Orleans, LA. In 1980 he relocated to his home state of Wisconsin where he currently resides in Milwaukee.

He is currently is a SSG Cavalry Scout in the USAR and has completed Reserve Tank Commanders Course 1984, active duty Drill Sgt.'s. School, Fort Jackson, SC 1986, attended Air Assault course 1989, completed Airborne Training, Fort Benning, GA 1991 and attended Pathfinder School in 1992. He has one daughter and two sons.

A.J. DONAHUE III, "AJ," was born on Sept. 8, 1941 in Stamford, CT and graduated from high school in 1959. He attended Georgetown University from 1959-61 and joined Class 35-62 in Pensacola in September 1962 MARCAD. He was designated a Naval Aviator and Second Lieutenant in the USMCR in June 1964. He was assigned to HMM-264 MCAS New River and was on a Med. Cruise aboard the USS *Boxer* and USS *San Marcos*.

He joined HMM-263 in Vietnam in December 1965 and had more than 400 missions. He most memorable experience was taking some force recon types to do a recon of Hills 881N & S and feeling that 10,000 Vietnamese were looking at them but not a shot was fired.

Donahue was awarded the DFC, Single Mission Air Medal and 19 Air Medals. He returned to CONUS in January 1967 and instructed at MCAS New River until discharged in September 1967 with the rank of Captain.

He flew with the HMM Reserve Squadron at Floyd Bennett Field until 1971. He is currently VP regional manager for Offshore Financial Corps., specializing in yacht financing. He now resides in Ft. Lauderdale, FL.

GERALD F. DOOLEY, Lieutenant Colonel USMC (Ret) was born on April 26, 1935 in Lowell, MA where he graduated from Keith Academy in 1952; graduated from Saddleback College with degrees in political science (1979) and liberal arts (1987). He enlisted in July 1953 and attained the rank of Sergeant after tours as a machine gun squad leader in WESTPAC, drill instructor at Parris Island and Sergeant of the guard at the Naval War College. A meritorious NCO selectee to OCS, he was commissioned a Second Lieutenant in June 1958. He completed basic school in February 1959 and flight training in September 1960.

Dooley deployed with HMM-362, the first Marine squadron in Vietnam, in April 1962. He returned to CONUS and deployed to the Cuban Missile Crisis as a flight leader with HMM-361 in September 1962. A "plank owner" of HMM-265, he was assigned as CH-46 Project Officer by COMNAVAIRLANT and was the first fleet aviator selected for factory pilot training. Captain Dooley volunteered for his second Vietnam tour, with HMM-265 in May 1966. Upon his return, he was assigned to HMX-1 where he was a Presidential Aircraft Commander and Executive Aircraft Maintenance Officer for Presidents Johnson and Nixon. In October 1970, Major Dooley deployed for his third Vietnam tour, with HMM-364 as a flight leader and A/C Maint. Officer.

He served as AMO and XO of HMT-301, 1971-74; was selected for the USAF Air Command and Staff College in 1974; Maintenance Mgmt. Officer, MAG-16 and 3rd MAW, 1975, and Plans and Electronic Warfare Officer I MAF, 1975-76. He completed USAF School of Applied Aerospace Sciences in 1976 and assumed command of H&HS-38 in September 1976 and MATCS-38 in June 1977.

He was forced down by enemy fire on four separate occasions, receiving the Purple Heart for wounds received at Con Thien, RVN in March 1967. Among Lieutenant Colonel Dooley's decorations are the Meritorious Service Medal, two Distinguished Flying Crosses, Purple Heart, five Single Mission Air Medals and 32 Strike/Flight Air Medals. During his career Dooley accumulated 5000 accident-free flight hours while piloting 19 different aircraft. He retired in December 1978.

He is a former Marine Corps light heavyweight boxing champion who also played and coached varsity baseball for six years.

He is now a facilities manager, residing with his wife, Eleanor, in Mission Viejo, CA. Their three grown children are Deborah Jean, Stephen and David and they reside nearby in Southern California.

LAWRENCE L. DOWNEY Colonel USMCR (Ret) was born on Sept. 5, 1937 in Cedar Rapids, IA. He holds a BS degree from Colorado State University and an MA degree from Cal State University, Hayward. He was commissioned a Second Lieutenant on Dec. 19, 1960 at Quantico, VA. He was designated a Naval Aviator on March 30, 1962 at Pensacola, FL. He served with HMM-263 at MCAF New River from April 1962 to April 1963.

During that time he participated in the Kennedy State visit to Costa Rica, the Cuban Missile Crisis and the civil uprising at Ole Miss operating from the decks of the USS *Wasp*, USS *Boxer*, USS *Thetis Bay* and the county airport at Oxford, MS. In May 1963, he was transferred to HMM-262 for duty in the Mediterranean aboard USS *Hermitage*. From December 1963 through February 1965, he served with 2nd ANGLICO at Camp Lejeune, NC. In March 1965, he transferred to VMO-6 at Camp Pendleton, CA. In August 1965 he departed CONUS for Vietnam with MAG-36 aboard USS *Princeton*. He returned to CONUS in October 1966 and was assigned to HMMT-302 at MCAF Santa Ana for duty as a flight instructor. In December 1967 he was released from active duty. From 1969 to 1980 he flew with the reserve squadron HMH-769 at NAS Alameda, CA. From 1981 through 1990, he held a variety of MTU/IRA command and staff positions, including establishing a reserve environmental and engineering support program at MCB Quantico, VA. He was transferred to the retired list on July 1, 1991.

As a civilian, he pursued a career as a forester with the U.S. Forest Service. He served at several National Forests and did a tour in the Forest Service Regional Headquarters in San Francisco. He retired from the U.S. Forest Service in 1985.

Larry Downey and his wife, Carlene, now reside in Elephant Butte, NM. They have two children and two grandchildren.

JOHN ANTHONY DOWNING, "J. ANIMAL ZERO," was born on May 1, 1943 in Brewton, AL. He enlisted in the USMC on Aug. 21, 1961 and reported to Navy Flight School at Pensacola, May 2, 1962 as a MARCAD. He was commissioned as a Second Lieutenant and designated a Naval Aviator on Jan. 3, 1964 at Ellyson Field. He served with HMM-363 at Santa Ana MCAF California from February 1964 to August 1965. He qualified in the UH-34D. He joined HMM-361 in Vietnam in September 1965 and flew 567 combat hours.

He left active duty serving with HMM-264 at New River, NC in December 1966. His awards and medals included 20 Air Medals, Vietnam Service and Campaign Ribbons and the Army Achievement Medal. He has a total of more than 23 years of service and more than 3500 flight hours.

He obtained a BS in aerospace engineering from the University of Alabama in 1971. He joined the Alabama National Guard in 1977 flying the UH-1H. He was Commanding Officer of the 1133rd Medical Detachment at Mobile, AL from November 1981 to October 1985.

He is currently a CWO-3 flying UH-1s with the 445th AVN DET in Mobile. As a civilian he serves as the county engineer for Escambia County, AL and hopes to retire in the near future. He is married and has two daughters. He is an active member of the USMC/Vietnam Helicopter Pilots and Aircrew Reunion, 1st MAW Association and VHPA.

JOHN CREE DUNCAN was born on Jan. 8, 1946 in Clovis, NM. He graduated from high school in 1964 in Seattle, WA. He enlisted in the Marine Corps on Nov. 22, 1965. After boot camp in San Diego and helicopter mechanics training in Memphis, he was assigned to MCAS New River, to a training squadron that was eventually formed into HMM-161. In April 1968, the entire fleet of HMM-161's CH-46 helos flew from New River NC to El Toro, CA. The squadron then flew aboard the USS *Princeton* which in turn transported HMM-161 to Vietnam.

John served with HMM-161 as a gunner and crewchief. He made two successful emergency recon extractions which resulted in two medals. The first was June 1968 when he was a gunner, and the second was August 1968 when he was a crewchief.

He was discharged as a Lance Corporal at El Toro in June 1969. He was awarded 2 Single Mission Air Medals and a total of 43 Air Medals.

In June of 1977, he graduated from the University of Washington with a BA in business administration with a degree in finance. He was a stock broker from mid-1977 to mid-1979. Since mid-1979 he has been employed by the Washington State Department of Revenue. He began as an exempt property auditor in the property tax section and since 1987, has been an auditor-appraiser. In 1987, he also became a landlord and has since acquired additional rental properties.

John and his wife, Gail, just celebrated their 24th wedding anniversary. They have been blessed with two fine children; Julie, age 22 and Chad, age 21.

RALPH E. DUNN, "RALPHIE," was born on May 24, 1941 in Oklahoma City, OK. He graduated Valley Forge Military Junior College in June 1961. He entered flight training, Pensacola in Class 28-61 as a MARCAD. He was commissioned and earned his wings in February 1963. He was first assigned to HMM-264 New River for a Caribbean cruise. In July 1963 he was with HMM-162.

HMM-162 launched across the big pond to Futema to begin its year rotation from June 1964 to June 1965 from "The Sand Piper," Laguna Beach. In November 1963 he saw HMM-162 in Hong Kong aboard the USS *Princeton* for "The Birthday Party." "'Twas the night of the party..."

Returning to New River in June 1965, Dunn was assigned as an ops duty officer for his last seven months in the Corps.

In January 1966 he joined Pan American World Airways and 26 years later, in November 1992, he transferred to Delta Airlines. He is married with two children and with 18,000 flight hours and a five handicap. He looks forward to his last seven years of "productive employment."

THOMAS A. DUNN, Senator, was born on Oct. 12, 1942 in Joliet, IL. Enlisted on Oct. 1, 1965 for two years as a private with a BA in political science. Using his degree, he was trained in the mysteries of supply. He was relieved of his command in short order and made a door gunner with VMO-2 from 1966-67.

He was stationed in-country at Marble Mountain and Dong Ha. His memorable experience is feeling lucky to get out of South East Asia alive as a door gunner.

He received Permanent Combat Wings, Air Medals, Presidential Citation and got his fair share of NVA and VC. He retired on Feb. 1, 1968 as a Corporal after two years of honorable service and moved to Mexico to drink and ride motorcycles, successfully mastering both.

He received a law degree in 1972. He re-entered government service in 1986 and is still serving as state senator. He also owns a restaurant. He quit riding motorcycles and is now married with three children. He resides in Joliet, IL.

MELVIN W. EAKINS, Lance Corporal, USMC was born to Mr. and Mrs. Warren Eakins on Dec. 2, 1945 in Cincinnati, OH. He served with distinction in-country with HMM-262 and died May 2, 1969 at the age of 23.

GENE T. ELLIOTT Master Sergeant USMC (Ret) was born on May 27, 1929 in Salisbury, MD. He graduated from high school in June 1946. He joined the Marine Corps in August 1946 and following recruit training at Parris Island was assigned to MCATS, Quantico for training. In February 1947, he was transferred to VMP-254, El Toro. In July 1948 he joined HMX-1, Quantico. He was a member of the original complement when helicopters were the "new kid on the block." Everyone thought you were crazy in "those frustrated palm trees."

In July 1950 he was in the first helicopter unit, VMO-6, to leave for Korea; frozen Chosin and return. He later taught mechanics in Memphis and Jacksonville. He joined HMM-364 in 1963 as squadron maintenance chief. He went to Okinawa and on to Vietnam in 1964. His squadron trained pilots and mechanics for the VNAF. He turned over 24 H-34s and equipment to the VNAF.

He retired July 1967 to Florida with the rank of Master Sergeant. He was awarded an Air Medal and a Navy Commendation Medal.

He attended two years college for computer science, then went into banking as head of a Trust Department and onward to his current position selling real estate. Today he resides in Lake Panasoffkee, FL with his wife and three children.

WILLIAM EMERSON, "BING," Capt USMCR was born in Concord, MA and attended elementary schools in Danvers, MA and Glastonbury, CT before entering Belmont Hill in 1956. He won letters in football, wrestling and crew, was captain of the wrestling team in 1960, won the Robert Hurlburt Memorial Football Medal in 1959 and the wrestling award in 1960. Bing graduated from Belmont Hill in 1960, and one of his teachers who knew him well wrote, "I think that what I admired most about Bing was the combination of toughness and sensitivity. He certainly threw himself into every outdoor activity with complete enthusiasm, but at the same time he appreciated the wonders of nature and beauty-a rare young man." Yes, he was. No number one son ever gave more pride and joy to grandparents, parents, brothers, sisters, cousins and a host of friends than he. From the day he arrived Dec. 14, 1941, until he died he lived with gay, adventurous enthusiasm which spread happiness wherever he was.

At Harvard he majored first in government, then English, and finally economics but his real specialties were friendships, mountains, motorcycles, skis and confrontations with the dean. After a year at Harvard he enlisted in the Marine Corps Reserve, returning in 1962 and graduating in 1965. He was commissioned a Second Lieutenant in the Marine Corps on March 18, 1966 at Quantico, VA and then went to Pensacola NAS, FL for pilot training.

On April 1, 1967, a day of joy and beauty, he married Suzanne Robertson at New Haven, CT and they returned to live in Pensacola where he won his wings on Aug. 18, 1967.

He loved people and of his service in Vietnam he wrote, "My rationale. . . for being here is. . .in the eyes and faces that look up at us in the cockpit when we come into the zone and pull them out of a hot spot. They, these kids, men, 18 or 19 may not know why they're here but they're really something. They'll even try to come up in the cockpit and kiss you., and believe me sometimes it isn't only sweat running down my cheeks. They need me and I can help them."

On Nov. 20, 1968, Captain William Emerson, USMCR was killed in action when his helicopter was shot down approaching a landing zone in Quang Nam, Vietnam. He had flown CH-46 helicopters with HMM-265, MAG-16, based at Marble Mountain near Da Nang and Phu Bai since his arrival in Vietnam in January 1968.

Services for him were held at the Harvard Memorial Church in Cambridge on Nov. 29, 1968, conducted by Rev. Lawson Willard of New Haven, a great friend of Bing and Suzanne, who officiated at their wedding.

Besides Suzanne, he left his parents, Mary and David; parents-in-law, Mr. and Mrs. Raynham Townshend (of New Haven); three brothers, Alexander F., Daniel and Raymond II; four sisters, Mrs. William (Margaret) Bancroft of Dedham, MA, Ellen, Lauran and Amelia; and his grandparents, Mr. and Mrs. Raymond Emerson and Mrs. Moncrieff M. Cochran, all of Concord, MA.

"As long as there are such sons, our Stars and Stripes will fly bright and high for all mankind."

ALAN WAYNE ERDMAN, "BIG E," was born on Sept. 5, 1950 in Baltimore, MD. He enlisted in the Marine Corps in May and reported for active duty on Sept. 3, 1968 and went through boot camp at Parris Island. He reported to Memphis for aviation training in fixed wing aircraft and helicopters. His first duty assignment was with HMM-365 in New River, NC.

After a transfer to HMM-261 and a Caribbean cruise he was sent to Vietnam and attached to HMM-262 at Marble Mountain as a crewchief. He served from September 1970 to May 1971. Shortly before the Marines left Vietnam in 1971 he was transferred to HMM-263 returning to Quantico, VA.

He was discharged Aug. 1, 1971. During his tour he was awarded the Air Medal with Bronze Numeral "42," Navy Unit Commendation awarded MAG-16, Meritorious Unit Commendation with Bronze Star awarded to MAG-16 and 3rd Marine Amphibious Brigade, FMFPac for service in Vietnam, Good Conduct Medal, National Defense Service Medal, Vietnam Service Medal with Bronze Star, Vietnam Meritorious Unit Citation (Gallantry Cross Color), Vietnam Meritorious Unit Citation (Civil Action Color, 1st Class), Vietnam Campaign Medal, Vietnam Cross of Gallantry with Palm, Combat Aircrew Insignia with three stars, Combat Action Ribbon, Rifle Sharpshooter Badge and Pistol Sharpshooter Badge.

After running the family business for several years he went to work for the Milwaukee Road Railroad as a machinist and held that job until 1980 when the railroad closed. Changing careers he enrolled in college and earned a BS in business administration and an associate degree in heating and air conditioning. He is currently working as a project manager/estimator for a mechanical contractor in Chicago and lives in West Allis, WI.

BARNEY ESPINOZA JR. Master Sergeant, USMC (Ret) was born in Austin, TX on April 18, 1939. He joined the Texas National Guard while still in school and went though Army basic training at Camp Chaffee, AR in the summer of 1955. In 1957 he joined the Marine Corps Reserve and after high school graduation in 1957 went through Marine boot camp in San Diego. He served at Camp Pendleton and then in February 1960 reported to NAS Memphis and attended the Aviation Operations School.

He served in the flight operations of VMCJ-3 from July 1960 until he deployed to Iwakuni, Japan where he served with VMCJ-1 from October 1962 through October 1963 through the Far East and aboard a number of ships. After returning to the U.S. as a Corporal he served in Flight Clearance, MCAS Santa Ana, CA until the summer of 1965, when he deployed as a Sergeant with MAG-36 S-3 aboard the USS *Princeton*, landing in Ky Ha, RVN. His assignments at Ky Ha included the Provisional Rifle CO, unloading the SS *Iberville*, laying matting for the helo field and building bunkers. In the spring of 1966, he became Ops Chief of HMM-364 and flew as a gunner aboard UH-34s until returning to the states in October 1966 as a Staff Sergeant. He next served with H&MS-32 and VMFA-312 in Beaufort, SC until October 1967. He served at Base Ops., MCAS Kaneohe Bay, HI from January 1967 to October 1969. He returned to Marble Mountain RVN where he served as Ops Chief, HMM-263 and flew as a gunner aboard CH-46s from December 1969 until December 1970. He flew more than 1000 hours and accumulated 700 combat missions during his two tours in-country.

He reported to HMH-461, New River, NC in January 1971 and served as Ops Chief. In February 1972 he deployed as Ops Chief, HMM-261 aboard the USS *Guadalcanal* for the Mediterranean, and returned in September 1972 as a Gunny. From February 1973 till September 1975 he served as Airfield Operations Chief, MCAS Futema, Okinawa. In September 1975 he was assigned as Group ops Chief, MWSG-37, El Toro, CA as a Master Sergeant participating in TRANSPAC OPS to the Far East. In September 1976 he was assigned as MAG-24 Ops Chief, MCAS Kaneohe Bay, HI.

The "TOP" retired from active duty in September 1979 with almost 25 years of service. He earned his Combat Air Crew Wings, 35 Air Medals and four Single Mission Air Medals, Navy Commendation with Combat "V," CAR, PUC, NUC, two MUC, two Vietnam Crosses of Gallantry, Vietnam Service Medals w/stars for seven offensive periods, Vietnam Campaign Ribbon and seven Good Conduct Medals.

He earned a bachelor of professional aeronautics degree from Embry-Riddle, Aeronautical University, Daytona Beach, FL. Upon returning to Texas, he attended post-graduate schools and worked for the State Comptroller's office for eight years. In 1987 he was appointed as a municipal judge and serves as a consultant for the Texas Education Agency.

WILLIAM T. ESTES, "WILD BILL," was born on Oct. 9, 1928 in LaGrange, GA son of Tommy "D" and Louise Estes. He enlisted in the USMC on Feb. 21, 1946, and attended boot camp at Parris Island. Duty stations were Cherry Point, Midway Island, MCAF New River, NC, NAS Glenview, IL and NATTC Memphis, TN.

While in-country he served as crewchief (CH-46s) in HMM-261, HMH-361 and also in VMO-2. His duty stations were Peiping, China, Orote Point, Guam and Marble Mountain, Vietnam.

Advancing in rank from Private to Chief Warrant Officer-3, he completed his 28 year career as the academic department head of the Basic Helicopter School, NATTC Memphis. After retiring on Sept. 1, 1974 he was employed by the Little Rock District as a teacher. His awards included Combat Aircrew Wings, Air Medal, Navy Commendation with Combat "V," USMC Good Conduct Medal and RVNAFUMUC Cross of Gallantry.

Presently he resides in Alexander, AR and is enjoying his retirement.

EARL P. EWING, "EVIL EARL," born in La Jolla, CA and joined the USMC on May 13, 1951 being assigned to HMM-262 and HMM-364 as an acting gunner, crewchief, avionics chief, maintenance chief achieving the rank of Master Gunnery Sergeant. His military locations included Parris Island, MB Clarksville BAS, TN, Miami, FL, 3rd MAW, MCAS El Toro, CA, Cherry Pt., NC, 2nd MAW, Beaufort, SC, Atsugi, Japan, Vietnam, 1st MAW, Marble Mountain, Chu Lai, Ky Ha, Kaneohe, HI, Memphis, TN, Jacksonville, FL, Marine Bks. and COMNAV, Norfolk, VA.

His awards included Good Conduct Medal, National Defense Service Medal, Republic of Vietnam Campaign Medal w/Device, Combat Action Ribbon, Presidential Unit Citation, Vietnam Service Medal, Combat Air Crew Wings, Vietnam Cross of Gallantry w/Palm. He was honorably discharged on March 28, 1981. He and his wife Lana resided in Bullhead City, AZ and have a step daughter, Leann, son, David, daughter, Cathy with grandchildren Dylan, Harley, and Louise. He passed away in 1994.

JERRY E. FARROW, CLUTCH, was born April 2, 1941 in Eldred, IL. Holds a B. S. in Business Administration from Southern Illinois University, 1964. Commissioned Quantico, December 1964 as a 2nd Lt. Designated naval aviator in May, 1966 at Pensacola. Served with HMMT-301 Tustin, CA from June 1966 - August 1966. Joined HMM-361 in Vietnam September 1966 flying the UH-34. Flew in excess of 1000 combat hours and more than 1000 combat missions. Awards include the DFC and 50 Air Medals plus the standard Vietnam service ribbons. Served MCAS Iwakuni, Japan October 1967 to April 1969. Released from active duty April 1969. Joined Marine Corps Reserve squadron VMR-234 at Minneapolis June 1969. VMR-234 transferred to NAS Glenview, IL in 1971. Military aircraft flown included the UH-34, C-119, C-131, and the KC-130. Left active reserve status in May 1977 after serving as Operations Officer for VMGR-252 at the rank of Major.

As a civilian, he was employed with Saudi Arabian Airlines, domiciled in Saudi Arabia, from June 1977 to August 1988 as a First officer on the B-707 and as Captain on the B-737, L-1011 and the B-747. Currently with Korean Airlines as Captain B-747 domiciled at Los Angeles International. He is married, with one daughter and one son.

LARRY W. FENTON, "PHANTOM," was born on Dec. 6, 1945. He grew up on a farm near Aurora, MO and graduated from Aurora High School in 1963 and the University of Missouri in 1967. He attended the PLC Program in the summer of 1964 at Camp Upsher and the summer of 1966 at Mainside in Quantico. He was commissioned June 5, 1967 and reported for flight training in Pensacola on July 10, 1967. He received his wings in August 1967. He reported to Santa Ana for predeployment training on the UH-34 but subsequently transferred to the CH-46 after the UH-34 pipeline was closed down.

He checked into HMM-262 March 2, 1969. He was medically evacuated on June 14, 1969 to Japan for recovery from a gunshot wound to the knee and spent four months in Yokosuka and Pensacola Naval Hospitals. He reported to Meridian, MS after returning to flying and spent his remaining active duty instructing in the T2 Buckeye.

He was discharged on June 18, 1971 with the rank of Captain. He was awarded the Silver Star, Purple Heart, a Single Mission Air Medal and 17 Air Medals.

For the next five and half years he worked various jobs most of which involved either commuter, instructor, or corporate flying. He was hired by Delta in 1976 where he presently flies as Captain on the MD-88. Today he is married with two children; a 22 year old daughter and a 16 year old son. They presently reside in Dunwoody, GA.

HAROLD T. FERGUS, "TOM," Lieutenant Colonel USMCR (Ret) was born on March 23, 1939 in Cleveland, OH. He graduated from Cleveland St. Ignatius High School in 1957 and earned a BS in business administration from George Washington University in 1971. He was a mustang from July 1958 and was commissioned a Second Lieutenant and a Naval Aviator in May 1963.

He flew in Vietnam from August 1965 through October 1966 with HMM-263 and HMM-161. He flew more than 1000 combat hours.

He ended his Marine Corps career as a Reserve Air Officer, Fleet Marine Forces, Atlantic. He was awarded the Distinguished Flying Cross, Single Mission Air Medals and other awards.

As a civilian, he retired from the FAA as an airman certification inspector, B-727, B-737, DC-9, BAC1-11 and CV-580. He was Captain at People Express Airlines and much later at Continental Airlines. He has been married since 1969 and has one daughter and two sons.

MICHAEL D. FERRIN was born on March 5, 1947 in Ogden, UT. He enlisted in the USMC on Jan. 6, 1966 and attended boot camp at San Diego. He attended Jet Engine School at NATTC, Memphis.

He volunteered for Vietnam, arrived in Da Nang on Oct. 15, 1966 and volunteered for helicopters. He was assigned to H&MS-16, MAG-16, at Marble Mountain for six months.

He then volunteered to go with HMM-164 (CH-46s) for shipboard assignment as NCOIC of the engine shop aboard the USS *Princeton* and served as aerial gunner. He became a crewchief upon deployment at Phu Bai. He flew missions to the Citadel in Hue (Tet), Khe Sanh (siege), A Shau Valley, DMZ, Ho Chi Minh trail, Laos, all of I Corps, throughout 1967 and 1968. He was assigned as flight line NCO, awards NCO and relief crewchief in 1969. He accumulated 508 hours combat flight time. He had extended his tours three times (35 months) and rotated home on Sept. 17, 1969.

He was discharged Sept. 29, 1969 with the rank of Sergeant. He was awarded Combat Aircrew Wings, 30 Strike/Flight Air Medals, one Single Mission Air Medal, Navy Commendation w/Combat "V," PUC, MUC, NUC and CAR.

Today he is married to Verda and they have four sons. They reside in Lyman, WY. He is a mechanic/welder for the Trona Plant in Wyoming. He was chairman of the Wyoming POW/MIA Association in 1993-94 and commander of VFW Post 7798 in Bridger Valley, WY.

STEVEN E. FIELD, "CRAZY HORSE," Colonel USMC (Ret) was born on Sept. 22, 1937 in Greenville, SC. He attended Georgia Institute of Technology from 1955 to 1958 Majoring in industrial management. He graduated from Navy Flight School on April 6, 1960 and reported for duty with HMR-261 flying UH-34s.

He first flew in Vietnam on April 18, 1962 with Operation Shu-fly. Next he went to Thailand flying out of Udorn and Vien Chien, Laos. He returned to MCAS New River flying the CH-37 with HMH-461. He was back again to Vietnam in August 1965 and flew UH-34s with HMM-361; CH-46s with HMM-165 and UH-1Es with VMO-6.

Next he flew CH-37s as OIC, DET 1, MAG-16 and returned to MCAS New River again in late 1967. He transitioned to CH-53s with HMH-461. He attended USNPGS in 1969 and was assigned as XO, NROTC University of West Florida in 1972. He was transferred to HMH-463, Kaneohe Bay, HI and served as both XO and then CO. He reported to 1st MAW (Okinawa) in August 1978 as director, Aviation Safety and Standardization. He was assigned in August 1979 as XO MCAS, Yuma, AZ and remained until promoted to Colonel and reassigned to COMCABEAST as director of Moral, Welfare and Recreation.

He retired on April 1, 1985 and moved back to Pensacola, FL and taught at the University of West Florida in the College of Business. His decorations received in the 26 plus years of military service include Legion of Merit, three Distinguished Flying Crosses, 23 Air Medals plus numerous other awards.

He returned to flying as a chief pilot for a Pensacola based corporation in 1988. His total flight time is now approaching 9200 hours in more than 34 different aircraft types. Today he is married, with two sons (one a Marine) and two daughters.

SAMUEL FLORES JR. was born on Oct. 7, 1947 in Raymondville, TX. He enlisted on June 16, 1967 and trained at Parris Island. He reported to Memphis for Aviation Supply training and then to MCAS Cherry Point.

In September 1969 he was ordered to Camp Pendleton and then to Vietnam. He was assigned to MAG-16 Forward at Phu Bai and then served with HMM-262's S-4 and as a door gunner.

Flores was awarded the Combat Aircrew Wings, seven Air Medals, CAR, Vietnamese Cross of Gallantry and the Vietnamese Civil Action Citation. In 1976 he was selected for Gunnery Sergeant and WO Program. In 1992 he was promoted to Lieutenant Colonel and assigned to 3D MAW as Wing Aviation Supply Officer.

He is a member of the M.C. Mustang Association, VFW, USMC/Vietnam Helicopter Pilots & Aircrew Reunion and the Aviation Supply Reunion West Coast. He is still married to the former Patricia Baker and has two daughters Kimberly and Lydia.

MAX R. FLOYD entered Vietnam in March 1967 and left in May 1968. He reported to H&MS-16 at Marble Mountain and was assigned to Sub Unit-1 maintenance chief on CH-37 with the call sign "Junkman." He flew in north sector to DMZ. He was reassigned to HMM-163 with the call sign "Ridge Runner."

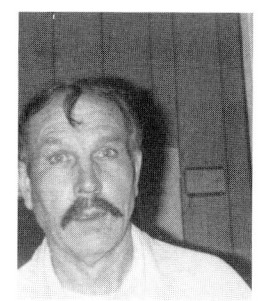

"Charlie hit the squadron on Aug. 1, 1967 and knocked out 22 helos, leaving only two that could fly. We worked night and day to make them ready to go on the carrier on Sept. 1, 1967. We operated off the carrier for one month at sea then off loaded at Quang Tri on Sept. 30, 1967."

"We took mortar fire and 122mm rocket fire and one mortar attack. We lost one pilot; one enlisted man was knocked out, but saved. I got hit in the left knee."

He retired on Jan. 1, 1971 with the rank of Master Sergeant. He received the Vietnam Service Medal w/three stars, Air Medal w/ five stars, Combat Aircrew Insignia with three stars, Purple Heart, Presidential Unit Citation, Combat Action Ribbon, Cross of Gallantry w/Palm and Frame and the National Defense Medal.

Today he is retired from working at the GE Nuclear Fuel Plant.

JAMES R. FOWLER, MC, Rear Admiral, USNR, was born on Feb. 4, 1937 in Douglasville, GA where he completed his primary and secondary education. He was commissioned an Ensign in 1959, and graduated from Emory University School of Medicine in 1964. Following an internship at Harborview Medical Center, University of Washington, Seattle, he reported to NAMI, Pensacola, FL in 1965 as a student Naval Flight Surgeon. Upon designation he was assigned to the 3rd Marine Aircraft Wing at El Toro. His active service in the States included VMT-2, El Toro and VMT-103 USMCAS, Yuma as a Naval Flight Surgeon until 1969.

While in-country he served with HMM-163, MAG-16, 1st MAW from 1966-67.

Rear Admiral Fowler's reserve career includes service with the 4th FSSG, Atlanta, GA; commanding officer Surgical Team 220, Salt Lake City, UT and 4th MAWMED MAG-42, Alameda and director of Health Services, REDCOM TWENTY San Francisco. Rear Admiral Fowler was the commanding officer, Fleet Hospital CBTZ-500, No. 9, REDCOM TWENTY-TWO, Seattle, when selected for Flag Rank in 1990.

Dr. Fowler is in the practice of surgery of the hand and upper extremity in Salt Lake City, UT. He is certified by the American Board of Plastic and Reconstructive Surgery and has a certificate of added qualification in surgery of the hand. He is assistant chief, Bureau of Medicine and Surgery and Surgeon General for Reserve Affairs, Washington, D.C. and was selected for his second star in 1993.

SAMUEL J. FULTON, "SAM," born Dec. 18, 1929 in Highlands, NC, and joined the USMC on Oct. 20, 1948.

His military locations include Parris Island; Aviation Electronics School, 1949; Naval Flight Training commissioned as 2nd Lieutenant January 16, 1952; assigned to VMF 223 and VMF 224 flying F9F-5 Panthers, 1953; 2D MAW as Wing Air Freight Officer, 1954; MAG-26, 1954-56; HMR-461, 1956-57; promoted to First Lieutenant, 1953 and Captain, 1955; with HMX-1, 1957-61; Regimental Training Officer for 9th Marine Regt., 1962-63; NAS Whiting, 1963-65, promoted to Major July 1, 1962; VT-6, 1965-66; VMO-1, 1966-67; promoted to Lieutenant Colonel, July 1, 1967; MAG-16, 1st MAW, MABS-16, VMO-2, 1967-68; flew 473 missions in UH-1; MACG-38, 1968-70; Asst. Comptroller 3rd MAW, 1970-71, Comptroller 4th MAW, 1971-75; promoted to Colonel, July 1, 1973; MAG-36, 1975-76; JUSMAG (K) Seoul, Korea, 1976-79; Comptroller MCDEC Quantico, VA, 1979-80.

Discharged Oct. 1, 1980. He married Lola on May 7, 1955. They have three daughters.

GARY NILES GARNER was born on May 9, 1944 in the San Diego Navy Hospital to Sgt. and Mrs. Harvey Garner USMC. He graduated from high school in Tampa, FL in 1962 and later graduated from the University of South Florida in 1970. In between, he was at Parris Island, Memphis Aviation Schools and Camp Pendleton VMO-6.

He was in VMO-2 Marble Mountain from 1965-66. His memorable experience of Vietnam was being overrun at Marble Mountain in October 1965.

Upon his return to Beaufort MCAS, he was discharged with the rank of Sergeant on May 1, 1967. He received Combat Aircrew Wings, 10 Air Medals, Navy Commendation with "V," Vietnam Cross of Gallantry with Palm and Vietnam Civil Action Medal with Palm.

He returned to college and graduated in June 1970. He has been married 27 years to Carol, the most forgiving woman God ever made, and the mother of their two girls. They reside in Tampa, FL. He has been a real estate broker for 23 years or whenever the fish quit biting or the birds wouldn't fly. By the way, Sergeant Harvey Garner (an Iwo Jima Marine) is 76 years young and can still whip his butt. Some things never change. *Semper Fi*.

JERALD B. GARTMAN, Colonel, was born in Stillwater, OK on July 27, 1942. He graduated from the University of Missouri and was assigned to TBS in June 1965.

He received his wings in February 1967 and was assigned to HMM-165, RVN in July 1967. In August 1968, he was assigned to H&MS-24. In July 1971 he was assigned to AWS; in January 1972 he was assigned to HMM-162 and in December 1972 he was assigned to HMM-165, RVN. In January 1973 he was Senior Marine VT-2 and in January 1976 he was in Naval Air Systems Command, H-46 Class Desk. He was in AFSC in January 1980 and in July 1980 he was in NARF at Cherry Point. In July 1982, he was in ICAF, in June 1983 he was CO of MABS-36 and XO 35th MAU. In July 1984 he was XO MAG-29. In June 1985 he was G-3, 2nd MAW. He was CO NARF (Naval Aviation Depot) at Cherry Point in March 1986.

He logged more than 4200 flight hours. His awards included Legion of Merit, Silver Star, Purple Heart, two Meritorious Service Medals, 21 Air Medals, two Navy Achievement Medals, two PUCs, two NUCs, CAR, SecDef-Productivity Excellence Award and the John A. LeJeune Award for Inspirational Leadership. He retired in September 1991.

BOB GAUNTNER, "BLAISE," was born on Feb. 5, 1943 in Cleveland, OH. He joined the USMC through the PLC Program in 1965. He received flight training at Pensacola from 1967 through 1968. He was then stationed at MCAS Tustin in 1968. He was also stationed at MCAS Beaufort from 1970-71.

He qualified as pilot and flew UH-34s for HMM-362 while stationed in Vietnam from 1968 to 1969. Everyday was a new experience for him.

He was discharged from active duty on May 15, 1971 with the rank of Major. He spent six years in the reserves at Selfridge AFB, Detroit and was discharged in May 1977. He was awarded the National Defense Medal, Vietnam Cross of Gallantry w/Silver Star, CAR, 35 Air Medals, Merit Unit Citation and Vietnam Service Medal.

He is married to Barb (Rogge) and they have son, Ryan, 18 and daughter, Stephanie, 20. He is V.P. in sales for Western U.S., Health Care Management, consulting ServiceMaster Company and resides in Mission Viejo, CA.

CHARLES E. GRUBBS, "BUD," was born on Feb. 8, 1945 in Los Angeles, CA. He graduated from Crespi High School in Encino, CA in 1963. He received a BS in accounting from San Diego State University in 1968. He attended the PLC Program at Quantico 1967 and was commissioned Second Lieutenant USMCR at MCRD San Diego on June 15, 1968. He was designated a Naval Aviator on Aug. 15, 1969 at Pensacola. He qualified in the CH-46 with "F Troop," HMMT-302 at MCAS (H) Santa Ana in September 1969 to March 1970.

He served with 1st MAW from March 1970 to March 1971 flying the CH-46A/D. He was a squadron pilot with HMM-164 at MCAS Futema, Okinawa and afloat with SLF (MAU 1970-71). ALO with the 1st Battalion, 9th Marines and he was squadron pilot with H&MS-36.

He served with HMM-163 at MCAS (H) Santa Ana from March 1971 to February 1973. He was certified PMIP and VIP pilot in the CH-46. He earned four "To PAR and Back" Awards and holds Vietnam Service and Campaign Ribbons. He completed active duty as a Captain with HMM-163 in February 1973 with 1100 flight hours.

He was a financial manager with Ford Aerospace in Newport Beach, CA from May 1973 to May 1988. He earned his MBA from Pepperdine University in 1975. He is president of his own management consulting firm. He is married with two daughters.

GORDON H. GUNNISS, "SCOOP," was born on Aug. 1, 1939 in Detroit, MI. He joined the USMC on Jan. 15, 1962 and was stationed in the States in Quantico, Santa Ana, Camp Pendleton, Pensacola, New Orleans, New River, Cherry Point and Hawaii.

He was stationed in Vietnam from 1966 to 1967 and then again in 1970 to 1971. His memories include the camaraderie that he shared with his former Marine buddies including, Sonny, an ex-grunt.

He was discharged on Feb. 1, 1982 and achieved the rank of Lieutenant Colonel. His awards included DFC, MSM, NCM, AM, Purple Heart, NAM, etc.

Today he is married with one son (former Marine) who is in college and one daughter who is also in college. He is an air conditioning contractor and resides in Gulf Breeze, FL.

MYRON L. HAAG, "MIKE," was born on March 14, 1938 in Columbus, OH. He graduated from Olterbeen College, Westerville, OH in 1959 and enlisted in the USMC in November 1961. He was commissioned as Second Lieutenant in the USMC at Quantico and entered the Naval Air Training Command at Pensacola, FL, where he earned his aviator wings.

Haag proudly served with HMM-362, HMM-163 and HMM-164 flying helicopters and combat missions in Vietnam from 1964 to 1965. He left the Marines in 1966 as a Captain.

He served as a pilot for American Airlines for 26 years, leaving American's service as a 767 Captain. He died on March 29, 1992. He is survived by his childhood sweetheart and wife of 29 years, Patricia, and their three children: Michele, Kristen and Michael (also a pilot).

JAMES G. HALL, "J.G.," GySgt USMC (Ret) was born on April 19, 1937 in Omaha, NE. He attended Omaha Tech High and joined the Marine Corps on June 22, 1955. Tours of duty in the States included W-2-9, 3rd Div., VMR-152, 1st MAW, MCB-29 Palms, CA, MCSC Albany, GA, Recruiting Kansas City, MO and I&I 4th MAW Chicago, IL.

While serving in-country from 1965-66 and in 1971-72, he was a port gunner in HMM-364 at Ky Ha, RVN flying UH-34s. He was also stationed at Camp Brooks, Red Beach, Da Nang, RVN and FSR Okinawa.

He was discharged on June 30, 1972 with the rank of Gunnery Sergeant. His awards included CACW, AM, CAR, PUC, NUC, MUC, NDS, GCM, Vietnam Service and Campaign Ribbons, Vietnam Cross of Gallantry and Vietnam PUC.

Since his retirement, he was worked as a job analyst for Mutual of Omaha and as a Veterans' Benefits counselor for Veterans' Affairs. He is a member of the American Legion, DAV, VFW, Forty & Eight, 1st MAW Association, USMC/Vietnam Helicopter Pilots & Aircrew and VVA. He is married to Judith and they have a daughter, Kym Hall Bybee, who is married to Thomas L. Bybee. He is now the owner of a retail store and resides on 10 acres of land with his wife, wildlife and trees in Ft. Calhoun, NE.

To all who served in "The Purple Fox, Give A Shit," "Hope your Strap Hanging is Peaceful!" To all those who flew their last mission with the "FOX": Womack, Chief, Johnson to name a few and to all Marine Aviators and Aircrews over the years, "Semper Fi!" You are not forgotten! Ingredients for Purple Fox Cooler: Good water (if you can find it) and Goofy Grape Kool-Aid. Mix in canteen cup with gin. Add ice (ha, ha!) Drink it down and get the Purple Fox Grin!

ORAMEL EUGENE HALL was born on June 29, 1946 in Flint, MI. He graduated high school in June 1964 in New Castle, IN. He enlisted in the Marine Corps in May 1965 with delayed entry. He started work with Western Electric in Indianapolis, IN in May 1965. He began a military leave of absence on Sept. 1, 1965 and went through boot camp at San Diego, CA. He reported to MCAF Santa Ana, CA to HMH-462 for aviation training in the UH-34 helicopter. The squadron was changed to HMMT-301 in January 1966.

In January 1967 he was transferred to Vietnam. From January 1967 to April 1967 he served with H&MS-36 and was assigned to the engine shop. From April 1967 to February 1968 he served with HMM-362 as crewchief. Duty locations in Vietnam were Ky Ha from January to June and USS *Okinawa* from June to September. While aboard ship, on a test hop, he received flight instructions and made a successful carrier landing in July 1967. From September 1967 to February 1968 he was at Phu Bai and while there he flew support for Khe Sanh and then Hue, Tet 1968.

In March 1968 he was assigned to HMH-461 at New River, NC. In October 1968 he was transferred to Millington, TN for jet school and was discharged in April 1969 as a Corporal at Millington, TN. While serving in Vietnam he was awarded Combat Aircrew Wings, 16 Air Medals, Purple Heart, PUC, two Vietnam Service Medals and Vietnam Campaign Medal.

He returned to work at Western Electric from military leave in April 1969 and worked a variety of production jobs and skilled trades. He attended Ball

State University in 1973 and graduated from Indiana Vocational Tech. College in April 1978 with an associate degree in industrial management. He completed tool & die-machinist apprenticeship in March 1981. In 1982-1983 he was commander of VFW Gold Star Post No. 1282 in New Castle, IN. The corporate name changed to AT&T, the then Indianapolis plant closed and he transferred to Phoenix, AZ in 1985.

He was married in August 1970 to Sharon. They now have three grown children, one daughter and two sons and three grandsons. He is a project trades planner for AT&T Network Systems and planning for early retirement. He and his family reside in Glendale, AZ.

RICHARD HARRY HALL, First Sergeant USMC (Ret) was born on March 2, 1937 in Portsmouth, VA. Carrying on the family tradition, as his father was also a Marine, Richard enlisted in the Marine Corps on March 27, 1954. After undergoing boot camp at Parris Island, he was transferred to Cherry Point to train on the F4U. He was subsequently transferred to New River in October 1956 to receive training at the H&MS-26 engine shop.

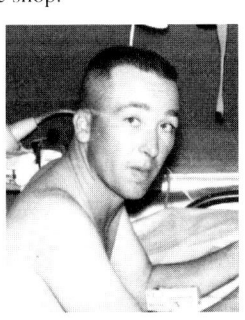

In June 1957, he was detailed to train with the British Commandos in the Mediterranean. He was then transferred to Oppama, Japan to join HMR-261 in December 1958. Thereafter, he returned to Quantico MCAS in August 1960 and was assigned to HMX-1. During this tenure he served three presidents. In December 1965, he was transferred to HMM-264 until December 1966, when he was detailed to the Caribbean for jungle training preparatory to serving in Vietnam. He served in Vietnam from December 1966 until March 1968 with HMM-361, as section leader.

During his tour, he was awarded the Marine Corps Aircrew Wings, 11 Air Medals, two Purple Hearts, CAR, NUC, PUC, MUC and the Vietnam Service Medal with three stars. He retired as First Sergeant on July 1, 1976 at Quantico.

First Sergeant Hall presently lives in Nathalie, VA on a 122 acre farm, where he spends great time and energy as a printer and Naval and Marine Corps historian. He works with military groups to commemorate the history of their units by creating commemorative envelopes post marked on the anniversaries of the historical events of the units. He is committed to ensuring that no one should ever forget the brave deeds accomplished by his fellow Marines.

DON E. HAMILTON, "HOLLYWOOD," was born on Dec. 2, 1939 in Brooklyn, NY. He attended Brown University and received a BA degree. He joined the service through the PLC Program in June 1959 and was commissioned Second Lieutenant in the USMC on June 7, 1961. While in the States he was stationed at MCAS El Toro and MCAS New River.

While serving in-country he flew UH-34Ds. He was stationed in Da Nang and Qui Nhon. His memorable experience was the strike of March 31, 1965 which was written up in *Life Magazine*..

He was discharged on March 15, 1966 and achieved the rank of Captain. He received the Silver Star and nine Air Medals.

He flew for United Air Lines and took an early retirement at age 50 on Jan. 1, 1990. He is also into international investing. He has one son, Scott, born on June 7, 1961.

DENNIS E. HAMMOND, "DUKE," was born on May 20, 1944 in Lockport, NY. He graduated from high school in 1962 in Newfane, NY and holds a BS degree from S.U.N.Y. Brookport and an MS from Virginia Tech. He joined the Corps in November 1964 and was commissioned Second Lieutenant on June 4, 1966. He was designated a Naval Aviator on Nov. 1, 1967 and was attached to HMM-263 Santa Ana from Dec. 1, 1967 to March 1, 1968. He qualified in UH-34s and CH-46s.

He joined HMM-265 at Marble Mountain, Vietnam in March 1968 and was then transferred to HMM-361, 363 and 362 as the UH-34s were decommissioned.

He was awarded 37 Air Medals, a Single Air Medal, Navy Commendation Medal with Combat "V" for rescuing wounded a pilot from his burning aircraft after a crash on medevac, Vietnamese Cross of Gallantry, CAR, NUC, MUC, Vietnam Service and Campaign Ribbon with Silver Star. He completed active duty with HMM-163 in Santa Ana as Captain in November 1970.

He was married to Candy Grassi on July 1, 1967 and they have four daughters. He is an environmental administrator with Florida Game and Fish. He is also a lay preacher involved in prison ministries.

J.B. DOIL HARVEY was born on May 22, 1937 in Sequatchie County, TN. He joined the service on Oct. 1, 1954 and was stationed at Parris Island, SC, Memphis, TN, Jacksonville, FL, Beaufort, SC, Jacksonville NC and Edenton, NC. He was also stationed aboard the USS *Boxer*, USS *Tarawa*, USS *Intrepid* and the USS *Tripoli*.

While serving in-country, he was a door gunner on UH-1Es in VMO-2 from 1966-67 at Marble Mountain, Vietnam. He remembers the installation of the first TAT-101 on a UH-1E gunship. Also the special night black box flights with Lieutenant Colonel W.F. Harrell, Commanding. His second tour of duty was from 1971 to 1972 with HMM-165 on station for the bombing of Haiphong Harbor.

Harvey was discharged on April 9, 1974 with the rank of Master Sergeant at New River, NC. He was awarded the Marine Corps Combat Aircrew Wings, two Air Medals, CAR, Vietnam Service and Campaign Ribbons.

In 1986 he joined the TDF Tennessee Defense Force as a Captain of Company (A) Commander promoted to Major and served as Bn XO, promoted to Lieutenant Colonel and served as Bn Commander until December 1993.

He has been married for 37 years and has two sons and two daughters and three grandchildren. Today he is an electrical and building contractor and does primary electrical work for the State of Tennessee.

VINCIL WAYNE HAZELBAKER, "WAYNE," Colonel USMC (Ret) began life 10 miles north of Grangeville, ID on Oct. 1, 1927. He enlisted in the USNR on May 6, 1945 for the duration and six months. During the Korean Conflict, he requested Naval Cadet Training and was commissioned a Second Lieutenant in the USMCR in September 1953.

He flew Corsairs with VMA-331 in Miami and married Nancy Kerr Scott in February 1954. He trained as a fighter pilot with VMF-312 in Cherry Point, NC. He served with VMF-115 in Korea during the winter of 1954-1955 and then returned to ranching in May 1956. They built a home and Nancy gave birth to Heather Ann, Brad Alan and Scott Wayne. He flew with Reserve Squadron VMF-216 at Spokane and Seattle.

He returned to active duty in October 1960 for the defense of South Vietnam and flew with VMA-211 during its WESTPAC deployment from 1962-1963. He was trained in helicopters in 1963-1964 and spent June 1966 through July 1967 at Marble Mountain, Vietnam attached to VMO-2. He flew 680 combat hours.

He received a regular commission in the USMC and a college degree in 1968. He was XO of the 8th Marines from 1968-1969. He was CO of HMM-261 from 1969-1970, graduated from Command and Staff College in 1971 and was assigned as CO of HMM-263.

He served at Nam Phong, Thailand from 1972-1973 and with FMFPAC Staff from 1973-1978. He retired in July 1979. Total service for pay was 34 years, two months; foreign and/or sea service was five years, 10 months. His awards included the Air Medal with 34 stars, Navy Cross Medal, VSM with three stars, PUC, CAR and the WWII Victory Medal.

Presently Nancy and he split their time between the ranch in Idaho and their home outside the gate at MCAS Kaneohe Bay, HI. They have been married more than 40 years.

RONALD JAMES HELLER, "RON," was born on April 14, 1937 in Milwaukee, WI. He enlisted in the Marine Corps on April 14, 1954 and went through boot camp at Parris Island. He spent the first six years in the infantry with the 2nd Marine Division and then transferred to the Air Wing in 1960. He was assigned to HMR(L)-261 at MCAF New River.

During the deployment to WESTPAC in 1961, he was TAD to HMM-362 during operation Shu-Fly at Soc Trang, RVN. On his return to the States he was assigned to HMM-362 at MCAF Santa Ana. He was deployed to RVN with MAG-36 in August 1965 and was stationed at Ky Ha with HMM-362 till October 1966. During this tour he served as crewchief and gunner on UH-34s.

He retired from the USMC in October 1974 as a

First Sergeant. He was awarded his Combat Aircrew Wings, seven Air Medals, CAR, MUC, NUC, VCG etc. Today he resides in Coloma WI.

RICHARD E. HENDRIE, "CLUTCH," was born in Connecticut in 1939. He grew up in Canada and graduated from high school in Toronto in 1958. He earned a BS degree from Ohio University in June 1963. He was commissioned at Quantico in December 1963 and designated a Naval Aviator in July 1965.

He joined VMO-2 in July 1966 flying UH-1Es out of Marble Mountain Airfield, Vietnam. He was awarded 22 Air Medals, PUC, Vietnam Service and Campaign Ribbons. He was released from active duty a Captain in 1968.

He flew helicopters with the Minnesota Army N.G. in 1970 and transferred to Air Force Office of Special Investigations as a special agent in 1981 and currently is an active ready reservist. He received an Air Force Commendation Medal in 1984. He was a D.E.A. and U.S. Customs Special Agent from 1969-1989. He supervised the narcotics group until retirement. He specialized in air smuggling investigations and accumulated 2500 hours of flight time.

He currently farms 360 acres near Askov, MN and has been married 29 years to the former Linda Miller of Sandusky, OH. He has two sons, Chad and Chris.

ROGER A. HERMAN, "DUKE," was born on June 13, 1944 in San Diego, CA. He joined the Marine Corps after his second year of college and went through flight training as a Marine Aviation Cadet (MARCAD). He received his wings and commission as a Second Lieutenant in August 1966.

In January 1967, he joined HMM-361 as the squadron was preparing to return to Vietnam. Based out of Marble Mountain and Dong Ha, he flew the UH-34. During the last part of his tour he was aide-de-camp to the Asst. Wing Commander for the 1st MAW. Three years later, after a tour of duty as a flight instructor in Pensacola, he returned to Vietnam, this time as a C-130 pilot based out of Okinawa.

His decorations include 26 Air Medals, the Navy Commendation Medal, the Navy Achievement Medal, the Purple Heart, the Vietnamese Service and Campaign Ribbons, the Combat Action Ribbon and the Navy and Presidential Unit Citations. He attended Amphibious Warfare School in 1972. He received his college degree shortly thereafter, attending night classes while serving as the Marine Liaison Officer for the Atlantic Fleet Weapons Range in Puerto Rico. He held different assignments that included NATOPS, Aviation Safety, Maintenance and Operations and after 20 years of active duty service in the Marine Corps and Navy, he retired in June 1985.

After retirement from the military, he began a second career as an airline pilot. He is married, with two grown sons and lives in San Diego, CA. In 1986, after locating a few old friends from his in-country squadron whom he hadn't seen in nearly 20 years, he founded what would later become the USMC/Vietnam Helicopter Pilots & Aircrew Reunion. He has served as president of the organization since its beginning.

HAROLD L. HERNANDEZ, "CRAZY CUBAN," was born on Dec. 16, 1942 in Tampa, FL. He took a military LOA from Pan American Airways, where he was employed as an A&P mechanic, to enlist in the Marine Corps in February 1966. He was the first employee to enlist since WWII. He attended boot camp at Parris Island at 24 years old and was the "old man" of the platoon. Prior helicopter experience landed him in Memphis, TN where he completed the Basic Helicopter course in November 1966. Then he was assigned to VMO-1 at New River, NC, where he made crewchief and was promoted to Corporal in April 1967. Upon discharge in February 1968 he returned to Pan Am with the rank of Corporal.

He is now a farmer and an active member with the Marine Corps League. He has been married for 30 years and has two sons in the Marine Corps both graduates of the U.S. Naval Academy in Annapolis. They reside in Venice, FL.

DAVID P. HETRICK, "SPIKE," was born on Jan. 19, 1948 in Wadsworth, OH. He enlisted in the Marine Corps on April 21, 1966, left for Parris Island on Aug. 10, 1966 and then was sent to Memphis for aviation training on helicopters. Next he was assigned to HMM-162 at MCAF New River.

He volunteered for overseas duty, arrived at Phu Bai, Vietnam on Jan. 15, 1968 and reported to HMM-164. In August of 1968 he was reassigned to HMM-364 and served until Feb. 21, 1969. He was stationed in-country at Phu Bai, Da Nang, and Atsugi, Japan. While in Vietnam, he also had the pleasure of flying the Bob Hope Christmas Show for four days in December 1968.

After leaving Vietnam, he reported back to New River where he served with HMM-162, HMMT-402 and then became an instructor with NAMTD at New River.

As a crewchief in both squadrons, he earned the Combat Aircrew Wings, three Single Mission Air Medals, 28 Air Medals, Navy Commendation Medal w/Combat "V," two Presidential Unit Citations, Combat Action Ribbon, Navy Unit Citation, Meritorious Unit Citation, Vietnam Service and Campaign Ribbons and the Vietnam Cross of Gallantry. He was discharged on Oct. 22, 1975 with the rank of Staff Sergeant.

He is presently living in Downingtown, PA with his wife, Susan, and their three daughters: Cindy, Kati and Amy and her husband, Matt. He has worked for Xerox Corp. since December 1975 and is presently a manager in the Wilmington-Delaware District.

JUDSON D. HILTON JR., "J-DOG," Lieutenant Colonel USMC (Ret) was born on July 8, 1944 in St. Paul, MN. He enlisted in the Marines on Oct. 30, 1964 and completed boot camp at Parris Island. He became a MARCAD in May of 1965 and was commissioned a Second Lieutenant with wings in December 1966.

He joined HMH-463 on Jan. 19, 1967 at MCAS Santa Ana and was deployed to Vietnam aboard the LPH *Tinian* in May. He became a ground FAC with BLT 2/4 in November 1967. He returned to Vietnam in May of 1973 with VMA-211 flying A-4s.

He was awarded the Silver Star, 18 Air Medals and Vietnamese Cross of Gallantry. He retired as the Commanding Officer of SOES at Cherry Point in November 1984.

He was a pilot for People Express Airlines and is now with Continental since their merger in 1987. He is married to the former Pamela R. Gerring and they have two sons, Judson III and Russell. They reside in Jupiter, FL.

ALLYN J. HINTON JR., "BIG AL," was born on Sept. 1, 1942 in Kansas City, MO and graduated from Soldan High School in St. Louis, MO in 1960. He holds a BS degree from Southeast Missouri and a master's degree from the University of Missouri, St. Louis.

From June 20, 1960 to June 19, 1963 he was a U.S. Army paratrooper. From Feb. 23, 1968 to June 30, 1980 he served as a U.S. Marine Corps Aviator. He served in Vietnam with HML-167 from December 1969 to December 1970. He piloted UH-1Es, T-2s, TA-4s, KC-130s, UH-1Hs and UH-60s.

He was discharged from the U.S. Marine Corps in 1980 with the rank of Captain and was awarded these medals during his military career: 46 Air Medals, NUC, MUC, VSM, ARCOM, AAM, SWASM and KLM.

Since his discharge from the Marine Corps in 1980, Hinton has been teaching American history and aviation at Parkway South High School, Manchester, MO. He has also been coaching football. As a Warrant Officer pilot with the 7th BN 158th Aviation Regt. He flew with his son who graduated from Army Pilot Training in June 1990 and went to operation Desert Storm where he flew 115 hours.

He is married and has one son, Allyn J. Hinton III, age 24, and one daughter, Elizabeth R. Martin, age 23 and one granddaughter, Hanah Martin, age 3. He is currently in the U.S. Army Reserve and holds the rank of CWO-4.

CHARLIE HIPSHIRE, served with HMM-265 for 18 months as a crewchief from approximately April, 1967 to November, 1968.

At the age of 24, he died in an automobile accident and was buried on February 21, 1971, in Morristown, TN, the year after his discharge. He was a member of the Morristown VFW and Moose Lodge.

He is survived by his parents, David and Lissie Hipshire, sisters, Mrs. Mary Lou Smith, and Mrs. Lois

White, and grandparents Rev. and Mrs. J.C. Hipshire and Mrs. Johnnie McGinnis. He was known and loved by all. He is sorely missed. *Submitted by Gary Spevacek.*

WILLIAM T. HOLMES "TEE" was born on April 13, 1943 in Dublin, GA. He joined the service in November 1963. He was stationed at New River in the States.

He was a co-pilot/HAC flying UH-34s while in-country and was stationed at Futema, Okinawa, Marble Mountain and Guantanamo Bay.

"One event that surely set some type of record occurred during a night medevac that Jim Van Gorder and I conducted at sea. We launched out of Da Nang on the proverbial 'dark and stormy night' to pick up a sailor who had been critically injured in a shipboard accident. We found the ship (which seemed to me to be about half way to Japan), and though quite large it had no helicopter pad or level spot for landing. What it did have was an array of antennae, radar domes and jagged pieces of steel just waiting to ruin the day and careers of a couple of bone weary, vertigo crazed UH-34 pilots.

"The deck crew, with whom we had direct radio contact, was having as much difficulty on the rolling and pitching deck trying to get the litter to a high point among the antennae and radar domes as we were having trying to hover over their position. A sailor finally managed to grab our hoist cable, but in an effort to free both hands to help his mates, he reached over and fastened the cable to the ship. Unfortunately, neither Jim nor I had been privy to the sailor's admirable intentions. All we knew was that suddenly our aircraft was being controlled, and rather abruptly I might add, by some giant invisible hand reaching up from the briny deep.

"To my knowledge this was the first time in Marine Corps history that one of its helicopters, while airborne, had been towed by a ship of the U.S. Navy. Attached to our tether, we found ourselves flying close (very close) night formation with the pitching stern of a U.S. Navy warship. Obviously, the alternating upward zooms and bone jarring jerks down provided us ample opportunity to dramatically overspeed and overboost the engine, and to exceed most of the other limitations of our trusty steed as well. Among those was the weight limit on the hoist cable which, though twanging like a banjo string, never sheared as advertised. We were only seconds from cutting the cable ourselves when the guy on deck finally got our message (relayed through the bridge) and unhooked us.

"After ruling out a gunnery run on the hapless sailor, we were eventually able to make the pickup and subsequently find our way home." Discharged in April 1968 with the rank of Captain and earned all the regular medals for a one tour "Dog Driver."

Today he is married to Peggy and they have one daughter, Carrie, age 16. They reside in Marietta, GA and he is a sales manager for Geographical Publishing Company.

EDWARD A. HORNE Lieutenant Colonel USMCR (Ret) was born on Aug. 8, 1945 in Cambridge, MA. He enlisted and entered boot camp in San Diego in 1966. He was ordered to OCS in 1968, graduating as platoon honorman and completed Officer Basic School as the distinguished academic graduate. He was designated a Naval Aviator in 1969 and attended the Army's Cobra Gunship School.

He was assigned to HML-367 Da Nang, Vietnam from January to December 1970 and flew more than 800 combat missions. He spent additional tours at Camp Lejeune, Philadelphia and New River. He was deployed in 1974 with HMM-264 to the Mediterranean and was transferred to USMCR in May 1976. He joined HMM-767 at Norfolk.

He returned to active duty in September 1981, to the Full Time Support program. He was promoted to Lieutenant Colonel and transferred to MAG-49, Atlanta as Executive Officer. He transitioned to the CH-53 and MAG-41, Dallas, as the Executive Officer and retired on Dec. 1, 1991.

He earned a BA in economics and graduated magna cum laude and he earned a MBA in management. He is currently an executive director for the Texas Association of Mortgage Brokers. He is also an active member of the USMC/Vietnam Helicopter Pilots & Aircrew Reunion and numerous Marine and Navy associations. He is married and lives in Arlington, TX.

ANSLEY S. HORTON Col USMCR (Ret) was born on Dec. 8, 1937 and attended high school in Daytona Beach, FL. He is the son of a USMC Master Sergeant. He enlisted in the USMCR in 1955-59 and obtained a communications degree at the University of Florida. He earned an MA in personal relations and an MA in management at Webster University. He was in the 25th Officers Candidate Course-platoon and was honorman of Honor Company in December 1959. He was designated a helicopter pilot in May 1961 at Ellyson Field. He was with HMM-262 at New River when the squadron was involved in the recovery of astronauts.

He was on Mediterranean and Caribbean cruises with VMO-1. He was involved in the Cuban Missile Flap (another long story) with HMM-162 and then deployed with 162 to the Republic of Vietnam in 1964-65. He was part of Operation Shu-Fly (considered combat advisors) until the Gulf of Tonkin brought American forces in-country.

He was an instructor at Ellyson from 1965-68 and Instructor of the Year. Released from active duty in 1968 to be an airline jock. He was laid off six times from Pan Am, World and TWA. He flew UH-34s and CH-53s in reserves at Alameda. He was in the reserves in New Orleans and CAT-6 Executive Officer in 1974-78 MARTD Glenview. Then he flew UH-1s and C-130s. He was CO 4th MAW DET A NOLA and OIC Mobilization Station NOLA. He has flown 13 different USMC aircraft and had approximately 4600 hours in the UH-34. After 35 years he had a retirement ceremony in the National Museum of Naval Aviation in Pensacola in June 1990.

Presently he is a T-34 Simulator instructor at Whiting and is also self-employed.

JEFFREY J. HOUSER, "DOC CD," HMCS(AC) USNR-R was born on March 31, 1948 in Los Angeles, CA and he enlisted in the Naval Reserve Hospital Corps in September 1966. He attended college, received military, hospital and combat medical training from the Navy and Marine Corps at NTC and Balboa Hospital in San Diego and Field Medical Service School at Camp Pendleton. He was activated in early 1968 for two years.

He served with the 3rd Marine Division in Vietnam as a platoon corpsman along the DMZ and in Laos. After recovering from wounds he was reassigned to an emergency medical facility near Dong Ha and also worked in outlying villages and orphanages with the Medical Civic Action Projects (MedCap). He then served with the MAG-16 south of Da Nang as a Medical Evacuation/Search and Rescue Aircrew corpsman on CH-46 helicopters; duty known as the "To Hell and Back Medevac."

As a reserve corpsman in a Marine support medical unit, he was again activated to full duty, from January 1991 to March 1993 for the Persian Gulf War as clinic supervisor, Marine Combat Training-School of Infantry, Camp Pendleton. He reaffiliated with his reserve medical unit upon release from active duty.

His awards included: two Bronze Stars, three Purple Hearts, seven Air Medals, NCM, NAM, CAR, MUC, Vietnam Service and Campaign Medals, Vietnam Gallantry Cross, Vietnam Civic Action Medal, seven NRMS Medals and two AFR Medals and Marine Corps Combat Aircrew Wings.

He is married with two children, and is a nature photographer residing in Southern California. He retired in 1995 with nearly 29 years combined military service.

EDWARD J. HUNNEYMAN, "EASY ED," was born on Aug. 26, 1939 in Oneida, NY. He graduated from Oneida High School and in 1962 he received a BA in economics from Guilford College, NC. While attending college he joined the U.S. Marines Corps Reserve which he served in for five years prior to entering on active duty in March 1963. He was commissioned as a Second Lieutenant upon completion of OCS and proceeded to NAS Pensacola where he subsequently received his wings as a Naval Aviator.

He arrived in the Republic of Vietnam in July 1966 and was assigned to HMM-364. When HMM-364 rotated back to CONUS, he transferred to HMM-362, with whom he remained until late spring 1967 when he transferred north to join HMM-363 at Dong Ha. Upon return to CONUS, he served with MAG-26 at MCAF New River where he transitioned to the CH-46. His memories include flying out of Khe Sanh, working with the Korean Marines, any night medevac's and recon extracts and inserts.

After a Caribbean cruise with the BLT, he returned to New River until his release from active duty in November 1968. He was awarded the Distinguished Flying Cross, 28 Air Medals, Vietnamese Cross of Gallantry and Vietnamese Service Ribbons. Units he served with were also awarded the Presidential Unit Citation and the Naval Unit Citation.

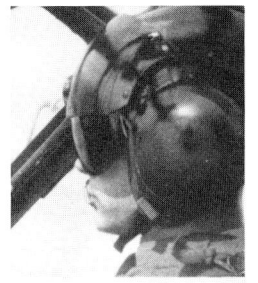

Following service in the USMC, Hunneyman began duty with the FBI as a Special Agent. After serving the FBI in Missouri, Texas and Alabama for 25 years, he retired in November 1993. Presently he is serving on the staff of the Alabama Attorney General and residing in Montgomery, AL. He is married and has a daughter and a son.

WILSON ROGER HURD, "BILL," was born on a ranch outside Monte Vista, CO on Nov. 4, 1941. He was graduated from Monte Vista High School in 1959. After one and one fourth years of college he joined the USMC. He went to boot camp at MCRD, San Diego in May 1961. He started as a MARCAD in April 1962 with Class 14-62 (Best Class Ever). CarQualed April 16, 1963 and was designated a Naval Aviator T-8485 on Oct. 4, 1963. He was assigned to MAG-36, MCAF Santa Ana, CA and joined HMM-462 and became a "Deuce Driver." He then went on to MABS-36 to become the TAFDS officer (Fun! Fun!).

He went to Vietnam to run the fuel farm at Ky Ha for MAG-36 in September 1965. He joined HMM-364 and spent the rest of tour of duty being a "Dog Driver."

Hurd was awarded 15 Air Medals and separated from active duty as a Captain upon return to the States in October 1966.

He hired on with United Air Lines in November 1966 and is still flying for a living. He has been married for the past 31 years to the same beautiful, tolerant wife, "Pokey." Present count is two daughters and two grandsons. He is living at the present time in South Fork, CO at the foot of Wolf Creek Pass (east side). He spends his time skiing, kayaking, bow-hunting for elk, fly-fishing, climbing fourteeners (14,000 foot mountains), motorcycling and punching cows. He is still a "Junkman" at heart.

SEPPO I. HURME, "FLYING FINN," Captain USMCR was born on Jan. 21, 1944 in Vyborg, Finland. He emigrated to the United States in 1960 and graduated from Venice High School in California in 1963. He attended UCLA for three years prior to entry into service. He was commissioned a Second Lieutenant and designated a Naval Aviator on June 9, 1967 in Pensacola. He served with HMMT-301 at MCAF Santa Ana from July 1967 to September 1967. He qualified in the UH-34D.

He joined with HMM-361 in Vietnam in October 1967. He flew 380 combat hours before the squadron was decommissioned in May 1968. He transferred to VMO-2 and then to HML-167. He flew the UH-1E for 95 hours before an AK-47 round to the right heel ended his tour on July 10, 1968 and he was medevac'd to Yokosuka, Japan.

He was awarded the Purple Heart and holds a total of 17 Air Medals plus Vietnam Service and Campaign Ribbons. He completed active duty as a First Lieutenant with HMM-561, HMH-363 and then H&MS-56.

He obtained a BS in engineering in 1972 and MBA in 1974 from UCLA He flew with the reserve squadrons HMM-764 and HMA-773 from April 1971 to August 1974.

As a civilian he has served in various contracts and marketing positions with several aerospace subcontractors. The aircraft programs have included B-1, E-2C, F-14, F-15, F-18, H-53, YF-117A and the Space Shuttle.

Currently he is a manufacturer's technical representative, providing marketing representation for several small businesses in the Southern California area. He is married with two daughters.

He is currently an active member of the USMC/ Vietnam Helicopter Pilots & Aircrew Reunion and Vietnam Helicopter Pilots Association.

ROBERT E. ILZHOEFER, "ILZ," was born on May 10, 1937 in Brooklyn, NY. He enlisted in the Marine Corps in 1956 and was transferred to the Naval Aviation Cadet Program in 1957 and was designated a Naval Aviator in 1959. He served with various UH-34 squadrons at MCAS New River.

He took part in Expeditionary Operations with HMM-264 in 1961 in the Belgian Congo. Later in 1961 he was deployed to the Mediterranean with HMM-262 and in 1962 he took part in the Cuban missile crisis with HMM-261. He was deployed to Vietnam in 1963 with HMM-261 and he flew the UH-34 in support of Operation Shu-Fly. After a flight instructor tour in Pensacola, he returned to Vietnam in 1967 to fly the UH-34 "Evil Eyes" with HMM-163 and completed his second tour flying the CH-46 with HMM-164. He returned to the Mediterranean in 1974 with HMM-162 taking part in the mine clearing operations in the Suez Canal and the evacuation of Nationals from Cyprus.

His awards include two Distinguished Flying Crosses, 36 Air Medals, two Presidential Unit Citations, two Navy Unit Citations, one Meritorious Unit Citation and the Vietnamese Cross of Gallantry. Major Ilzhoefer retired in 1976 with a total of more than 5900 accident-free hours.

He returned to college and earned a degree in education and has been a middle school teacher in Onslow County for the past 15 years. He is married, has two sons and resides with his wife, Christine, in Jacksonville, NC.

REED JACKSON, "TWEED," was born on Dec. 15, 1942 in Mexia, TX and graduated in May of 1961 from Mexia High School, where he was a member of the National Honor Society, an All-State trombonist and captain of the golf team. In 1966, he received a bachelor of business administration from Southern Methodist University where he was initiated in Phi Mu Alpha and Kappa Alpha Order. After a short stint with the Internal Revenue Service, he entered the Marine OCS in March of 1966, followed by flight training at NATC Pensacola where he received his wings in August 1968.

After training in the CH-46 with HMM-162 at MCAS New River, he reported to HMM-265 at Hue Phi Bai, RVN in January 1969. When HMM-265 returned stateside aboard the USS *Iwo Jima* in September 1969, he was assigned to H&MS-36 at MCAS Futema, Okinawa until returning from his tour and reporting to HMM-163 at MCAS (H) Santa Ana in February 1970.

His last active duty billet was as assistant Special Services officer for the 3rd Marine Air Wing at MCAS El Toro, where he was released from active duty on Nov. 15, 1971 and assigned to HMM(later HMH)-777 at NAS Dallas, driving CH-53s until transferring to a Judge Advocate's position at MRTC Houston. He received 19 Air Medals, Vietnamese Service, Cross of Gallantry and a Campaign Ribbon.

While with HMH-777 he completed law school at Baylor University and since being licensed in November 1974, he has engaged in "defending the weak and the oppressed." He has lived and worked in Fairfield, TX since the spring of 1975, where he at various times has served on the library board, hospital board, city council and youth baseball/softball association board. Politically he has served both Major parties, having been sent as a delegate to the Texas Democratic Convention in 1978, and the Texas Republican Conventions in 1980, 1984 and 1986. He was chairman of the Freestone County Republican Party from 1980-84 and served as secretary of the Texas Republican County Chairman's Association from 1982-84. He is licensed to practice before the Texas Supreme Court, the United States District Courts for the Northern, Eastern, Southern and Western District of Texas, and the U.S. Court of Military Appeals. He has served as president of the Bi-Stone Bar Association and on the Lawyer Referral Committee of the State Bar of Texas and is chairman of the State Bar District 2-C Grievance Committee. He was inducted as a Fellow of the State Bar of Texas in 1993. In addition to his private practice he serves as city attorney for the city of Fairfield.

He is married to Stacy and they have three children: Brian, Lauren and Jordan. They reside in Fairfield, TX.

ROBERT W. JOHANNESEN, JR., "PACK RAT," was born on July 29, 1946 in Schnectady, NY. He enlisted in the USMC in 1967 (Honorman) and served with F-2-2 on a Med cruise. He was next assigned to Reconnaissance School at Camp Pendleton and later commissioned in November 1968. He attended basic school in May 1969 and Army Flight School and was designated a Naval Aviator in 1970.

He flew CH-53s in HMH-461, UH-1Es with VMO-6 and HML-167 in Vietnam. Later he flew UH-1Ns in HML-167 and CH-46Ds in HMM-774.

He was discharged in 1973 with the rank of Captain.

He finished a BS in law enforcement in 1975.

In 1973 he applied to Bureau of Narcotics and Dangerous Drugs (BNDD) and in September 1975 was trained as a special agent with the Drug Enforcement Administration (DEA) and has been stationed in Wilmington, NC; Newark, NJ: Washington, D.C. and Dallas, TX. He has worked undercover assignments in the U.S., Europe, Mexico, South America, Canada and the Caribbean. Johannesen has worked as a "street agent," technical operations agent, special agent pilot and supervisory special agent. He is now the Aviation Maintenance supervisor for DEA's fleet of 124 aircraft at Alliance Airport in Ft. Worth, TX. With his ATP, CFI and A&P ratings, he is a senior member of DEA's Aircraft Accident Investigation Team which was trained by NTSB, University of Southern California and the USAF.

He is married and has two teenagers.

BRUCE W. JOHNSON "B.W." was born March 28, 1948 in Miami, FL. and graduated from high school in June 1966. He joined the Marines on April 26, 1966 and attended boot camp at Parris Island, S.C. After receiving helicopter training at NATTC, Memphis, TN, he was then transferred to MCAF New River, N.C. After a Caribbean cruise from August 1967 to December 1967 he was transferred to Vietnam. He served with HMM-265 in-country as a crewchief and machine gunner April 3, 1968 to Nov. 27, 1968. During his tour of duty he was awarded the Purple Heart, Combat Aircrew Wings, Single Strike Flight Award with Bronze Star, PUC, MUC, CAR, Vietnamese Cross of Gallantry, Vietnam Civil Action Citation, Vietnam Service and Campaign ribbons.

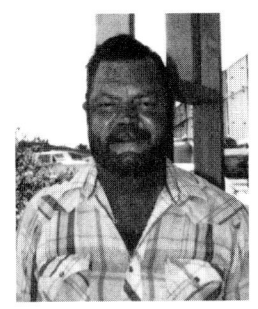

He was medically discharged on Aug. 22, 1969 with the rank of Lance Corporal. He later attended college and earned an AA degree in business administration. He is now finishing his bachelor's degree in Psychology.

GERALD "JERRY" W. JOHNSON, "RAG-ARM," Lieutenant Colonel USMCR (Ret) was born on July 9, 1939 in the suburbs of Vanduser, MO (pop. 281). He graduated fourth in his class from Vanduser High School in 1956, (only eight graduated). He was a four year letterman in basketball and fast-pitch softball (everyone had to play). He holds a BS from the Southeast Missouri State University (1962) and a BA from the University of West Florida (1974). He was commissioned as a Second Lieutenant on Dec. 21, 1962 at Quantico, VA. Then he went on to TBS for John Wayne Day (wounded by flying grenade fragment) and the Three Day Way (had to leave early to practice softball). He received his orders to Pensacola for Naval Flight School and was designated a Naval Aviator on Feb. 2, 1965.

He received a great set of orders to join HMM-161 at Kaneohe which were canceled after arriving in SFO for transportation to the islands. He was then assigned to HMM-363 at Santa Ana, CA. He was deployed in August 1965 with HMM-363 aboard the USS *Princeton* to the Republic of Vietnam. He transferred to HMM-361 in Da Nang on Sept. 12, 1965 and completed his tour in-country with HMM-361 at Chu Lai on Sept. 21, 1966.

His assignments included: Flt. Equip. O, HMM-363; Pers. O, Assist. Admin. O, HMM-361; Flt. Inst., HT-8; OPS O, HMM-767; XO, HMM-768; XO, MARTD New Orleans; Reserve Aviation Plans & Programs O, HQMC; and XO & CO of HMH-772.

His many memorable experiences range from the medevac of Gunnery Sergeant Jimmy Howard's recon team from Hill 488; five flights into Hiep Duc with the VC chained to their 50 cal.; leaving country with WW and Swamp to represent MAG-36 in 1st MAW softball tournament in Okinawa (Swamp never saw the ball and WW lost his shoe); living with LQ, BT, J-Animal and Cochise when they were attacked at Marble Mountain and when the tent left the hard frame about 3:00 a.m. in a driving rain and wind storm; volunteering for SAR North for a day of R&R, only to be launched with BJ and wondering what all those red marks were on our map (didn't know a UH-34 could go that high); having the privilege of being picked to instruct the Apollo astronauts in the TH-13M, simulating the moon landing; playing a lot of softball with McSouth at Ellyson Field, winning the 6th Naval Dist. tournament and inches from winning the regional; map reading course from Black Bart in N.Y.C. while he was driving (said UH-34 drivers couldn't read a @#$%^&&*?# map); planning the going away party for Lieutenant Colonel "P" in N.O. where he received a special plaque and gift to show him our appreciation and love; many fun filled weekends and ATD's in the reserves with the Gorilla, carrier quals (boots and helmets); someone mooning the flag at Quantico, buying paneling at New River; FIVE out and back from Willow Grove and Yuma; setting the record for nacho eating in Yuma for a squadron; but most of all just serving with the best group of guys in the world, USMC Helo Pilots and Aircrews!!!

Personal decorations include: DFC, 20 Air Medals, PUC w/star, NUC, MUC w/star, Vietnam Campaign Medal, Vietnam Service Medal, National Defense Medal, OMCR Medal and Vietnamese Cross of Gallantry.

He is currently retired from the USMCR and pitching softball (except for hitting practice for his daughter's high school team). He is now employed by United Airlines as a Captain on the 737-300/500, based at O'Hare Intl. Airport in Chicago, IL. He lives in Sleepy Hollow, IL and is married to the former Susanna Bryant (UAL F/A) and they have two sons, Brad, age 24 and Chad, age 22, not living with them, and one daughter at home, Marti, age 16.

LOUIS E. KAISER, "GUNNER," was born on Oct. 5, 1944 in Hazel Green, WI. He graduated from Cuba City High School in 1962 and joined the USMC in Dubuque, IA on July 9, 1962. He attended boot camp at San Diego and went on to Memphis, TN for more training. He was stationed at MCAF New River with HMM-263.

He had an extended tour of duty to go to Vietnam in 1965 and was attached to HMM-263 at Marble Mountain. He was a gunner/crew member in HMM-263 in UH-34Ds.

As the attack began on Marble Mountain, RVN by insurgent communist guerrillas (Viet Cong) on the night of Oct. 27 1965, Corporal Kaiser was in his quarters approximately one half mile from the mat. He was ordered to proceed with four other Marines to the flight line to assist in damage control. On arrival he found the office tents infested with Viet Cong. In the ensuing action, Corporal Kaiser aggressively attacked hidden enemy positions in the line shack, the ready room and in the face of automatic weapons fire and grenades, proceeded from tent to tent to flush out Viet Cong. After all the enemy had been killed or captured, he volunteered to locate and mark the numerous unexploded grenades and satchel charges to prevent their accidental firing, even though the conditions, stability and state of fusing were completely unreliable.

He was discharged on Jan. 8, 1967 with the rank of Sergeant from Willow Grove NAS. He was awarded the Combat Aircrew Insignia with three stars, Navy Commendation Medal with "V," Vietnam Cross of Gallantry, Air Medal with three stars and recommendation for a Bronze Star.

He broke his back in 1967 and has been confined to a wheelchair since. He served as city councilman for 11 years and served six years as mayor of Cuba City. He is chairman of the Grant County Veterans Service Commission and has served as vice-chairperson on the Grant County Community Option Committee. He has three children and three grandchildren and resides in Cuba City, WI.

WAYNE KARLIN was born on June 13, 1945 and enlisted in the Marine Corps in 1963.

He was sent to Vietnam in 1966 and served in Utilities and in the security platoon of MACS-7 at Ky Ha. After volunteering for and completing aerial gunner training, he was reassigned to the administration section of HMM-164 at Marble Mountain, HMM-161 at Phu Bai and HMM-263 at Marble Mountain. He flew combat missions in CH-46s for HMM-164 and in UH-34Ds for HMM-263.

He received Air Medal Awards, Combat Aircrew Wings and a Letter of Commendation. He was discharged with the rank of Sergeant in 1967.

After the war, he completed his BA and MA degrees, worked as a journalist abroad and is currently a writer, who has published four novels. He is also a professor of language and literature at Charles County Community College, LaPlata, MD. He is married and has one son.

WESLEY D. KEITH, "FLUFF," was born on Sept. 20, 1944 in Los Angeles, CA. He graduated from high school in Napa, CA in 1962 and received a BA from San Francisco State University in 1966. He entered active duty on Aug. 15, 1966 at OCS, Quantico, VA. He was designated a Naval Aviator in November 1968 in Pensacola (10 month delay at HT-8 due to an accident). He qualified in the UH-1E with HML-267 at Camp Pendleton from November 1968 to April 1969 and qualified in the AH-1G at Cobra Hall, Hunter Army Air Field, Savannah, GA in May 1969.

He was attached to VMO-2 at Marble Mountain

in June 1969 flying AH-1Gs. HML-367 later became the designation of the Cobra (AH-1G) squadron. In January 1970 he was attached to Da Nang DASC as a helicopter coordinator.

He was awarded the Distinguished Flying Cross, Single Mission Air Medal and the Navy Commendation Medal. He completed active duty with the rank of Captain at HT-8 as a flight instructor in February 1972. He flew with the Arkansas National Guard until 1981.

He received his MS in administration from the University of Arkansas. Since 1974 he has been a federal employee with SSA. He was ordained a Southern Baptist minister in 1982 and has been active as a bi-vocation pastor since then. His wife, Katie, and he have three grown children. In 1987 he founded the Vietnam Veterans for Christ Ministries as an outreach ministry to help veterans and family members who are having difficulties coping with their Vietnam experiences. This includes mailing out a quarterly newsletter and speaking at various churches, prisons and other organizations and one-on-one counseling. He has published one book entitled *Victories: Stories of Christian Vietnam Veterans*. Currently, he is active in several service organizations, including serving as chaplain for the Arkansas VVA State Council and serving two terms as president of his local VVA Chapter.

JOHN PATRICK "PAT" KENNY, "SWIFT CHUCK," was born on July 18, 1944 in St. Louis, MO. He joined the USMC on June 5, 1968 and had duty stations in the States in Pensacola, FL, MCAS, Santa Ana, Tustin, CA and Camp Pendleton, CA.

While serving in-country he was stationed at Marble Mountain Air Facility, Vietnam and with the Fleet Air Western Pacific, Atsugi, Japan.

His memorable experiences include: HMM-364 Purple Foxes, second flight in RVN when Dave "Frenchie" Legas and Kenny tried to land on the German hospital ship *Helgoland* in Da Nang thinking it was the USS *Sanctuary*;" You're a real Charlie Lindbergh, FNG, from now on we call you Swift Chuck"; flying night medevac or anything else for that matter with J.D. Bell especially when he was in the mood to stir up trouble; getting chewing-out by Lieutenant Colonel Scaglione for "stealing" one of "his" helicopters to deliver beer and oranges (a full sling load) to Bob "Uncle Norman" Stocker, who was on FAC duty at An Hoa and was soon to be overrun by Vietnamese; touch and go's on Highway 1, with "Worm" Wise the day he made HAC.

He was awarded 29 Strike/Flight and misc. other stuff. He was discharged on May 31, 1974.

Today he's married to wife, Nancy, (who has put up with all of the trouble over the years and four daughters who thank God weren't around during the old man's grungier years.) His present occupation is as a manager/grain merchant for Cargill, Inc. They reside in Richland, WA.

TIMOTHY R. KIRKMAN, "Chief", was born March 9, 1943 in Dinuba, CA. He graduated from the U.S. Naval Academy and was comissioned a 2nd Lt., June 1965. After attending Basic School in Quantico, VA he went to flight training in Pensacola, FL where he earned his wings in 1967. After combat readiness training in Santa Ana, CA he was transferred to Vietnam, where he served as a UH-34 pilot with HMM-361 from August 1967 to March 1968 when he was wounded flying a medevac. After recovering from his wounds in 1969, he was assigned as a flight instructor with VT-5 in Pensacola, FL training flight students in T-28 carrier qualifications.

He was discharged as a Captain and awarded a 40% V.A. disability in 1970. He has more than 40 Air Medals, a Purple Heart, PUC, NUC, Vietnamese Cross of Gallantry, and campaign ribbons. He began his civilian career in computer technology with EDS as a Systems Engineer. He earned his MBA in business management from the University of Dallas in 1977. He has held several executive management positions in MIS with American Express, Peer Services and is currently a Director in MIS with Burger King Corp. in Miami, FL. He is married to the former Marilyn Stewart of Pensacola, FL and has two sons in the Marine Corps and a daughter with two grandchildren in Ft. Worth, TX.

CLAUDE E. KING was born on Aug. 25, 1944 in Washington, D.C. He graduated from The Citadel in 1966 then went to the Basic School, winning the Top Military Skills Award. He married his high school sweetheart, Nancy, and took her to Pensacola. He attended Flight School in 1967 and was assigned to VMO-1 at New River.

Then he went to HML-367 at Phu Bai. From June 1968 to July 1969 he flew more than 1100 hours of gunship missions as "Scarface-33." He remembers being shot down and the aircrew all being rescued without being captured, even though NVA soldiers were very close.

He was stationed at Camp Pendleton from 1969 to 1972 and transitioned to OV-10, and his first child, Melody was born. He was discharged on March 10, 1972 with the rank of Captain. He was awarded three Distinguished Flying Crosses, 52 Air Medals, a Purple Heart and three rows of various other ribbons.

After separation from active duty, he joined the FBI. After agent training, he was assigned to Newark, NJ where his son, Brady was born. He helped develop FBI air operations flying more than 7000 hours in helicopters and airplanes. He opted for his latest assignment as a "street agent" in Savannah, GA, but still flies the UH-1H for the FBI. You may have seen his helo circling at Waco.

PETER G. KNESE was born on April 18, 1942 in St. Louis, MO. He graduated Chaminade College Prep in 1960, and earned a BA at St. Mary's University, San Antonio, TX in 1965. He was commissioned a Second Lieutenant USMC at Quantico, VA in December 1965 and was designated a Naval Aviator in 1967. In November 1967 he joined HMM-262, then went on to Okinawa. In January 1968 the squadron rotated to Quang Tri, South Vietnam until January 1969.

"Assigned to HMM-262 from November 1967 to January 1969 my tour was exciting and rewarding. Had I not recently been married, I probably would have extended. As your basic line pilot, the most rewarding mission was being able to extract a wounded Marine and fly him within minutes to the best medical attention. I have yet been able to equal the sense of satisfaction, responsibility and accomplishment as I did while in my 20s, during that one tour in Vietnam."

Returning as a Captain to MCAS New River, NC, he survived a loop in a CH-46 with Captain James Darrell Purdy and crewchief Corporal John J. Murphy. In January 1971 he joined the Federal Bureau of Investigation. In 1979 he earned an MS from the University of Louisville.

He is currently a special agent for the FBI and a surveillance pilot for the St. Louis Special Operations Group. He has been married to the former Janet (Janie) Dobson since April 1967. They have two children, Peter II, who married Melanie Connelly (January 1994) and Mary Brittany.

ERNEST JOHN KUN, "EJ," was born on June 14, 1940 in Palmerton, PA. He graduated from high school in 1958 and attended Steven State Tech and then Iowa State University obtaining a BS in 1965. He was commissioned on Dec. 17, 1965 as a Second Lieutenant from Quantico. He was designated a Naval Aviator on May 26, 1967.

He was assigned to HMM-364 "Purple Foxes" and as a squadron went to Vietnam in October 1967. While in Vietnam, he was assigned to 1st Battalion 5th Marines as a forward air controller.

He was awarded the Distinguished Flying Cross, 16 Air Medals, Combat Action Ribbon, Vietnam Campaign and Service Medals and the National Defense Service Medal. He completed duty as a Captain with HMM-264 at New River, NC.

As a civilian, he is a U.S. Secret Service Agent. He is currently the special agent in charge of the Philadelphia District. He is married and has a son and daughter.

GREGORY W. KUSKE, "CUSS," was born on Sept. 22, 1946 in New Richmond, WI. Enlisted in January 1964 on the delayed program. He was graduated from New Richmond High School and went to boot in San Diego in July 1964. Avionics School was next at Millington NAS and from there he went to MAMS-37 at El Toro. He was transferred to HMM-165 in January 1966.

He loaded on the *Valley Forge* LPH-8 in the fall of 1966 and went in-country. He stayed with HMM-165 until due for discharge. He flew as gunner on CH-46s, UH-34s and UH-1Es. Awarded the Purple Heart and the Silver Star. Discharged at El Toro as a Sergeant in May 1968.

75

He turned down a job as a field representative for Boeing Vertol and went to work for GE, the builders of that awesome T-58 engine. He married in 1970 and has three great kids. He is still with GE and with the same wife. They reside in Limestone, TN.

NORMAN S. LAFOUNTAINE, "FRENCHY," was born on Oct. 10, 1947 on Nantucket Island, MA. He enlisted in the Marine Corps on Sept. 13, 1965, completed boot camp at MCRD San Diego and ITR at Camp Pendleton. He graduated from Basic Helicopter "C" School, NATTC Memphis, TN in June of 1966; Class 624B. He was assigned to HMM-162 (UH-34D), MCAF, New River.

He was deployed on a Caribbean cruise in April 1966 and on June 23, 1967 he was injured in a mid-air collision between a UH1E and a CH-53A. He returned to limited duty in February 1968 with HMM-162 (CH-46D). He transferred to H&MS-26 jet shop when HMM-162 deployed on a Caribbean cruise. He transferred out of H&MS-26 in late spring of 1968. He arrived at MMAF, MAG-16 and was assigned to HML-167 as a mechanic/gunner and then qualified as a crewchief on a UH-1E gunship.

He was honorably discharged on Nov. 23, 1969 with the rank of Corporal and was awarded the Combat Aircrew Wings, 33 Air Medals, NUC, MUC and Vietnamese Gallantry Cross.

He graduated from Fisher College with an AS degree in business management and completed flight school, King Aviation, Taunton, MA. He became a commercial, instrument airplane pilot. He took up sport parachuting and scuba diving, and was certified as an Advanced Open Water Scuba instructor.

He entered a sales management training program with a national company in 1974 and four years later, started a manufacturers' representative business. He contracted out to companies throughout Africa and the Middle East as a procurement specialist.

Currently he is testing for a USCG commercial Captain's license to run a charter boat with his son. They plan to buy or build a marina (Yankee Marine Enterprises) somewhere on the southeast coast. He is active in many veteran organizations and projects. He has been married since 1970 to Johnelene, and they have one son, Jonathan, age 19. He first came aboard as a volunteer recruiter/propaganda SLJO for the "REUNION" in late 1988 and hospitality HOOTCH SLJO at the 1992 Reunion. He became the first aircrewman to sign on as a corporate officer when the reunion went "official" in 1993 and evolved into the USMC/Vietnam Helicopter Pilots & Aircrew Reunion.

BRUCE R. LAKE was born on Feb. 16, 1947 at Camp Lejeune, NC. He enlisted in the Marine Corps after high school and completed boot camp at Parris Island and Advanced Infantry Training at Camp Geiger before being selected for the MARCAD Program. He was commissioned a Second Lieutenant and designated a Naval Aviator on June 2, 1967 and the following day married Claudia Hackler of Keene, NH.

He served with HMM-261 in New River and on a Caribbean Cruise in February 1967 aboard the USS *Guadalcanal* and served as an CH-46 pilot with HMM-265 in Vietnam from March 1968 to April 1969. The squadron was headquartered at Marble Mountain, and later at Phu Bai and aboard the USS *Tripoli*. He also served as an CH-46 post maintenance test pilot in Atsugi, Japan. He was discharged as a Captain in 1974. He was awarded the Silver Star, 42 Air Medals and the Vietnamese Cross of Gallantry with Bronze Star.

Bruce graduated from the University of New Hampshire in 1980 and managed the university's horticultural research farm. He served several years as an officer of the National Association of ASCS County Office Employees and presently works for the U.S. Department of Agriculture in Maine, New Hampshire and Vermont. Bruce is also the author of the book, *1500 Feet Over Vietnam*.

Bruce and Claudia live in New Hampshire and have one daughter.

HAROLD B. LAMB, "HANK," was born in Salt Lake City, UT on Jan. 31, 1946. He joined the USMC in September 1966. He attended boot camp at MCRD San Diego. He was also stationed at OCS Quantico and Army Helicopter Flight School. In August 1977 he joined the USN (Medical Corps).

While in-country he was a pilot flying CH-46Ds for HMM-364. It was the highlight of his military career. He also served in HMM-264, HMM-365, HMM-163 and HMMT-301.

He was discharged from the USMC with the rank of Captain in 1974 and he retired from the USN with the rank of Captain on July 1, 1994. He was awarded two DFCs, two Single Mission Air Medals and 53 Strike/Flight Air Medals for his service during Vietnam, 1969-70. For his service during the Persian Gulf War, 1991, he was awarded the CAR and NUC.

He is married to Martha Dickerson, a retired Navy commander, and has a son, Harold Lamb IV. They reside in Salt Lake City, UT. He is chairman of the FHP/Utah Department of Anesthesiology.

DENNIS LANDWEHR, "POPEYE," was born on Jan. 11, 1946 in Ft. Wright, KY (near Cincinnati). He enlisted in the USMC on Jan. 11, 1964 and graduated from Covington Catholic High School on May 27, 1964. He then left for boot camp in San Diego on May 28, 1964, went to Memphis for aviation training in structural maintenance and his first assignment was Oahu, HI for a 24 month tour. He spent one day in Hawaii.

He then went on to Japan for 11 months. He volunteered to go south from Japan to Okinawa for additional training in helicopters. He joined HMM-263 stationed at Marble Mountain/Dong Ha in 1966, where he served as an aerial gunner. After Vietnam, he returned to New River and Cherry Point followed by a Caribbean cruise.

During his tour he was awarded the USMC Combat Aircrew Wings, two Air Medals and a Naval Commendation Medal with Combat V. He was discharged on May 28, 1968 as a Sergeant and returned to Northern Kentucky and the University of Kentucky. In 1969 he went to work with a former WWII leatherneck in the photography industry.

He has owned his own commercial photography business in Cincinnati, OH, Corporate PhotoGroup, since 1976. He resides in Lakeside Park, KY and is married with two children.

DAVID F. LAVIOLETTE, "FALCON," was born on Oct. 12, 1941 in Appleton, WI. He graduated from high school where he was captain of the track team, a starter on the 1960 championship basketball team and All-Conference in football. Laviolette graduated from the University of Wisconsin-EauClaire in 1964 with BS degree. He was all conference halfback in both 1962 and 1963 and served as captain on the unbeaten 1963 squad. On Oct. 22, 1989 LaViolette was inducted into the University of Wisconsin-EauClaire's "Hall of Fame." He was commissioned in December 1964 as a Second Lieutenant from Quantico and designated a Naval Aviator on April 8, 1966 in Pensacola. He was promoted to Captain in March 1967.

He was with HMM-363 in Vietnam from August 1966 until September 1967. He returned to Pensacola as a flight instructor at HT-18 from November 1967 until August 1970. He was with HML-167 in Vietnam from October 1970 until June 1971. In both tours LaViolette flew a total of 1,366 combat hours.

He has a total of 53 Air Medals plus the Distinguished Flying Cross, Single Mission Air Medal, Vietnam Service Medal with four stars, MUC, CAR, PUC, Republic of Vietnam Meritorious Unit Citation, Cross of Gallantry with Palm, Air Cross of Gallantry and NUC. In his 10 years in the USMC, he compiled a total of 3,824 flight hours.

Since leaving the USMC in 1975, LaViolette has been a Florida real estate broker and owner of the Real Estate House in Gulf Breeze, FL.

MICHAEL R. LAYMAN, Major, USMC (Ret) was born in Portland, OR on April 14, 1942. He was graduated from high school in 1960 and joined the USMC in June of that year.

After boot camp and aviation school he joined

HMM-361 at Santa Ana. Went to the nuclear tests at Johnston Island with HMM-364 in 1962 as a UH-34 crewchief. Joined HMM-365 in 1964 and went to RVN, where he served as UH-34 crewchief, and was awarded two Single Mission Air Medals, Navy Commendation Medal and 17 Air Medals.

Sent to OCS in late 1966 and commissioned as a Second Lieutenant. Studied Vietnamese at Defense Language Institute and returned to RVN in 1969 as a tank officer. Awarded the Bronze Star. Attended a college degree program in 1975 and earned a BA. In 1979 returned to language school for Greek. From 1980-83 he served as OPS Officer for Joint US Military Air Group in Athens.

He retired from the USMC in 1983 as Major. He is a licensed pilot for hot air balloons and has a commercial balloon ride business know as Teenie Weenie Airlines.

A. MICHAEL LEAHY, "POGUE," was born on Jan. 19, 1933 in Cambridge, MA. He enlisted on his 17th birthday. After Parris Island and NATTC Memphis in 1950, he served as a Sergeant, crewchief on HTLs, Marine Experimental Helicopter Squadron One, (HMX-1) Quantico, VA through 1952. He entered Naval Flight School as NAVCAD and graduated in March 1954. As a helicopter pilot, he served with HMR-363 Santa Ana, CA, HMR-162 Oppama, Japan, HMR-161 DMZ and Munsan-ni, Korea; VMO-1, HMR-261, HMR(M)-461 (Plank Owner), New River, NC, and also in the HMX-1 Executive Flight Detachment, Quantico, VA and HMR-772 Willow Grove, PA. He attended the Philadelphia College of Art from 1960-64 and graduated with a BFA in graphics.

As a Major, executive officer Marine Corps Combat Art Program (1967-69), he completed two assignments to Vietnam to administer the program and produce combat art involving ground/flight operations. He flew 73 missions as a helicopter machine gunner for VMO-2 on UH-1Es, for HMM-363 on UH-34s and for HMM-161 on CH-46s.

He was awarded the Presidential Service Commendation, the Combat Action Ribbon and Bronze Star with Combat "V." He was selected as Combat Artist of Year in 1969 by Marine Corps Combat Correspondents Association. Leahy retired from the USMCR in 1980 as a Lieutenant Colonel and then served as a Navy civilian info manager.

He was dispatched to Grenada in 1983 by the Navy shortly after Operation Urgent Fury to artistically reconstruct the combat operations of all Armed Forces. He provided documentary art for two all-services trans-Atlantic REFORGER exercises. He provided Watergate courtroom art for ABC-TV Network News and the *Washington Post* and aircraft hijack art for ABC-TV's "Nightline." He was featured in a Smithsonian Traveling Exhibit; one-man show "Air Expo," Dulles Airport. He is an artist member of the American Society of Aviation Artists. He has art displayed internationally and in numerous publications, e.g. *Vietnam Magazine* and *VFW Magazine*. Leahy won an award in the first National Naval Aviation Museum contest in Pensacola, FL in 1993. He is Vice President of the USMC/Vietnam Helicopter Pilots & Aircrew Reunion.

He has been married to Ann for 42 years, and they have five children and seven grandchildren. He is painting full-time.

DONAT J. LeBLANC, "DAN" Cpl USMC (Ret) was born on April 10, 1946 in New Bedford, MA. He graduated from high school in 1964. He attended SMU in N. Dartmouth, MA. He enlisted in the USMC in 1965, went to boot camp at Parris Island and attended helicopter school in Memphis, TN.

He was assigned to HMM-363 in Vietnam in April 1966 and was wounded on Sept. 15, 1966.

He was discharged in 1967 and was awarded four Air Medals, Combat Aircrew Wings, the Purple Heart, Expert Rifleman Badge and the Presidential Unit Citation.

He joined the post office in 1968 and then joined the VA in 1972 as a prosthetic representative. This is a position he holds today having served in nine different locations. He now serves as acting, staff assistant to the director, VAMC, Brockton/West Roxbury, MA.

He is married with three sons and a daughter and is a life member of the VFW, DAV, 3rd MAR DIV, USMC/VHP & Aircrew Association and the MOPH.

FITZHUGH B. LEE, "BAR," Lieutenant Colonel USMC (Ret) was born on Feb. 2, 1947 and continued his family tradition of military service by enlisting in the Marine Corps in December 1965. Meritoriously promoted out of Parris Island, he served as regimental adjutant's assistant awaiting orders to Officer Candidates School. He graduated as platoon honorman and was commissioned as a Second Lieutenant in 1967. Shortly thereafter he began flight training in Pensacola, FL and received his wings in December of 1968.

Subsequently, he served with HMM-162 and deployed to the Caribbean with HMM-365. He received a designation as helicopter aircraft commander (CH-46) from that unit and reported to HMM-161 in Quang Tri, Vietnam, September 1969 where he served for one year amassing some 800 combat missions.

He was awarded two Single Mission Air Medals, 39 Air Medals and the Distinguished Flying Cross.

Returning to the States, he served various billets in HMM-265 (later redesignated HMM-161) and subsequently attended Amphibious Warfare School returning to HMMT-301 in March 1973. In July of 1975 he served as assistant operations officer MAG-36, Okinawa, Japan while flying with HMM-164.

Upon returning to the United States, he attended Pepperdine University, graduating with honors in psychology before being assigned to Camp Pendleton, CA as a regimental air liaison officer with the 1st Marines. In March 1979, he served as executive officer for the newly commissioned HMM-268 and operations officer for their first deployment from MCAS(H) Tustin, CA. Upon return from Okinawa he was assigned as the executive officer of HMM-161 and subsequently took command in May of 1982. In May of 1983 he reported to Commander Training Air Wing Five, NAS Whiting Field, Milton, FL, as senior Marine and Operations officer until his retirement in October 1986.

Currently he is employed as a contract flight simulator instructor and lives with his wife, Sally, and children, Fitz (ENS USN), Mason (student at Brown University) and Kathy (student at Washington High School) in Pensacola, FL.

JAMES D. LEJEUNE, "SWAMPY," was born on March 12, 1948 in Minneapolis, MN. He graduated from Sacred Heart Military Academy in 1962 and DeLaSalle High School in 1966. In September 1966, in the example of his great uncle, General John A. Lejeune he enlisted in the Marine Corps.

He received WESTPAC orders in September 1967 and was assigned to Chu Lai, providing prime seating for the Tet fireworks display provided by the locals. In December he was sent to Quang Tri for service as both a door gunner on VMO-6 UH-1E gunships and safety and survival equipment tech on the new OV-10A Broncos. Lejeune left Vietnam in July 1969 and drew orders to H&MS at El Toro. While preparing for an appointment to OCS he was in a severe car crash and spent his last eight months in the Corps recuperating at naval hospitals in Long Beach, CA and Jacksonville, FL. The end result was a medical retirement.

Lejeune's personal awards include Marine Corps Combat Aircrew Wings, a Single Mission Air Medal and three Strike/Flight Medals plus the usual "Been There's," "Done That's," "Me Too's" and a genuine "By God" Good Conduct Medal.

His three children are grown (twin daughters and a son) and on Sept. 10, 1994 he married for a second time. His wife, Donna, is a juvenile court judge and Lejeune is president of Favrand & Lejeune Ltd., a company manufacturing limited edition and presentation grade sidearms.

JOHN J. LEONARD JR., "J.J.," was born on Oct. 30, 1946 at Bar Harbor, ME. Enlistment in the Marine Corps began on July 29, 1965 and he trained with PLT-259 at Parris Island. After boot camp he went to ITR training at Camp Geiger, NC and on to Memphis, TN for aviation school.

Leonard was next assigned to staging BN, Camp Pendleton and was in to Vietnam from May 1966 to July 1967. He served as a gunner and crewchief with HMM-161, HMM-263, HMM-361 and H&MS-16.

He was awarded Combat Aircrew Wings, two Purple Hearts and 14 Air Medals and was discharged a Sergeant at New River, NC on May 1, 1969.

In February 1972 he joined the Maine Army National Guard as a crewchief in an air ambulance company. In 1984 he graduated from the U.S. Army Ser-

77

geant Majors Academy. Leonard is in his sixth year serving as the Maine State Command Sergeant Major.

He was activated on Nov. 17, 1990, spent four months in Saudi Arabia and was discharged from the U.S. Army at Ft. Devens, MA on Dec. 21, 1991.

Leonard and his wife reside in Glenburn, ME and their two daughters attend college.

AL LEVIN, "FLIGHT QUACK," was born on Jan. 12, 1938. He was drafted into the USNR-Med. Corps in January 1966.

He was a flight surgeon for HMM-265 while serving in-country from January 1967 to February 1968. He had many memorable experiences too numerous to put to paper.

He was discharged in 1969 with the rank of Lieutenant (MC) in the USNR. He has a number of awards and medals.

Today he is married with no children. He resides in San Francisco, CA and is practicing medicine.

GUY W. LOBDELL, "MAGNET ASS," was born on Jan. 19, 1948 in Aberdeen, WA and graduated from high school in June 1966 in Richland, WA. He reported to MCRD San Diego for boot camp in July 1966 and further trained at Camp Pendleton, CA and the Naval Air Training Station at Jacksonville, FL. He reported to his first "duty station," the Marine Air Facility, Santa Ana, CA and was assigned to HMM-364 the "Flying Purple Foxes."

In October 1967 he deployed to S.E. Asia with HMM-364. He was wounded in May 1968 and spent several months in quaint little military hospitals. He was eventually re-assigned to HMM-263 (the Go-fer Brokes) and deployed to S.E Asia with that squadron in January 1969 and was wounded again in March 1969. He flew as a gunner for both squadrons.

He earned 14 Strike/Flight Air Medals, one Single Mission Air Medal, Campaign Medal with a Silver Star device, two Purple Hearts, a PUC, two NUCs, the Vietnamese Cross of Gallantry/Unit Citation and the regular ones. He retired on Feb. 1, 1970 with the rank of Sergeant.

After 13 years in law enforcement, he eventually became the administrator for Pre-employment Investigations for the Westinghouse Hanford Company in Richland, WA. Lobdell has been married for 25 years and has one daughter and two sons.

WESLEY R. LOGAN "RANDY" was born on Dec. 8, 1943 in Anacortes, WA. He earned a BA in German literature at Northwestern University in March 1966. He was commissioned a Second Lieutenant at Quantico, VA in October 1966 and was designated a Naval Aviator at Pensacola, FL in December 1967. He served with VMO-5 at Camp Pendleton, CA from January 1968 to May 1968.

From May 1968 to June 1969, he was with VMO-2 at Marble Mountain, RVN flying UH-1E gunships. "One of my more memorable moments was when my hootch mate, Jack Hogan, decided to test the rockets over MMAF at 0600 while on an HML-167 maintenance test flight, (It was on the test card.) Rather than do the test firing over the water as called for, Jack unleashed a rocket ripple from directly overhead MMAF. MMAF and Da Nang Main both went on full rocket alert. There were no serious injuries, just a few bunker burns, and some spilled coffee. I am told Jack did one hell of a rug dance from the squadron, through Group, and all the way to 1st Wing."

From June 1969 to December 1970 he was stationed at New River, NC with VMO-1 (UH-1E, OV-10). He was released from active duty in December 1970. He was awarded the Single Mission Air Medal, 35 Air Medals for 700 combat missions, three Distinguished Flying Crosses, a Silver Star, Navy Marine Corps Medal and a Purple Heart.

From 1970 to 1974 he was at El Toro, CA with VMO-8 (OV-10) and left the reserves as a Captain.

He flew helicopters for the United States Forest Service in forest fire suppression during fire seasons from 1970 through 1973. He attended law school at the University of New Mexico and has practiced law in Albuquerque, NM since 1979. He is married and has three children; two daughters and a son. They reside in Placitas, NM.

VERNON FRANCIS LOUD JR. was born on Oct. 26, 1947 in Long Branch, NJ. As a boy he enjoyed the outdoors and spent many happy hours fishing, crabbing and boating on the Shrewsbury River in Fair Haven, NJ where he lived. Loud graduated from Rumson-Fair Haven Regional High School in June 1966 and enlisted in the Marine Corps on Aug. 8, 1966 at the age of 19, for a four-year term which included a 13 month tour of Vietnam. His other military locations included Parris Island, SC, Cherry Point, NC, Memphis, TN and Kansas City, MO.

Loud served as an aerial gunner in Vietnam in VMO-2, VMO-3 and HMM-164.

Medals awarded to Loud include the National Defense Service Medal, Presidential Unit Citation, Vietnam Service Medal with two stars, Republic of Vietnam Campaign Medal, Presidential Unit Citation, Navy Unit Citation, Air Medal with star, and Combat Aircrew Insignia with three stars.

Loud achieved the rank of Lance Corporal but was demoted to PFC because he went to New Jersey to visit his fiancee, was caught in a severe snowstorm and returned to his base late. He received an honorable discharge on March 9, 1970.

After he was discharged, he attended Brookdale College part-time from 1976-1986 where he earned an associate's degree in applied science as well as certificates in automotive mechanics and computer training. He was self-employed as an animal control specialist.

In 1973 Loud purchased a 1920 Herreshoff Cutter, a 36 foot sailboat, christened it the Aquarius, obtained his Captain's license and sailed it from Maine to the Bahamas. He was devastated when this beautiful sailboat was destroyed by the "Noreaster of 1992."

Loud was very active in the Vietnam Veterans of America, Chapter 12 and held certificates of appreciation for his efforts. He made frequent trips to The Wall in Washington, D.C. and met two Vietnam buddies he hadn't seen since he served with them in Vietnam. This was in September 1992 and he was so excited about seeing them and sharing memories from the past.

Loud passed away on Jan. 2, 1993 of a service-connected illness. He was 45 years old and was survived by his parents, a brother and half-sister, a fiancee, and two sons, Orion and Jason Loud.

The Marine Corps accorded Loud full military honors at his funeral and burial. "I'm sure "Vern" was very proud!" *Written by Lorene T. Loud, Vernon Loud Jr.'s mother, on April 25, 1994.*

MORRIS W. LUTES, "MOOSE," was born in Blytheville, AR. He entered the NAVCAD Program on Jan. 5, 1955. After completion of Jet and Helicopter School he was assigned to HMR-263 at New River, NC.

In 1958-59 he served with HMR-163 at Oppama, Japan and Okinawa. He returned to the West Coast and was an air traffic controller and served with HMM-364. He participated in "Operation Dominic," a nuclear test program with HMM-364. He transferred to Hawaii in 1962. He transitioned into the F8 Crusader in Hawaii and made his first WESTPAC Vietnam Tour aboard the USS *Oriskany* in 1965. He was assigned to the Advanced Training Command in Kingsville, TX until 1968. He transferred to HMH-363 for training and then to HMH-463 at Marble Mountain, Vietnam. Half of this tour was assigned as air liaison officer to 1st Marine Division at Hill 55 south of Da Nang.

He returned to CONUS and completed the "Boot Strap" college degree program at the University of Miami and was assigned to HMH-363 as executive officer until 1974. Lutes returned to the Far East and was assigned as the Executive Officer of MAG-16 at Futema, Okinawa. He served on General Richard Carey's staff aboard the *Blue Ridge* command ship for the evacuation of Saigon and then returned to CONUS and was assigned to MAG-29 at New River. He served as the Commanding Officer of HML-167 for his last year of service.

Lutes retired in August of 1977 with the rank of Lieutenant Colonel. His personal awards include: the DFC, Bronze Star with "V", Air Medal with Gold Star and Numeral 24, Navy Comm., CAR, Vietnamese Cross of Gallantry with two Gold Stars and the Meritorious Service Medal.

Lutes is married with five children: David, Rebecca, Brian, Walter and Matthew. He currently is the vice-president of Finance for Life College in Marietta, GA.

ROBERT D. MABEY was born in Chicago, IL on Nov. 18, 1941. He graduated from high school in Bountiful, UT in 1961. He earned a BS at the University of Utah in 1965. He was commissioned a Second Lieutenant September 1966 and designated a Naval Aviator in December 1967.

He served in Vietnam from May 1968 to May 1969 with HML-367 (Scarface) out of Phu Bai and Quang Tri.

He joined the Utah Army National Guard in February 1971. He commanded the 396th Aviation Company (Attack Helicopter) and 211th Aviation Group (Attack). Mabey began working for the Utah National Guard in August of 1971 as a flight instructor. He became State Army Aviation officer in 1984. Currently he is the director of Aviation and Safety, UTARNG.

Decorations: three DFCs, 53 Air Medals and the

Vietnamese Cross of Gallantry. He is a master Army Aviator with more than 8,000 flying hours and 1,053 hours of combat time in helicopter gunships.

Mabey resides in Bountiful, UT with wife, Pat. One son, Sean, is an Army aviator and Second Lieutenant Apache pilot in the 1st Battalion, 211th Aviation Regiment, Utah National Guard.

KEN MACKIE born Aug. 15, 1940 in No. Salem, NH, and enlisted in the USMC on Oct. 20, 1960. After assignment to the training command 10-60/6-62, he went to MCAF New River, assigned to HMM-265 and then HMM-261 in Vietnam in 1963, HT-8, 1964-67, and HMM-165, 1967-68. His most memorable experience was his overseas tour with HMM-265, Vietnam, on LPHs *Iwo Jima* and *Okinawa*. His awards include the DFC (Vietnam) and Air Medals. He was honorably discharged in August, 1968 with the rank of Captain. He divorced in 1983 and has two daughters, Kathy who is in college, and Dee who is married with one son. He flies Bell 212s at Abudhabi Aviation in the United Arab Emirates.

CHARLES P. MADDOCKS, "MAD-DOG," was born on May 7, 1946 and enlisted in the Marine Corps in August 1965. He went to boot camp at San Diego (e.g.) a "Hollywood Marine" and then reported to Memphis for aviation training in helicopters.

He was assigned to FMF, 1st MAW MAG-16 VMO-2 in the Republic of Vietnam. He served as a crewchief/aerial gunner aboard UH-1E gunships.

During his two tours overseas he was awarded 38 Air Medals, two Purple Hearts, PUC and two Single Mission Air Medals and had 1536 combat hours. He was awarded a "combat promotion" to Sergeant meritoriously. He was discharged at El Toro, CA in 1968.

He recently retired from a teaching and coaching career and as an assistant principal at a junior high school. He also served as a part-time sheriff's deputy in Terrebonne Parish, LA. Presently, he is an investigator for a law firm in Houston, TX. He is divorced and has one daughter.

DAVID J. MAESTRINI, "WOP," was born on Aug. 23, 1945 in Meriden, CT. He went through boot camp at Parris Island in July 1963, after which he reported to Memphis for aviation training. He was then assigned to VMO-6 at Camp Pendleton.

He arrived in Ky Ha with VMO-6 in September 1965, flying in UH-1E helicopters.

During his tour he was awarded both Vietnam Service Medals, Marine Corps Combat Aircrew Wings, Air Medals and the Navy Unit Commendation Medal. He returned to VMA (AW)-224 in Cherry Point where he was discharged as a Corporal.

He is a Sergeant with the Connecticut Department of Motor Vehicles and is an active member of the USMC/Vietnam Helicopter Pilots & Aircrew Reunion, Vietnam Veterans of America and the Veterans of the Vietnam War. He is a marathon runner, having run six Marine Corps Marathons. He is married with two children, and continues to reside in Meriden, CT.

CLIFFORD HUDSON MANNING, "HUT/HUD," was born on Sept. 6, 1942 in Birmingham, AL. He graduated from high school in Walden, CO in 1959 and attended the University of New Mexico. He left in 1963 for Pensacola. He spent 25 years in the USMCR, starting as a MARCAD. He completed CNATRA in 1965, then left for HMM-363 MCAF Santa Ana. He spent time with HMM-164 and became a "Plank Owner" with HMM-165.

He went with the advance party to RVN, awaiting the arrival of HMM-165. When the CH-46s couldn't fly, he flew UH-34s and UH-1Es for other squadrons.

While in-country, he earned three DFCs, 29 Air Medals, Vietnam Cross of Gallantry, PUC, NUC and other awards.

Upon returning to CONUS, he joined HMM-763, later to be HMA-765, in Atlanta to fly UH-34s, UH-1Es, AH-1G&Js, S-2Fs and OV-10s, serving as maintenance, operations and executive officer. He spent one year with the 4th FSSG as air liaison officer. His twilight tour was flying the C-9B with SOES at Cherry Point. He retired in September 1988.

Civilian duties included 22 years at Eastern Airlines as S/O and F/0 on the B-727, F/O on the 757 and F/0 and Captain on the DC-9. He obtained a BS cum laude in aeronautical science at Embry-Riddle Aeronautical University and started a company, Automaster Precision P&B, Inc., in 1986. Currently, he resides with wife, Frankie, and sons, Mike and Patrick in Kennesaw, GA. Manning is presently employed by United Airlines as a first officer on B-737/300 based in Chicago.

EDWARD CHRISTIAN MARTENS JR., "RAGIN' CAJUN," Gunnery Sergeant USMC (Ret) was born on Oct. 30, 1939 in New Orleans, LA. He graduated from high school in 1959 in New Orleans, LA and enlisted in the Marine Corps on Aug. 26, 1959. He went through boot camp at San Diego, CA and graduated Sea School at the same depot. He was assigned to the USS *Canberra* (CAG-2) at Norfolk, VA for a world cruise in 1960. He reported to CO "C"; 2nd Anti-Tank BN, Camp Lejeune, NC in 1962 and was classified for aviation training to NATTC in Memphis, TN in 1963.

He was assigned to HMM-261 at MCAF New River, NC and departed for Vietnam in 1965. He served as crewchief. In 1966 he was assigned to the Sea Air Rescue Unit at MCAS Beaufort, SC. The second tour in-country came with HMM-265 in 1969. He returned to New River and HMM-264 and was assigned to Drill Instructor's School at Parris Island, SC from 1972-74. He had a tour with HMM-164 in Okinawa, Japan from 1977-78 and a twilight tour with HMM-165 and H&MS-24 at Kaneohe Bay, HI from 1978-80.

Martens graduated from Embry-Riddle Aeronautical University with BPA in 1980 and retired in 1981. He was awarded the Marine Corps Combat Aircrew Wings, 17 Air Medals, two Single Mission Air Medals, CAR, VCAC, PUC, NUC, VSM, VCM, NDR AFEM and the VCG.

As a civilian, he served as a quality assurance specialist for naval weapon systems at NADEP in Pensacola, FL. He is a member of the National Association of Government Inspectors, and he completed graduate degrees in public administration and management. He is married to the former Miss Katherine R. Valentine. His current plans are to locate to Tucson, AZ as CEO of his company. He is an avid supporter of veterans organizations.

JAMES M. MARTIN, "JAY EM," Major, USMCR (Ret) was born on March 3, 1945 in San Pedro, CA. Nothing much else happened until January 1967 when he enlisted in the USMC and attended his first of three boot camps. Finishing boot camp in September of 1968, Sgt. Martin arrived at Quantico with an "attitude." In his 10th week (of 10) he was invited to attend yet another 10 weeks of OCS or be discharged and go home. Ten weeks later Lt Martin went to TBS. Having once flown on a Boeing 707, he was found qualified for flight school. He and 75 of his fellow Lieutenants crossed over to the U.S. Army Rotary Wing Aviation School and suffered through 48 weeks of full per-diem and American Airlines untrained "stews" at Love Field, Dallas, TX.

Designated both an Army and Naval aviator, he went from HMMT-302 to the following: HMM-265/HMM-263 (RVN) H&MS-16 (RVN), HMM-262/SOES MCAS Yuma, 43rd Artillery BN (USAR)/MARTD Glenview/HML-776/HML-770/MTU WA-41.

He retired in January 1991. During his service he managed to earn three rows of ribbons but the ones with meaning were made possible by both Ho Chi Minh and Jane Fonda.

KURT MASON, "SMOKE," Major USMCR (Ret) was born on April 5, 1934 in Seattle, WA. He enlisted in the U.S. Navy in September 1952 and was an Air Controlman 1st in October 1958 when he entered Flight Training as a NAVCAD. Commissioned and designated as a Naval Aviator on May 11, 1960, he served with HMR(L)-261 and HMM-362, flying UH-34s.

He arrived in Vietnam in September 1965 with MATCU 68 at Marble Mountain. As a "groupie," he flew UH-1Es with VMO-2. He was assigned to VMO-2 from January 1966 to August 1966. He went back to the States for duty with the 1st Marine Brigade in Kaneohe. Once again, he went to Vietnam in December 1968, joining VMO-6 at Quang Tri for one month and then VMO-2 at Marble Mountain in January 1969, flying UH-1Es, AH-1Gs and OV-10s. He had more than 1,000 combat hours.

He was awarded the DFC (two), Single Mission Air Medal, 38 Strike/Flight Air Medals, Navy Commendation Medal, CAR, NUC, PUC and MUC. Mason retired in June 1973 while serving as a weapons instructor (Low & Slow) with MAWTULANT.

79

As a civilian, he became a licensed aircraft mechanic, worked as a general aviation mechanic and taught aircraft maintenance technology at a community college. He worked as an instructor with Alaska Airlines and is currently with the Boeing Company as an aircraft maintenance instructor. He owns and flies a Cessna 172, logging more than 800 hours in light aircraft since retirement. He has been married since 1960 and has three children and four grandchildren.

HAROLD E. MASSEY, "GENE," was born on June 12, 1943 in Flora, IL and moved later to Oklahoma. He graduated from high school in 1961 and received a BS from East Central University in Ada, in 1965. He joined the Marine Corps in January 1966, completed 39th OCS as part of the "third herd" and was commissioned a Second Lieutenant on March 17, 1966. He began flight school in Pensacola in May 1966 and was designated a Naval Aviator on June 16, 1967. He joined HMM-161 at New River in June 1967 which was commanded by Lieutenant Colonel Niesen.

Massey flew with the squadron to Long Beach to board USS *Princeton* for transport to Quang Tri, MAG-39. The highlight of his time in the Corps was serving with the men of HMM-161.

He was awarded the DFC, two Single Mission Air Medals, 33 Air Medals, NUC, MUC, Vietnam Service and Campaign Ribbons. He returned to Camp Lejeune in June 1969 and completed active duty as a Captain with Headquarters Bn 2nd Marine Div.

Massey has been a minister in service to the Lord Jesus Christ since 1973. He and his wife, Sally, live in Kingfisher, OK and have three children.

JOHN T. MAXWELL, "THE RED MAX," was born on March 15, 1932 in Pittsburgh, PA. He joined the service in 1953.

He returned from Korea as a Sergeant USMC and attended Flight School. He received his wings in 1956 and joined HMM-265 in 1965. He went to Vietnam with the squadron flying CH-46As in 1966. His aircraft were damaged by enemy fire 10 times and he was shot down once. There must be a better way to make a living! He had a three year tour at Kaneohe Bay after Nam and four unaccompanied overseas tours plus a Med Cruise.

Maxwell had 5,400 flight hours and was awarded the DFC, three Single Air Medals, 11 Strike/Flight Air Medals, a Navy Commendation and two Navy Commendations for achievement. He retired in 1974 as a Major.

In post retirement he completed college and served as a counselor/program director at a local correctional facility. He is married to Barbara and has a son, John T. III, and a daughter, Julie. They now reside in Marietta, GA.

LEN MAZON was born on March 2, 1943 in Kansas City, MO. He attended the University of Hawaii and Idaho State University. He enlisted in the USMC on Aug. 10, 1961. He became a Marine Aviation Cadet after serving two and a half years as a USMC air traffic controller at Kaneohe Bay, HI. He received a Marine Corps commission as a Second Lieutenant and aviation wings in Pensacola, FL on Aug. 10, 1965.

Mazon participated in Vietnam service from April 1966 to May 1967 as a squadron pilot with the HMM-161 "Pineapples" at Hue Phu Bai and with the HMM-263 "Screaming Eagles" at Marble Mountain, RVN.

He later served as a logistics officer for HMHT-301 at MCAS(H) Tustin, CA and for HMH-462 at Futema, Okinawa flying the Sikorsky CH-53 aircraft. He later served as officer-in-charge of TME-31 at El Toro, CA. He was discharged from active duty as a Captain in June 1975. He flew 416 missions with more than 900 combat flight hours in the Sikorsky UH-34 helicopter and was awarded 20 Air Medals along with the standard Vietnam stuff.

Mazon has more than 18 years experience in real estate development, management and sales. He currently serves as a sales representative for CENTEX Homes Corp. in Rancho Cucamonga, CA. He is an active member of the USMC/Vietnam Helicopter Pilots and Aircrew Reunion. He is active as a member of the Church of Jesus Christ of Latter-Day Saints. He has been active in civic, community, local school PTA and Boy Scout leadership. He has been married to LaWavna for 29 years and they have two daughters, one son and five grandchildren.

PETER A. MAZURAK, "DOC," was born on Sept. 6, 1942 in Berkeley, CA. He graduated from Lincoln High School in Lincoln, NE in 1960. He earned a BS in chemistry in 1964 from the University of Nebraska in Lincoln, NE and a MS in chemistry from the University of Wisconsin in Madison, WI. He was commissioned on Jan. 21, 1968 in the 47th OCC Quantico and was designated a Naval Aviator on April 1, 1969 at Pensacola.

He served with HMM-365 in New River which included a three month Caribbean cruise. He joined with HMM-162 in Okinawa on Oct. 5, 1969 and was transferred to HMM-263 at Marble Mountain on Nov. 28, 1969. He served until Sept. 30, 1970. He served as a flight instructor in HT-8 and HT-18 at Pensacola until Oct. 15, 1972.

"I would guess my most meaningful memory of my Vietnam tour would be the Christmas/New Year's Holiday Season of 1969. Christmas Day or thereabouts a bunch of Lieutenants, feeling no pain, drug the Christmas tree out of the O Club and set it on fire. New Years eve was quite an experience. At midnight everybody cut loose with small arms fire. The outposts on Marble Mountain, just to the south of the Marble Mountain Air Field set off about 1000 pop up flares which was spectacular. The small arms fire increased to the point where all aircraft at Da Nang and Marble had to orbit out to sea for about an hour. About 30 some aircraft were hit by friendly fire. After a few minutes a friend and I decided it would be safer to retire to our bunker. The time I spent in the Marine Corps and Vietnam was probably the most meaningful period of my life." Discharged on Oct. 30, 1972 with the rank of Captain.

Mazurak worked with Air America flying UH-34s, UH-1As and UH-1Hs in Laos, Cambodia and S. Vietnam. He returned to graduate school and received Ph.D. (chemistry) in 1979. He served with the Wisconsin Army National Guard from 1977-1983 as a UH-1B, C, D, H and M model pilot.

For the last 15 years He has worked for Kimberly Clark Corporation in Neenah, WI (currently he is the director of technical services for the Neenah Paper Division of Kimberly-Clark).

He is still married to his second wife, Elizabeth. They have a son, age 12 and a daughter, age 10. He has two grown children from his first marriage and one grandchild.

RONALD W. McAMIS, "MAC," Cdr USN (DC) (Ret) was born on Feb. 17, 1946 in Orange, CA. He graduated from high school in 1964 in Liberal, KS and received his BS in biology on June 5, 1968 from Kansas State College at Pittsburg, KS. He was commissioned as a Second Lieutenant in the USMCR through the PLC Program on same date. He received his Navy wings on Jan. 7, 1969 at Pensacola, FL. He trained in CH-46s in HMM-261 from July 15, 1969 to Oct. 15, 1969 at New River, NC.

McAmis served with HMM-262 in RVN from Nov. 16, 1969 to Aug. 5, 1970. He was medevaced to USNH Yokosuka, Japan in August 1970. After meeting a beautiful Navy nurse, he managed to extend his stay in Japan to two months. He married her a year later at Camp Pendleton. He was stationed at MCAF Tustin, CA from November 1970 to March 1972 and augmented to regular during that time. He was a flight instructor at VT-1, Pensacola until August 1973. He resigned his commission as a Captain and became an Ensign in the Navy Dental Corps, (worst day of his life!!). He had four years of dental school on a Navy Scholarship at Kansas City, MO, two years at NRTC Great Lakes and a tour in Okinawa with 3rd FSSG.

He had one year at back at NRTC and then graduated from Command and Staff College, Quantico in 1982. He was in Oral and Maxillofacial Surgery Residency at the University of Missouri at Kansas City from 1982 to 1985 and was head of the Dental Dept., onboard CVN 70 from 1985 to 1987. He was head of the Dental Dept. and medical director of Beaufort Naval Hospital, SC from August 1987 to November 1989 when he retired. McAmis was awarded the Meritorious Service Medal, 35 Air Medals, Vietnam Service and Campaign and Vietnamese Cross of Gallantry, Navy Achievement with Combat "V," NUC and Navy Comm.

Presently he is in private practice as an oral and maxillofacial surgeon in Kansas City and on the teaching staff at the University of Missouri Medical Center and the Kansas University Medical Center. He is married to Flo, the Navy nurse, and they have one son, Conor, age 13.

NORIS L. McCALL, "LYN," was born on Jan. 19, 1946 in Bloomington, IN. He graduated from Belleville, IL high school in 1964 and he enlisted in the USMC on Oct. 23, 1966. He reached the rank of Corporal and graduated from OCS in January 1968, TBS in June 1968 and was designated a Naval Aviator in April 1969.

While in-country he served with VMO-1 and HML-367 at MMAF from June 1969 to July 1970. He had 1,150 combat hours and 788 missions.

His post Vietnam assignments were with First ANGLICO, HMH-363, VT-9, VT-19, VMAT-102, MWHS-3, HMH-361, Naval Safety Center, MAG-36, H&MS-36, HMH-362, HQMC and OSD. His squadron billets included: Embark O, S-4, Flight Line O, S-1, AMO, S-3, DSS, XO and CO. His non-flying billets included FAC, platoon commander, Aircraft Mishap Analyst, HQMC Branch Head and OSD military assistant.

He was awarded DSSM, MSM, DFC, SMAM (two), S/FAM (39) and NCM. He retired as a Colonel in November 1992.

He completed a BA in general studies (chemistry) and master's degrees in international relations (Soviet Studies), National Policy and Strategy and resource management. He is a graduate of AWS, USMC C&SC, USAF C&SC, ICAF and NWC and became a Harvard Senior Executive Fellow in 1991.

One wife (Vickie age 29) and three kids (Luke, age 22, Liz, age 19, and Rug, age 17). They are currently living in Herndon, VA and he is working with Standard Technology, Inc., in Rockville, MD.

ROBERT A. McCLELLAN, "MAC ATTACK,"

Lieutenant Colonel USMCR (Ret) was born on Sept. 10, 1942 in Pascagoula, MS. He graduated from high school in Smyrna, GA in 1960. He received a BS (Summa Cum Laude) from Embry-Riddle Aeronautical University in 1973. He entered the MARCAD Program in October 1963, was commissioned and designated a Naval Aviator June 1965. He served with HMM-264 at New River.

His RVN service was with HMM-161 and VMO-2 from 1966-67.

McClellan was awarded the Navy Commendation in lieu of DFC, 26 Air Medals, Combat Action Ribbon and the usual.

He transitioned into the A-4 prior to RelActDu. He served 15 years with VMF(A)-142 at NAS JAX and Cecil, from SLJO to CO. Deployments including SATS, CARQUALS (two), Red Flag Aggressor, CAX's (two), USAF Mini-Dart Aggressor (many) and Three-Man Lifts (several). He was awarded a life membership in Lieutenants Protective Association and designated an honorary Second Lieutenant for accomplishing 20 years in the cockpit; always in a squadron. He also holds a Moose Rolfe Tact & Couth Award for exemplary relations with regular Marine component. Currently he is a member of USMC/VHP & Aircrew and LPA.

As a civilian he was employed for 21 years with Eastern Airlines from 1968-89. Presently he is self-employed (will consider any venture that is as much fun as flying). He has two sons and has been a Florida resident since 1968.

FRED McCORKLE, "ASSASSIN,"

Major General, USMC was born on Nov. 9, 1944 in San Francisco, CA. He joined the service on March 27, 1967. Major General McCorkle is the Commanding General, 3rd Marine Aircraft Wing, MCAS El Toro, CA.

He was designated a Naval Aviator in January 1969.

He served in Vietnam with HMM-262 during 1969 and 1970 and flew more than 1500 combat missions. Throughout his career he has accumulated more than 5500 flight hours to include several hundred combat flight hours.

Operational assignments have included billets as Commanding Officer, Marine Aviation Weapons and Tactics Squadron One (MAWTS-1) and as Commanding Officer, Marine Aircraft Group 29 (MAG-29). He also served as Commander, Marine Corps Bases Eastern Area, and Commanding General, MCAS, Cherry Point, NC.

His personal decorations include: the Legion of Merit with three Gold Stars in lieu of second through fourth awards; the Distinguished Flying Cross with Gold Star in lieu of second award; the Purple Heart and the Air Medal with Single Mission Award and 76 Strike/Flight Awards, Navy Commendation Medal w/ combat "V," and the Navy Achievement Medal.

Major General McCorkle is married to the former Katherine Schwartz of Johnson City, TN.

DANIEL B. McDYRE, "SAVAGE,"

born Nov. 9, 1941 in Philadelphia, PA. He graduated from LaSalle University in 1963 and commissioned via PLC. He went to flight school in 1965, designated NA in April, 1966 being assigned for one year to HMM-261 in New River flying CH-46s. Then sent to HMM-165 in Ky Ha, RVN in April, 1967, for thirteen months flying 500+ missions, 1,000+ hours later he shucked "magnet ass" reputation and returned to CONUS. He joined HMX-1 for three years as White House Liaison and Presidential Pilot. He completed AWS and was sent to WESTPAC as G-3A (Helo) at 1st MAW HQ. He was Marine rep in Hanoi (Operation Homecoming) then transferred to VT-1 (Saufley) as Senior Marine/Ops officer. After DIS and language school he was assigned as Assistant Naval Attaché, the Netherlands, 1976-79.

He completed C&SC then went back to WESTPAC as XO, MAG-36 Futema, 1980-81; then XO, MAG-46 El Toro/1981-83; Air War College, Maxwell AFB; CO, MAG-49 Willow Grove, 1984-86; HQMC, Branch Head ASL, 1986-89; Chief of Staff, 2MAW, June, 1989; AWC/CofS until retirement November, 1991; Commander, 2MAW during Desert Shield/Storm, January-May, 1991. Awarded LOM (2), DFC, PH, DSM, MSM, AM (27), CAR and Presidential Service Badge. Currently, he is the General Manager, Mercury Aviation & Reno Jet Center in Reno, NV.

MICHAEL M. McELWEE, "MIKE,"

Lieutenant Colonel USMC (Ret) was born on Jan. 14, 1946 in Cumberland, MD. He graduated from high school in West Hollywood, FL in 1964 and enlisted in 1965. He qualified for the MARCAD Program while in boot camp and went on to receive his wings and commission in June 1967.

His first squadron was HMM-161 and he deployed with them to Quang Tri AB, RVN in May 1968. He completed his Vietnam tour as FAC with 2nd Bn, 26th Marines. After CH-46s, he flew AH-1s from 1972-77.

After flight instruction duties in VT-6, and completion of the college degree program at Pensacola, he returned to the CH-46 and was CO of HMM-265 (1st Marine Brigade) from August 1982 to August 1984. He retired in 1987. Since 1989 he has been employed as an Emb-120 pilot with Comair, Inc. out of Orlando, FL.

QUINTEN R. MELAND, "GOOSE,"

was born in Bismarck, ND. He graduated from high school in Oceanside, CA and from college at San Diego State University. He entered the USMC as a PLC candidate in 1962 while attending college. Meland was commissioned in 1964 and went to Pensacola for flight training. He won his Wings of Gold in June 1965 and was transferred to MCAS(H) Tustin where he became the first helicopter aircraft commander to be graduated from HMMT-301.

He joined HMM-361 in Ky Ha, Vietnam in May 1966 and flew more than 500 missions. He was discharged from active duty in October 1968 at El Toro, CA.

He joined HMM-769 in Alameda, CA in 1969 and held various positions until being selected as commanding officer of HMH-769 (CH-53) in October 1980. He was promoted to Colonel in June 1985 and retired after more than 29 years of service in 1991. He holds the Distinguished Flying Cross, 25 Air Medals and various other decorations.

Meland and his wife, Shirley, have three grown children. They reside in Soquel, CA, where they own an insurance brokerage business. He is also a TransWorld Airlines Captain based in St. Louis, MO.

MICHAEL STERLING MELIN, "MUSH,"

was born in St. Paul, MN on Feb. 29, 1940 (making him one of the "youngest" Marines ever to serve since he's only had 13 birthdays). He enlisted in the Marine Corps in 1957 and attended boot camp in March 1958, followed by the PLC Program in 1962. He graduated with a BA in 1963 from San Jose State University where he played football and served as president of his fraternity, SAE. He received his Second Lieutenant commission in June 1963.

He served in Vietnam from 1965-66 where he flew HAC in the UH-34. His claim to fame is crashing three UH-34s in the ocean while in Vietnam. In 1967 he was stationed in Taipei.

He left active duty in November 1967, having last served in the reserve squadron VMR-216 NAS Whidbey Island in 1973 where he flew PC in the C-119. He was awarded 23 Air Medals.

He joined United Air Lines as a flight officer in December 1967. He is currently a B-727 Captain, based in San Francisco, and is a member of the Air Line Pilot Association. He is married, with one daughter and one son. He is also the proud owner of an English bulldog named, "Oorah" in honor of the USMC.

WILLIAM H. MILES, "BULLETHEAD,"

Major USMCR, was born in New York City on March 15, 1940. He graduated from High School of Commerce on June 25, 1957 and enlisted in the USMCR on June 27, 1957. He was commissioned a Second Lieutenant on Jan. 1, 1966 and designated a Naval Aviator on May 20, 1966. Undergraduate studies included: BS program in psychology, SUNY Stony Brook; BA program in marketing, Dowling College, Long Island. His graduate studies included: MA program in communications and the arts, Regent University, Virginia Beach.. Boot Camp at Parris Island from

June 27, 1957 to Sept. 30, 1957; Aviation Prep School at NAS Jacksonville, FL; and "A" School, structural mechanics (hydraulics) at NAS Memphis, TN.

His first FMF assignment was to H&MS-24 MCAS Cherry Point, NC. He transferred to the State Department Security School, HQMC and was assigned to the U.S. Embassy in Managua, Nicaragua. He returned to the FMF with VMA-331 at MCAS Beaufort, SC. He was deployed with VMA-331 to Guantanamo Bay, Cuba during the Bay of Pigs invasion. Miles re-enlisted and was assigned to Hotel Battery, 3rd BN. 10th Marines in May 1963.

He was accepted to Naval Flight School, Pensacola, FL as a MARCAD on Aug. 14, 1964. After commissioning and designation, he was assigned to VMO-1, MCAS New River, NC for transition to UH-1 helicopters. He joined VMO-6, MAG-36 at Ky Ha, RVN and was WIA on May 18, 1967 in Operation Hickory-Beaucharger. Was reassigned to 2nd BN, 3rd Marines as Air Liaison Officer. His last FMF tour with HML-267 was at MCAF Camp Pendleton, CA

Miles left active service as a Captain on Nov. 30, 1977. He was awarded 12 Air Medals, two Purple Hearts, Marine Corps Expeditionary Medal (Cuba 1961), National Defense Medal, CAR, PUC, NUC, Vietnamese Service Medal, Vietnamese Campaign Medal, Vietnamese Cross of Gallantry w/Palm and would you believe, two Good Conduct Medals!

He flew with HML-767, NAS South Weymouth, MA from 1972-1973. He reentered active service on May 14, 1974 as contact team officer at HQ, 1st MCD Garden City, NY and was promoted to Major on Aug. 1, 1975. Miles was discharged on Aug. 31, 1977.

He is director of Membership Services with the Christian Coalition, a conservative public policy and lobbying organization. Previously he was: president, the InterMedia Group-marketing and advertising PR; director, marketing and sales-CBN Founders Inn & Conference Center; regional director of sales-Cruise International; international product manager-Subsidiary Sumitomo Heavy Industries; manager, marketing and sales-Subsidiary Monsanto Company.

Married to Margaret Arleen LeGere (Aug. 2, 1968) with three daughters, Amy (1970), Hillary (1974) and Leslie (1976). Amy and Leslie are in missionary work, (Ukraine, Botswana and Ghana) and Hillary is in franchise management. Semper Fi.

HARRY R. MILLS, "BOB" "SIOUX-CROW/ APACHE", joined as a MARCAD in February, 1960, and was commissioned on October 18, 1961. He participated in two Mediterranean and one Caribbean cruise, one tour of Gitmo flying the fence, two tours in Okinawa, Vietnam, 1962-63 with HMM-163 then HMM-265, 1966-67. He had 651 day/night shipboard landings, command tours included CO H&HS, 1972-74, XO HMM-264, 1974-75, XO 2nd Anglico, 1975-76 and XO HMM-164, 1977-78. He retired USMC in April, 1981 at the rank of Major.

His most memorable experience was with HMM-265, "Bonnie Sue," in 1967 when awakened by mortar fire. He jumped out of the rack in his skivvies and ran outside, tripped over a tent wire, then jumped to his feet to dive under a deuce and a-half. While peering up at the gasoline tanks, he decided that another place was preferable. He dove under a track as the incoming ceased. He slunk back to the tent embarrassed for his fear and fell asleep immediately.

LAWRENCE C. MONSERRATE, (LARRY) "PAPPY," Colonel USMCR (Ret) was born on Aug. 7, 1941 in Helena, MT. He graduated from Wilson High School, American University, Washington, D.C. and he received a master's degree from University of West Florida, Pensacola. He was commissioned in September 1966 and designated a Naval Aviator in June 1968. He transitioned to CH-46 at MCAS New River from June 1968 to September 1968.

He served with HMM-265 (Highboy) in Phu Bai/ USS *Iwo Jima,* RVN from September 1968 to September 1969. While flying as a CH-46 HAC with HMM-265, he acquired 650 combat hours.

One of Monserrate's memorable experiences occurred when he was flying as an H-46 HAC with HMM-265. "I was fragged to find a Marine squad that had been separated from its main unit south of Da Nang. Once in the area, the following call came in over the Fox mike: 'Bird (aircraft) in the sky, bird in the sky, this is Bravo 1-2, your at my 12 o'clock.' Since I had no idea which direction this young Marine was looking (his 12 o'clock position could be in any direction) the following radio transmission was made in response to his original call, 'Grunt on the ground, grunt on the ground, this is bird in the sky, please say which compass direction is at your 12 o'clock.' With that, the Marine answered with a very definitive and correct identification of his squad's position and we quickly left to pick them up. Needless to say, we could have flown around in "Charlie Territory" for a long time trying to get into position to be at 'his' 12 o'clock!" He completed active duty as a VT-1 flight instructor in June 1971 and was awarded two Single Mission Air Medals, 30 SF, CAR, etc.

He joined HMM-767/8, MAG-46, NAS New Orleans in October 1971. He served in 4th MAW from June 1971 to March 1994. As a reserve pilot for 20 consecutive years, he has more than 3,600 hours. His last USMCR assignment (March 1992-March 1994) was at El Toro, Base Closure (BRAC staff.) He has a total of 28 years in the reserves.

Monserrate has been married for 30 years to Gayle Monserrate and they have two daughters and three grandchildren. His civilian occupation is that of director of Public Projects, Development and Environmental Planning for the City of San Diego, CA.

FRANK A. MOORE, "COLORADO," was born on March 14, 1944 in Denver, CO. He graduated from Regis College, Denver, CO in 1966 with a BS in mathematics. He was commissioned on June 8, 1966 as a Second Lieutenant and designated a Naval Aviator on July 14, 1967. He served with HMMT-301 at the Santa Ana Air Facility from August to November 1967.

He served in Vietnam from December 1967 to August 1969 while flying the UH-34 with HMM-363 and H&MS-16 and the UH-1E with HML-167. He flew more than 1,400 hours in-country.

Moore was awarded the Single Mission Air Medal, 55 Air Medals, NDS Medal, Vietnam Service and Campaign Medals, MUC and the Vietnam Cross of Gallantry. After Vietnam he was assigned to HMM-261 (CH-46) at New River, NC and completed active duty in September 1970.

He flew with the reserve squadron HMH-769 (CH-53) in Alameda, CA from July 1972 to August 1974.

Moore joined the FAA in June 1973 and has worked in California, Alaska, Oklahoma and Nevada. He is an aviation safety inspector with the Las Vegas FSDO. He is married and has a son and a daughter.

PAUL MOORE, was born on Feb. 20, 1925 in Ogdin, WV. He enlisted in the USMC on April 6, 1942 and attended AMM School for six months in Jacksonville, FL. and spent six months in NAP training, MARTD Dallas, TX. He volunteered as crew member on VMTB 232 in the Pacific Campaign, Caroline Islands and Okinawa Campaigns, WWII from 1944-1945. He was awarded three Battle Stars, two Presidential Citations and Occupation Medal.

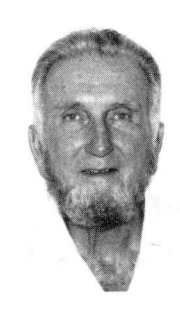

He received helicopter training in 1950 in Lieutenant Colonel K. B. McCutcheon's squadron, HMX-1 at Quantico. Served in Korean War from 1953-54 with VMO-6. Awarded the following: Presidential Citation, Korean PUC, United Nations Medal and three campaign stars.

He was next assigned to H&MS-16 Okinawa 1962-63 where as QA Chief he delivered overhauled UH-34Ds to Da Nang, completed assembly and flew with Ridgerunners.

He was transferred to Fleet Reserves in October 1963 and accepted flight orders with Air Force Advisory Group to train VNAF pilots in UH-34D, C & G helicopters from 1964-1968. He flew 900 hours at Da Nang, Nha Trang, Saigon and Can Tho with VNAF squadrons.

While assigned as Sikorsky Rep on Okinawa served on LPH Okinawa with HMH-462 in Operation "Frequent Wind" - evacuation of Saigon Embassy in April 1975. After 14 years as Sikorsky rep on Okinawa with deployments to Korea, Philippines and one to Oman to prepare helicopters for Iran prisoner rescue, he retired as a senior representative for Sikorsky Helicopters.

Moore works part time as aviation consultant with assignments in Dallas, TX; Cairo, Egypt; and Geneva, Switzerland. He and wife Lieu have seven children.

RONALD A. MORIN, "MOE," was born on Jan. 22, 1940 in Pawtucket, RI. He joined the USMCR in February 1957 and attended boot camp at Parris Island in 1958. He attended Naval Flight Training at Pensacola FL from October 1963 to May 1965 and was commissioned Second Lieutenant in May 1965. He was stationed at MCAF New River, NC from May 1965 to June 1966 and again from August 1967 to May 1968. While in-country he was helicopter aircraft commander on UH-34s and flew with HMM-363 in Vietnam in I Corps from June 1966 to July 1967. He was stationed at Ky Ha, on the USS *Iwo Jima*, and at Dong Ha.

Awarded two Distinguished Flying Crosses, one Single Mission Air Medal and a total of 22 Air Medals with more than 900 combat hours flown. He was discharged on June 30, 1976 with the rank of Captain.

He was married in August 1969 to Cynthia and they have a daughter, Kimberly (1970) and a son, Michael (1973). He is an airline Captain with Delta Airlines. He resides at Manchester-By-The-Sea, MA.

MICHAEL "MIKE/M&M" K. MORRISON

was born on July 4, 1942 in Waltham, MA. He was raised in South Boston, MA and graduated from high school in 1960 in Burlington, MA. He completed boot camp at Parris Island, 1961; Avionics training at NAS Memphis; and served at NAS Willow Grove, PA until selected for the commissioning program. He attended the Naval Academy Prep School at NTC Bainbridge, MD. He joined NROTC unit at the University of Nebraska where he earned a BS in math and was commissioned as a Second Lieutenant in January 1967. He was designated a Naval Aviator in January 1969 and trained as a CH-46D pilot with HMM-264 at MCAF New River and after three months in the Caribbean, he joined HMM-365 to complete his training.

Morrison departed CONUS for WESTPAC in December 1969 and joined HMM-161 in RVN.

He served as a flight instructor at Ellyson Field from 1971-1972; attended Communications Officer School, 1974; served as an instructor at MarCor Extension School; and then ended his career with the Communications Company of the 1st Marine Division in Okinawa.

He was discharged on March 31, 1976 with the rank of Captain. He received the Silver Star, 21 Air Medals, Meritorious Unit Citation, Good Conduct, etc.

He trained in computer programming techniques at the University of West Florida and works as a "software engineer" at Westinghouse near Baltimore, MD.

WILLIAM B. MOTTER, "TANK,"

born on Dec. 5, 1942 in Aurora, IL. Joined on Feb. 3, 1964 and had boot camp at San Diego, CA Platoon No. 316. He schooled at NAS Millington Aviation Training MOS 6418. Reported in January 1965 to HMM-164, MAG-36, 3rd MAW MCAS Santa Ana, CA.

He embarked on board the USS *Princeton* on Feb. 16, 1966 with HMM-164 for Vietnam. He arrived at Marble Mountain on March 8, 1966 and served as ground crew and gunner until he qualified as crewchief. In January 1967 he was transferred to HMM-265, MAG-16, 1st MAW after it arrived in-country. He was wounded on Jan. 26, 1967 near Khe Sanh trying to extract a six man reconnaissance team who were surrounded by a force of approximately 150 North Vietnamese Army Regulars. While going into a landing zone, his helicopter sustained extensive damage and had to be abandoned. Assisted by the pilot and the co-pilot, they moved to the reconnaissance team's defensive position. After a very exciting night in the jungle, everybody was air lifted out the following day. He was medevac'd to Great Lakes Hospital and placed on temporary retired list in September 1967.

Motter was honorably discharged on Oct. 1, 1972 by reason of permanent physical disability with the rank of Corporal. He was awarded (two) Navy Commendation Medals with combat "V," (first one September 1966 with HMM-164 and the other Jan. 26, 1967 with HMM-265), Combat Aircrew with three stars, Air Medal with five stars and the Purple Heart.

He was employed by the city of Mendota Police Department serving as a juvenile officer. He retired from the police department in 1993 and is now working for a local area newspaper.

He is married to Chere and they have three sons. All three are serving or have served in the Marine Corps; Jeff-Avionics, Chris-tanker having served in the Gulf War; Jason-serving Presidential Guard duty with HMX-1.

JAMES P. MULROY,

born June 27, 1940 in Chicago, IL. He graduated from North Illinois University in 1963 with a B.S. Commissioned a 2nd Lieutenant in June 1963, he was assigned to Pensacola, FL and designated naval aviator on Jan. 8, 1965. He served with HMM-363, MCAF Santa Ana, CA.

Embarked for Vietnam August 1965, USS *Princeton* with MAG-36 to Da Nang; reassigned HMM-361, MAG-16, Marble Mountain, Chu-Lai. Transitioned to CH-46, November 1966, MCAF New River, NC, HMM-161 (Paul W. "Tiny" Niesen, CO) and served as personnel officer and instructor pilot, while embarking a full complement of manned A/C from New River to Vietnam, April 1968.

Awarded single mission Air Medal (Hiep Duc 1965). He has 16 Air Medals plus various unit and campaign ribbons.

Graduated from Indiana University School of Law in January 1971 and is a corporate law partner in a 450 man international law firm. He is married and has two children and one grandson.

JOHN T. MURPHY, "MURPH,"

Lieutenant Colonel USAF (Ret) was born on Sept. 18, 1941. He was commissioned as a Second Lieutenant via Marine Aviation Cadet (MARCAD) Program in 1963.

He served with HMM-161 MCAS Kaneohe Bay from November 1963 to March 1965. "Mounted out" with HMM-161 to Vietnam and transferred to HMM-363 half-way through the tour from March 1965 to April 1966. He was a T-28 instructor at VT-3 NAS Whiting from 1966-68. His second RVN tour was flying the OV-10A with VMO-2 at Marble Mountain from July 1968 to August 1969. Final USMC tour OV-10 instructor with HML-267 at Pendleton from September 1969 to January 1971.

Murphy's awards include: DFC (two), Single Mission Air Medal, 37 Air Medals, Navy Commendation, PUC, NUC and CAR.

From 1972-1991 he flew fighters with the Ohio Air Guard and logged more than 2,000 hours in the F-100 and A-7. While in the Guard, he was promoted to Lieutenant Colonel and served as the flight squadron commander. He has 29 years of service and more than 6,000 hours in military aircraft.

He lives with his wife in Swanton, OH where he owns a Ben Franklin Craft & Custom Frame store.

JOHN SHERMAN NEALY

was born on May 2, 1947 in Piqua, OH. He enlisted in the Marine Corps on May 31, 1965 and went through boot camp at Parris Island receiving a meritorious promotion to PFC at graduation. He reported to Memphis for aviation training and was next assigned to Cherry Point, NC. He attended NCO School. Afterwards he went to Vietnam and served with HMM-361 as an aircrew member until October 1967.

He was then stationed at NAS Willow Grove, PA, serving as a drill instructor for Navy recruits and was assigned to Casualty Assistance Notification until his discharge as a Staff Sergeant on April 11, 1969.

During his tour he was awarded the Marine Corps Combat Aircrew Wings, Purple Heart, Air Medals, PUC, NUC, Vietnam Service and Campaign Ribbons.

As a civilian he presently works as an applications engineer with an instrumentation company in Pennsylvania. He is currently a member of the USMC/Vietnam Helicopter Pilots and Aircrew Reunion and VFW. He is married with one daughter.

RUDOLF M. NEBEL, "THE BLUE MAX,"

was born in Camden, NJ on Nov. 22, 1935. He joined the Marine Corps on Feb. 19, 1954 as an enlisted man. He finished boot camp at Parris Island and went on to aviation schools for Air Traffic Control. He worked in GCA primarily until July 1957 when he was ordered to Flight School as a Sergeant. He finished NAVCAD flight training in January 1959 and reported to MCAF(H) New River.

Nebel served with HMM-162 there and on Okinawa and many interesting places around the Pacific Rim. He was a flight instructor in T-28s then went on to Yuma for a tour with a MATCU. He served with HMM-363 in-country from June 1966 through July 1967. He came back to New River and joined HMH-461. He then helped form HMH-361 with 53s. "Got 'em up" and flew all 18 to San Diego in the summer of 1969. He went back to the MATCU at Marble Mountain until the summer of 1970. He went to Quantico, basic school staff and later was OIC of the O Club. He was at the Naval War College and then went back to Okinawa for staff as S-4 for the MAG. He helped plan and execute the evacuations of Phnom Penh and Saigon. Nebel returned to New River and commanded MABS-29 and then HML-268 until its disestablishment.

He retired as a Lieutenant Colonel on Sept. 1, 1978. He was awarded the DFC, 30 or so Air Medals, Purple Heart, PUC, NUC and a Good Conduct Ribbon, plus the normal range of Vietnamese service and National Service type awards.

Since retirement he has managed restaurants, supervised a steak house chain, managed a "prestigious" yacht club and is a substitute teacher.

WALTER T. NEIGHBOR, "WALT" was born on Aug. 18, 1943 in Chariton, IA. He joined the USMC on July 7, 1965. He was stationed at Camp Lejeune, NC with HQCO Force Troops.

From November 1967 to October 1968 he was with MAG-36 and was stationed at Phu Bai, South Vietnam while in-country.

He was discharged on Nov. 5, 1968 with the rank of Corporal.

Today he and his wife, Marcia are living in Ankeny, IA where he is a pharmacist.

BILL NELSON, "FUELS," was born on Sept. 7, 1944 in Chicago, IL. He was commissioned in March 1967.

He joined HMM-262 in Quang Tri in February 1969. He moved with HMM-262 to Phu Bai and Da Nang and flew more than 900 hours in-country. He was a T-28 formation flight instructor for Training Squadron 2 at Whiting Field from April 1970 to June 1971.

While flying a CH-46 in Vietnam, he will never forget his first flight in-country which was an emergency extraction of remaining members of a recon team off a very much less than secure mountain top in Laos. He will also never forget being caught on top of clouds chasing two solo students in T-28s over Luverne, AL. All three planes were running on fumes and there was no daylight left. The students with no real instrument or night time training; found a hole and landed at Luverne Airport, though the "airport" was a street of a new housing development when they initially started their descent. The only damage sustained was to the instructor's reputation.

Nelson was awarded 51 Air Medals and the Distinguished Flying Cross. He was discharged as Captain in June 1971 at Pensacola.

He has been married to the same girl, Bonnie for 23 years. They reside in Litchfield, MN. Their oldest son, Ryan (19) is attending the U of M in Duluth. He flies on skis. Their youngest son Monty, (15) flies on hockey skates. Bill has been working in construction management and engineering since leaving the Marine Corps.

RALPH W. NICHOLAS, "NICK," GySgt USMC (Ret) was born in Louisville, KY on Dec. 2, 1925. He enlisted in the USMC on March 11, 1944. He went through boot camp in San Diego and transferred to Camp Pendleton for advance infantry training as a hard charging young B.A.R man.

He proceeded on to the Pacific and joined 1-3-6 2nd Div. in Saipan. He participated in the Okinawa invasion and the initial occupation of Japan. He was discharged on Feb. 18, 1947 but re-entered on Oct. 19, 1950. He served at Cherry Point, Quantico and in HMR-161 in Korea for all of 1953. He also served at MCAS Miami and at MCAF New River in the following squadrons HMR-261, H&MS-26, HMR-461, HMR-162 for a Far East tour 60 and 61, HMM-262 and HMH-461. He served as a Quality Control Chief in HMM-163 from July 1966 to July 1967 in Vietnam and then returned to H&MS-26.

He retired on April 30, 1968. He was awarded the Air Medal, Navy Unit Commendation, Good Conduct with four stars, Asia Pacific Campaign with one star, World War II Victory, Navy Occupation Service, National Defense Service with one star, Korean Service with two stars, United Nations Service, Korean Presidential Unit Citation, Vietnamese Service with one star, Vietnamese Campaign Medals and Combat Aircrew Insignia with three stars.

After retirement he worked as chief mechanic for Doan Helicopter, Inc. in Daytona Beach, FL restoring War Birds and converting H-34s to S-58 Helicopters. Nicholas was in charge of the restoration of the only F4U-1 birdcage canopy Corsair known to be in existence. He also had a hand in the re-work of a B-25, C-54, T-6 and DC-3. Now fully retired, he is rearing his third set of children.

PAUL W. NIESEN, Colonel, USMC, born Feb. 25, 1930 Racine, Wisconsin. Enlisted July 8, 1948, assigned Parris Island recruit training, afterwards served as a guard at Marine Barracks, Charleston, SC through July 1949. Underwent flight training at NAS Whiting Field, Milton, FL, and NAS Corpus Christi, TX for jet training, and designated Naval Aviator March 1, 1951. Served with VMA-251 and VMF-214, MAG-13, MCAS El Toro, CA through March 1952. Ordered to Korea, where he served with VMA-121, MAG-12. Returning to the States December 1952, he was a flight instructor at NAS Corpus Christi, TX. Transferred to MCAS El Toro, he served with VMCJ-3.

After helicopter training at NAS Pensacola, he was transferred in 1956 to Santa Ana, where he flew with HMM-363, HMM-362, and HMM-462 of MAG-36. In 4/59, he became Aviation Safety officer with H&HS MCAS Beaufort, SC, then assigned as Asst. Air Ops 0 with the MAD, USS *Boxer*. In March 1963, he was selected to serve with the Executive Flight Detachment, Marine Helicopter Squadron One, MCAS Quantico as Ops 0. From February 1967, he served as S-4, MAG-26.

He was selected in April 1968 as CO HMM-161, at MCAF New River NC, and took HMM-161's CH-46s to Vietnam as a full squadron unit. The outfit flew cross country, and embarked aboard ship for Vietnam. In country, he earned the Purple Heart, Silver Star, and his 2nd through 25th Air Medals. He then served as XO, ProvMAG-39 at Quang Tri.

He returned home and was assigned as Head, Air Assault Support Section, Aviation Weapon Systems Branch, DCS(AIR), at HQMC. During 1969, he received Marine Corps Aviation's most distinctive honor when he was selected as Marine Aviator of the Year. From 1970 through 1972, he served at Quantico MCDEC's Air Branch, when as a Special Project Officer, he took a complement of AH-1 Cobras to Vietnam from 1/71 through 4/71 for evaluation. From 1973 through 1975 he served as G-4, 2nd MAW at MCAS Cherry Point NC. After retirement from the Marine Corps in March 1975, he worked as a deputy division director with Bell Helicopters in Teheran, Iran through June 1977. Col Neisen died on April 3, 1990.

ROBERT B. NORCOTT, "BOBBY," was born on Sept. 16, 1947 and grew up in Milton, MA. He enlisted in the Marine Corps in November 1966 on the "120 day delay" program, while in his first semester at the University of Massachusetts. He went active on Jan. 20, 1967, completed recruit training at MCRD Parris Island, SC and went on to attend electronics schools at NATTC Memphis, TN, graduating as a radar/navigation technician in April 1968.

Norcott was attached to HMM-162 for training in CH-46s at New River, NC and subsequently completed a tour in the Caribbean with them aboard the USS *Boxer* (LPH-4), returning to CONUS just in time for orders to FMF WESTPAC. In October 1968, Norcott joined HMM-262 in Quang Tri, Republic of Vietnam. He spent his entire tour with "262," assisting in their move to the south to Phu Bai in September 1969. Corporal Norcott was awarded 11 Air Medals and Combat Aircrew Wings.

Once back in CONUS, Norcott was promoted to Sergeant and completed service with H&MS-26 and HMM-365 joining the latter on a Med cruise aboard the USS *Guam* (LPH-9). While aboard, he supervised the "radio shop" as the Quality Control Inspector.

Sergeant Norcott was discharged from active duty on Jan. 20, 1971. After discharge he returned directly to the University of Massachusetts, where he completed a bachelor's degree in English. For the past 17 years, he has been a teacher for the Commonwealth of Massachusetts in institutional settings. He is currently teaching at-risk juvenile offenders and recently completed his master's degree in education. Norcott has been married to Kathie for 15 years. They have a 13 year-old son, Rob, and live in North Attleboro, MA, 30 miles south of Boston.

DR. JOHN F. NORMAN, "HAWG," was born on Jan. 25, 1946 in Coleman, TX. He joined the service on Aug. 15, 1965 at Parris Island. He graduated from Class 10-66 at NAS Pensacola,

While in-country he was a pilot flying UH-34s and UH-1Es from 1968 to 1969. He was also stationed at VT-6 NAS Pensacola as an instructor from 1969 to 1970.

Dr. Norman was discharged in October 1970 with the rank of Captain. He was awarded 44 Air Medals, Vietnam Cross of Gallantry and the usual "I've been there medals."

Today he is living in Bradenton, FL and is a dentist. He is divorced with no children.

AVERY C. NORRIS Gunnery Sergeant (Ret) arrived in Ware County, GA on Aug. 29, 1929. He departed for the big MCAF in the sky on June 17, 1994. Sometimes called "Cuz," sometimes "Reb" and frequently that #@#* SOB by those who did not measure up to his standards for Marines.

Norris enlisted in the U.S. Army Air Force on Dec. 5, 1946 and was discharged on Oct. 7, 1949. He enlisted in the USMC on Nov. 10, 1949 and served continuously until retirement on Feb. 27, 1967. All the time he served as a helicopter mechanic and flight crew except for approximately two years on AD Skyraiders and one tour as an instructor at Memphis.

He worked on HO3S, Bell 47, HO5S, HRS, HR2S, HUS and CH-46 helicopters. He participated in the Berlin air lift as a crew member on a C-54 aircraft. He served two tours in Korea, one with VMO-6 and one with HMR-161. Norris served one tour in Vietnam from 1966 to 1967 with HMM-265. He participated in assorted Panic Button emergencies, Carib cruises and overseas tours in between.

Norris was awarded the following medals: Army Occupation, Japan and Germany World War II Victory; Berlin Device; U.N. Ribbon; Korean Service with three stars; Presidential Unit Citation, Army Distinguished Unit Emblem; Good Conduct with four stars; National Defense Service 2nd Award; Vietnam Service; Vietnam Campaign; Air Medal and Combat Aircrew Insignia with three stars.

He was married to Patricia Dramble Norms for one day shy of 38 years. They have three daughters and three grandsons.

After retirement he worked as a shop foreman at Jet Power, Inc. in Miami, FL and later as director of maintenance at Jet Cargo, Inc. all over the Mid-East.

May he always be remembered as one hell of a Marine, husband, father and my best friend. RWN

JOSEPH A. NOVAK, "MOMS," born on Oct. 14, 1941 in Milford, CT. He enlisted on Nov. 24, 1959 and went to boot camp at Parris Island. He saw duty at MCSD Albany, GA from March 1960 to March 1961. He then reported for retraining in aviation to NAS Jacksonville, FL, where he attended Aviation Electricians "A" School. He then reported to HMR(L)-262, (H-34) at New River from May 1962 to April 1964.

He transferred to HMM-161 (H-34) for deployment for Vietnam, returning in May 1965. After returning, he was assigned to Aviation Electricians "B" School, NAS JAX, FLA. In March 1966 he reported to HMH-461 at New River and was sent to TAD to Sikorsky Aircraft Company for training on the CH-53A. In April 1968 he was transferred to HMM-161 (H-46) for deployment to Vietnam. In June 1968, he was transferred to HMH-463 (H-53A) where he served through June 1969. He then returned to New River and served with HMH-461 until his discharge on Nov. 26, 1969 as a Staff Sergeant.

During his service he was awarded Combat Aircrew Wings, an Air Medal, Vietnam Service and Campaign Medals, Combat Action Ribbon, Good Conduct Medal, Presidential Unit Citation, Rifle and Pistol Expert Badges and other awards.

As a civilian he has worked for Union Carbide and Ionics Incorporated as a sales and service engineer. He is presently employed by Leeds and Northrup as the Southern regional analytical manager.

He has been married to Marilyn (Montambault) since 1966 and they have a daughter Kimberly.

ROBERT M. NYE JR., "BOB," Lieutenant Colonel USMC (Ret) enlisted on July 18, 1964 and graduated from the Marine Aviation Cadet program as a Second Lieutenant on April 21, 1966.

He was assigned to UH-34s and joined HMM-163 at Futema, Okinawa. He flew aboard the USS Iwo Jima at Phu Bai and Dong Ha during 1966-1967.

He then served as air liaison officer with the 1st Marine Division. Transitioning to KC-130s in 1969, he flew with VMGR-352 and VMGR-152. During 1971-1973 he was the first director of the MCAS El Toro Staff NCO Academy. He was assigned to Washington, D.C. and flew the Commandant's VC-118 from 1977-1981.

After retiring in 1984, he wrote for NBC television and flew for Jet America Airlines. Currently he is an MD-80 Captain for Alaska Airlines living in Big Bear City, CA with his wife, Shirley.

JOHN R. ODOM, III, "BOB," was born on Feb. 16, 1945 in Miami, FL. He graduated in 1966 with a BA in history from VMI and joined the service in July 1966. He earned his wings on Jan. 23, 1968. He was stationed at MCAS New River and at MCAS Cherry Point.

From May 1968 to June 1969 he was stationed in-country at Quang Tri and Futema, Okinawa with HMM-161. He was an aircraft commander flying CH-46s. Was discharged in January 1974 with the rank of Captain. Awarded the DFC, four Single Mission Air Medals, 65 Strike/Flight Awards, two Meritorious Commendations, USCG Commendation, USAF Commendation, two USCG Achievements and USCG Comdt. Letter of Commendation Ribbons.

In June 1975 he entered the U.S. Coast Guard and was stationed at Elizabeth City, NC, Sacramento, CA, McChord AFB WA, Kodiak, AK and RAF Woodbridge, UK.

He was executive officer for USCG Air Station in Clearwater, FL. He was selected for Captain in July 1994. He is also an aircraft commander on HC-130s with a total of 8,800 total flight hours.

He has been married to Natalie Gregory since April 1967 and they have two daughters, Elizabeth and Chandler. They reside in Clearwater, FL.

GLENN A. OLSON, "ANDY," Lieutenant Colonel USMC (Ret) was born on July 4, 1936 at Monmouth, IL. He enlisted on June 14, 1954. His education consisted of Monmouth College, San Jose State, bachelor of science; and U.S. Naval Postgraduate School, master of science in management. He was a graduate of U.S. Naval Flight Training, Pensacola, FL, Kingsville, TX from 1957 to 1958; and Amphibious Warfare School, Quantico, VA in 1966.

His service highlights include 123rd Field Artillery Battalion, 33 Div., Illinois National Guard, Monmouth, IL (Corporal) 1954-56; U.S. Navy Flight Training Pensacola, FL/Kingsville, TX. NAVCAD, 1957-58; commissioned Second Lieutenant May 1958; VMR-263, VMR-363, R4Qs) MCAS Cherry Point, NC 1958-1959; VMR-352, First Lieutenant, (R5Ds) MCAS El Toro 1959-1961; VMR-253, MCAS Iwakuni, Japan, 1961-1962; MARTD Los Alamitos, Alameda, (Captain), (R5Ds) 1962-1963; HMR-362, HMR-363, MCAF Santa Ana, (H-34s) 1963-1964; aide-de-camp to CG Third MAW, 1964-1965; Amphibious Warfare School, 1965; VMO-1, MCAS New River, (UH-1Es) 1966; VMO-6, Aviation Safety Officer, Ky Ha, RVN, 1966-1967; U.S. Naval Postgraduate School, (Major), 1967-1968; Data Systems Division, HQMC, 1968; Special Assistant Secretary of Navy's Office, Senior Systems Engineer 1969-1970; HML-257, OV-10 transition, Camp Pendleton, CA 1970; VMO-2, Executive Officer, Da Nang, RVN (OV-10s), 1970; First MAW. Assist. Chief of Staff-Management, Da Nang, RVN/Iwakuni, Japan, 1971; HML-257, Executive Officer, Commanding Officer, Camp Pendleton, CA (OV-1Os), (UH-1Es), 1970-1973; MCTSSA, MTACCS, Camp Pendleton, CA (Lieutenant Colonel) 1973-1975; Naval Air Systems Command, London Office Program Manager-Harrier, London, England (AV-8As), 1975-1977; and Naval Air Systems Command, Deputy Program Manager, Washington, DC, AV-8B, 1977-1978.

He was discharged on June 1, 1978 with the rank of Lieutenant Colonel. His awards include: the Distinguished Flying Cross, (three awards), Bronze Star w/Combat "V," 28 Air Medals, Presidential Unit Citation, Navy Unit Commendation and other citations.

He is currently president of Global Initiative and resides in Lake Oswego, OR.

KENNETH R. OLSON Colonel USMCR (Ret) was born on June 23, 1945 in Hobbs, NM. He attended San Diego State College. He enlisted in the Marine Corps Reserve in August 1965 and was selected for

the Marine Aviation Cadet program. He was designated a Naval Aviator and commissioned a Second Lieutenant in June 1967. He joined HMM-161 at MCAS New River for training in CH-46s. He was stationed at Santa Ana, Yuma, New Orleans and Alameda.

He was sent TAD to HMM-365 for a Caribbean cruise from December 1967 to February 1968, rejoined HMM-161 and was deployed to Quang Tri, RVN in May 1968. He transferred to HMM-265 at Phu Bai in December 1968 and returned to CONUS in May 1969. He flew more than 800 combat hours.

Olson joined HMM-764 as a reserve Captain in May 1972 and served in various positions with MAG-46 while transitioning to the OV-10 and the A-4. He returned to active duty as an operations officer MAG-42 Det. A from January 1978 to November 1982. He was OIC MABS-49 Det. C from April 1984 to May 1985, Commanding Officer VMA-142 from May 1985 to May 1987, assistant G-3 4th MAW from June 1987 to October 1989 and ACE/Deputy Group Commander MAG-42 from November 1989 to December 1991.

Olson retired in September 1993. His awards include: MSM, the Single Mission Air Medal, 46 Air Medals, CAR, RVN CG, NDSM, VNCM, NUC, VSM, MUC, SMCRM and AFRM.

He lives in Cave Creek, AZ and is married with one son. He is a Captain with Southwest Airlines.

JARDO OPOCENSKY JR., "SKI," was born on Aug. 7, 1947 in Leuthershausen, Germany. The family moved to Glastonbury, CT in 1956 where he graduated high school in 1965. He enlisted in the Marine Corps on Oct. 19, 1965. After completing Recip and Helicopter School, he was assigned to VMO-1 MCAF New River.

Volunteering for Vietnam, he served with VMO-2 as a crewchief on UH-1E gunships and for LtGen Cushman, commander of III MAF.

During his tour from December 1966 to January 1968 he was awarded Combat Aircrew Wings, the Navy and Marine Corps Medal, 13 Air Medals, PUC, the Vietnamese Cross of Gallantry and the Vietnam Civil Action. He was discharged as a Sergeant at New River on Oct. 18, 1969.

He graduated from college in 1973, where he earned his pilot wings. He pursued an engineering career and later sales and marketing. He is presently a sales rep. for Wyman-Gordon Investment Castings selling casting to the aerospace market. He resides with his wife and son in Hebron, CT.

DOUGLAS R. ORAHOOD, "CAPT. AMERICA," was born on Oct. 10, 1946 in Crawfordsville, IN. He graduated from Ball State University in 1968. He was designated a Naval Aviator in July 1969 and transitioned to CH-46s in HMMT-302.

He departed for RVN in December 1969 reporting to HMM-364, "The Famous Purple Foxes." Incountry time he flew with "Swift" and "Scarface." He departed RVN in December 1970 with 1,000 hours of "Red Ink" time. High points of RVN were sleeping on plastic wrapped mattresses in the medevac hooch, bologna sandwiches and warm milk, taking a crap while sitting on a cold, wet plywood seat and being awakened by the "bee bonk" of Da Nang DASC breaking squelch at o'dark-thirty for an emergency evac. "It ain't no . . . picnic." "My best to all you guys and our buds on The Wall. 46's-ooh-all!"

He spent a lifetime with 2nd ANGLICO (12 months) at Camp Swampy and finished with HMM-162. He was awarded a DFC, 44 Air Medals and other "standard trimmings." Orahood returned to Indiana to teach "shop," now industrial technology, farm and raise a family. He retired from the Indiana Guard with more than 27 years service and a lot of holes bored in the Midwest skies.

RONALD G. OSBORNE was born on May 10, 1939 in Auburn, IL. He enlisted in the USMC on March 3, 1958. He completed boot at MCRD San Diego and was assigned to the 7th Marines, 1st MAR DIV at Camp Pendleton. He was selected for NESEP in 1960, completed college at Ole Miss and was commissioned in August 1963. After TBS Quantico, he was ordered to flight training and designated a Naval Aviator on Oct. 9, 1965. He trained in the UH-1E at VMO-1 New River and reported to VMO-2 at MMAF in late August 1966.

He was the recipient of 18 Air Medals and the MC Good Conduct. He flew as an instructor pilot at VMO-5/HML-267 and then was assigned to grad school at NPS Monterey. He re-trained in the AH-1J at New River and was assigned to H&MS Sub-unit 1 at Futema. He became CO of that unit and then first CO of HMA-369 when the squadron was designated in April 1972. He left the USMC in January 1974.

His most gratifying mission was pulling a recon team off "the shelf" near Khe Sanh after they had been stuck for five extra days due to bad weather. The hairiest mission was in the month of February 1967 flying Army SOG out of Khe Sanh. He was assigned to Wing G-3 for the last five months of his tour.

Special memory or mission assignment was the recon extraction mentioned above. "The team was on "the shelf" area near Khe Sanh in late December ?? 1966. Area had been fogged in for days and team could not be extracted on schedule. Daily flybys for radio contact said they were running out of food and water. Finally found a hole in the fog on the fifth day and got them on board a UH-1E slick. As we came out of the zone and dropped over the cliff, my crewchief said on the ICS, "Sir, that son-of-a-@#$%* kissed me!"

GEORGE V. OTTO, Major, USMCR. was born on March 26, 1943 in Beaumont, TX. He received a BS in mathematics from Lamar University and in January 1966, he was the first graduate of Lamar to be commissioned through the PLC program.

Served in Vietnam from November 1967 through December 1968 with HMM-265, 262 and 165, flying the CH-46. Instructed in Flight 18 of VT-1, then went back to the CH-46 in Reserve Squadron HMM-767.

Otto was awarded 28 Air Medals, the Presidential Unit Citation, Combat Action Ribbons, Vietnam Service and Campaign Ribbons.

Otto and his wife, Carolyn, were married December 1965 and have one son, Mark, born December 1969. Carolyn teaches third grade and Mark is a graduate student at the University of Florida. He has been employed with Prudential Insurance for 24 years and currently works as a division manager in Jacksonville, FL. Otto continues his flying as a volunteer with the Civil Air Patrol.

JOHN ANTHONY OUBRE was born on Aug. 3, 1946 in Pensacola, FL. He enlisted in the Marine Corps on June 15, 1964 and went though Parris Island, SC. He reported to Camp Lejeune, NC, 3rd BN 6th Marines and participated in the Dominican Republic Conflict. He re-enlisted in 1967 and was assigned to HMM-361 Marble Mountain, RVN (1967-1968). His second tour was at Marble Mountain, MABS-16, (1969-70) and his third tour was at Bien Hoa, VMA-311, (1972-73).

Oubre's awards include two Purple Hearts, Navy Commendation with Combat V, three Air Medals, two Combat Action Ribbons, three Good Conduct, Armed Forces Expeditionary, NUC, MUC, Vietnamese Civic Action Color with Palm/Frame, Combat Aircrew Wings and the Vietnamese Service and Campaign Ribbons. He was discharged as a Sergeant at Treasure Island, CA on June 11, 1973.

Oubre served with the Florida National Guard from 1986 to 1992. He has been with the U.S. Postal Service since 1975 and is currently the postmaster in Kissimmee, FL.

DOUGLAS B. PAGE, "WINDY," was born on Aug. 23, 1943 in San Jose, CA. He graduated from Lassen High School in Susanville, CA in 1961 and attended San Jose State College. He enlisted in the USMC in February 1962, attended boot camp in San Diego, Aviation Electronics School, in NAS Millington, Memphis, TN and made Corporal. He went to MARCAD for flight training wings in 1964.

He joined VMO-6 at Camp Pendleton and then went to VMO-6 and VMO-2 in RVN at Marble Mountain and Ky Ha from 1965-66. He came back to Pensacola as a T-28 flight instructor and then went back to Pendleton in VMO-5/HML-167 in UH-1E/

OV-10A. He went back to RVN at Quang Tri with VMO-6 tour as ALO with 1st BN, 4th MAR from 1968-69.

Memorable experiences: "The night I almost made a rocket run on a moonbeam that I thought was a VC searchlight? A downed USAF pilot rescue out of North Vietnam and four months as a 'grunt' as ALO in-country."

He went back to MCAS Cherry Point, NC as a weapons instructor in UH-1E/OV-10, flew with VMO-1 and transitioned to the A-4 Skyhawk in 1973. He was stationed with VMA-331, MWSG-17, VMA-223, H&MS-32, 2nd MAW and MAL-32. He served variously as S-2, S-3, NATOPS, Safety, S-4, XO and CO at Sqdn., Group and Wing.

He received a bachelor's degree in human resources management. He retired as a Major in March 1982. He is NATOPS qualified in the UH-1E, OH-43, T-28, OV-10, O-1E, C-117 and A-4, with a total of more than 5,700 hours and 1,300 plus combat missions.

His awards and decorations include: a Silver Star, Distinguished Flying Cross, 55 Air Medals (Strike/Flight), two Single Mission Air Medals, Bronze Star, Meritorious Service Medal, Navy Commendation Medal, Navy Achievement Medal, two Purple Hearts, Combat Action Ribbon, Presidential Unit Citation, Navy Unit Citation, Meritorious Unit Citation, Vietnamese Air Gallantry Medal, Vietnamese Cross of Gallantry, two Vietnam Unit Citations and various service ribbons/medals.

Page is currently flying a DeHaviland DHC-6 Twin Otter and Bell Jet Ranger for the police department. He has been a sworn law enforcement officer for the last 10 years. He is married to Paula and the have one son, Chris, age 19. They reside in Garden Valley, CA.

GREGORY JOE PARR, "JOE," was born on July 20, 1948 in Anderson, IN. He joined the service on Jan. 2, 1968. He was stationed at MCRD San Diego, Camp Pendleton (two times) MCAS Kaneohe, HI, Ft. Walters, TX and Savannah, GA for Army flight training.

He was stationed in-country at Marble Mountain, RVN and was a pilot flying the AH-1G.

He was discharged in January 1974 with the rank of Captain. His awards include two Single Mission Air Medals and 45 Air Medals.

Today he is a pilot supervisor in Maui and for Papillon Helicopters in Hawaii. He is married with three children and they reside in Kihei, HI.

FRED D. PATTERSON, "PAT," was born on Feb. 8, 1933 in Maitland, MO. He enlisted in the NAVCAD program in 1952, and was commissioned in the USMC in 1954. He joined VMF-235 and soon was at NAS Atsugi. He then went to 2nd MAW Headquarters prior to Pensacola for helicopter training. He served in MAG-26, HMM-263 before heading west in 1959, to the Pacific Fleet Air Intelligence Training Center, NAS Alameda. He again went to Japan to MAG-11 at NAS Iwakuni. He was also stationed at MCAS El Toro, CA, MCAS Cherry Point, MCAF New River, NAS Alameda and MCAS Santa Ana.

He flew in HMH-461 from 1961-1964, when he became an advisor to the Chinese Marine Corps at Tsoying, Taiwan. He met and married a Navy nurse while in Taiwan and after leave in Japan he reported to 1st MAW, Da Nang, Vietnam on the 2nd or 3rd of January 1966. After five months with the Green Berets in I Corps, he joined HMM-361 in May of that year at Okinawa and came back in-country with the squadron. He flew out of Chu Lai until September 1966 when he was shot while leading a re-supply flight into a hot zone and was medevac'd to the USNH Yokosuka.

Patterson retired with the rank of Major from Quantico, VA in October 1967. He was awarded Joint Comm. Ribbon, Air Medals and a Purple Heart.

Presently living on a farm in Amherst, VA, he now has a master's degree in counseling and works for Central Virginia Community Service Board as an addictions disease counselor. His son attended Old Dominion University in Norfolk, VA.

DOUGLAS H. PAWLING, "DOUG," was born on Sept. 30, 1938 in West Chester, PA. He joined the service on Feb. 2, 1960 and was commissioned in December 1960. He earned his wings in June 1962 and was stationed at MCAS New River and NAS Whiting.

He went on a Med cruise on the LSD-21 *Fort Mandon,* with HMM-162 Sub-Unit 2 from fall 1962 to spring 1963. He was in service in Vietnam with HMM-162 from June 1964 to June 1965. He was stationed in-country at Da Nang, Okinawa and Futema.

He was discharged on Jan. 15, 1966 with the rank of First Lieutenant and was awarded six Air Medals.

Pawling joined Pan American World Airways on Jan. 31, 1966 and retired on Feb. 1, 1991. He is type rated on B-707, 727, 747, L1011. He was a Captain in Miami on the B-727 when he retired. His total flying time was approximately 20,000 hours.

Today he resides with his wife in Inverness, FL. He and his wife have three children, ages 18, 21 and 32.

COURTNEY B. PAYNE, "COURT," was born on Nov. 13, 1934 in Pikeville, KY. He joined the service on June 13, 1951.

Vietnam 1962-1963: He arrived by passport at Soc Trang in IV Corps in July 1962 as part of the 7th Fleet Task Unit, Operation "Shu-fly"; joined HMM-163 which had H-34s at Soc Trang; served as a crewchief and flew as a crewchief on the command's C-117 at night in support of the Shu-fly operation. He rotated to Japan in January 1963 and returned to Vietnam twice in early 1963 to support Shu-fly flying "Red Line" cargo comprising munitions, explosives and other ordinance material. His rank was Staff Sergeant and the reason for his arrival in-country in 1962 was as an "advisor."

Vietnam 1968-69: arrived at MAG-36 at Phu Bai in August 1968; and was assigned to HMM-364, the "Purple Foxes," as maintenance control officer. During this time he cobbled together a recovery team from the maintenance department to recover aircraft downed from enemy fire or which had maintenance related problems. He participated in recoveries during "Mead River," "Taylor Common" and "Dewey Canyon" plus operations of day to day support for forces in I Corps. He received a permanent commission in July 1969 and was promoted to Captain in December 1969.

Most memorial experience: "Having the privilege to serve in HMM-364 during my 1968-69 tour. Seldom if ever, is a man given the opportunity to associate with such a gallant, dedicated, professional and delightful group of men. To be a Purple Fox, to head our Recovery Team, live the experience of emotional highs, then plummet into despair at the loss of a crew, then, sober up the next morning, and go again, was the most unforgettable time of my life. The honesty, humor and outrageous conduct displayed by Purple Foxes taxed our leaders, but what leaders!! The "Six Actual" Gene, "OC," Ernie, Bob, Clay (the quack) and of course, the gang in the ready room. All remembered by the "Poet." During my 27 year career as a Marine, the absolute pinnacle of this career was the day in 1968 when at Phu Bai, I became a Purple Fox."

Payne was retired on Nov. 28, 1978 with the rank of Major. He was awarded two Bronze Stars and a Navy Commendation Medal for action involving recoveries of aircraft shot down while operating out of Phu Bai and Marble Mountain. He also received a Meritorious Service Medal, Korean Service, United Nations Service Medal, Armed Forces Expeditionary Medal, Vietnam Service Medal, Combat Action Ribbon, Vietnam Cross of Gallantry and a Good Conduct Medal.

After retirement, he joined Sikorsky Aircraft and is currently assigned as a project engineer, Heavy Lift Programs, Sikorsky Development Flight Center, West Palm Beach, FL. He is married to Donna Ann, a former airline hostess for Western Airlines. They have three sons and one grandson.

FRED PENNING, "FRIAR," was born on March 30, 1944 in Granville, IA. He was drafted into the Army on Dec. 15, 1965 and volunteered for Marine Corps OCS on December 10. He completed flight school in June 1967 and reported to MCAF New River.

He took a Caribbean cruise with HMM-365 and rotated to Vietnam with HMM-161 in May 1968. He was stationed at MAG-39 Quang Tri the entire tour as a pilot flying H-46s. They were the last load out of Khe Sanh.

Penning reported to MCAS El Toro in July 1969 and was discharged to the reserves in June 1970 as a Captain. His awards were Navy Comm. w/Combat "V," Single Mission Air Medal, Vietnam Campaign Medal and 39 Air Medals.

After active duty, he attended the University of Idaho Law School for one year, then spent 1972 through 1975 helicopter logging in the Pacific Northwest. From 1976 through 1984, he spent his time operating a wheat/cattle ranch near Grand Coulee, WA. He returned to school from 1984 to 1985 to earn a teacher's certificate. He taught in Spokane, WA until he began working for Waste Management in 1989. He started with them in Dallas, then became general manager in Fort Worth and transferred to San Diego in 1991. He is married to Jane and they have three children, Ben, Nicole and Josh. He resides in San Diego, CA.

MIKE PEPPLE, "PEP," was born on Sept. 24, 1942 in Columbus, OH. He joined the USMC in October 1963 and was stationed in Pensacola and Santa Ana.

He was stationed in Vietnam, and, while in-country remembers all the great times with his good buddies that are too numerous to mention.

He was discharged in June 1968 with the rank of Captain and was awarded various medals.

Today he is retired from TWA and is married with two children. He resides in Bluejay, CA.

CLARENCE ROBERT PERRY, "BOB," was born on June 25, 1931 in Detroit, MI. He graduated from high school in Indianapolis, IN and attended Butler University. He enlisted in the Navy in January 1951. He became an aviation photographer's mate 3rd when he was appointed to USNA in 1952. He was commissioned in the USMC in 1956; went to Basic School in March 1956; and took flight training from 1957-58.

He was with HMR(L)-162 (HUS) in New River from 1958-1960; in the Far East from 1960-61; P.G. School earning a meteorology degree in 1961-63; El Toro Aerology O. and SAR pilot (HRS) 1963-65; and HMM-165 Santa Ana (CH-46) 1965-66. He went to Vietnam from 1966-67, Ky Ha and then to Cherry Point in 1967-70 (KC-130). He retrained on the CH-46 for return to Vietnam in 1970-71 at Da Nang and Wing Staff. He joined HMM-263 at Quantico and went on a Med Cruise in 1971-74. He then joined MARTD CO with HMM-766 and VMO in 1974-1977.

Perry completed more then 26 years of service and had more than 4,700 flight hours in 15 types of aircraft. He was awarded two DFCs, 27 Air Medals, two Vietnamese Crosses of Gallantry, CAR, NUC, MUC, PUC, Vietnam Service and Campaign Ribbons. He completed active services as a Lieutenant Colonel in 1977.

Perry obtained a law degree in 1981 and is now in general law practice in Michigan. He is married with one son and three grandsons.

JIM MARVIN PERRYMAN, JR. was born Feb. 7, 1933 in Washington D.C. Enlisted in the USMCR in 1949. Reserve duty as F6F/F8F PFC/Cpl plane Captain at NAS Anacostia DC. Entered USNA in 1951. Commissioned 2nd Lt. June 1955 and married Frances McClooney of Baltimore, MD. TBS Quantico, VA and then Platoon Leader with Engineer and Infantry Battalions Camp Pendleton, CA 1956-57. Son, Jeffrey, born April 1956.

Flight School Pensacola, FL 1957-58. HMM-361 at Santa Ana, CA 1959-60 (UH-34). Son, Glenn, born April 1960. Then HMM-362, "Archie's Angels," serving in RVN 1962 (UH-34). SAR at Yuma, AZ 1962-64 (H-2/H-19). VMO-6 Cam Pen and RVN 1965-66 (UH-1E). AWS Q-town 1967. HMX-1 1967-69 (VH-1/VH-3/CH-46/CH-53). HMM-262/H&MS-16 RVN 1970-71 (CH-46). Cumulative RVN awards: Silver Star, DFC, Air medals (64). HMX-1 XO/CO 1972-74 Presidential Helicopter Pilot. Naval War College 1975. H-1/H-3/H-46/H-53 Program Manager, Naval Air Systems Command DC 1976-81. Retired January 1981. Analyst PRC Inc. 1981 to present supporting V-22/UH-1N/AH-1W Programs.

DAVID M. PETTEYS, "DELTA MIKE," was born on April 6, 1939 in Delano, CA. He graduated from Berkeley High School in June 1957. He then graduated from the University of California, Berkeley in June 1962. He entered the Marine Corps OCS in September 1962 and was commissioned in December 1962. He began the U.S. Naval Flight Program in January 1963 and earned his wings in June 1964.

He flew in HMM-363 and was one of the original officers in HMM-165 starting in 1965. He was deployed to Vietnam in August 1966 and returned in September 1967. While in-country, he was stationed at Ky Ha, Marble Mountain, Phu Bai, Dong Ha and Khe Sanh.

He was discharged on Sept. 27, 1967 with the rank of Captain. He was awarded the Silver Star, DFC and 28 Air Medals.

Petteys joined United Airlines in January 1968 where he remains to the present. He is married and has four children. They reside in Conifer, CO.

WADE A. PETTIS, "PET," was born on Feb. 17, 1949 in Bonifay, FL. He enlisted in the Marine Corps on Aug. 23, 1967 and went through boot camp at Parris Island. He attended the AR-15/M-16 test fire in Panama Canal. He reported to Memphis for helicopter training (CH-46).

He transferred to Vietnam in January 1969 and was there until February 1970. He served in-country with HMM-263 as a crewchief and door gunner. He flew recon insertion, extraction; night and day medevacs and vertical troop insertion. He was transferred to HMMT-402 training squadron and then to H&MS-40 as a quality assurance inspector.

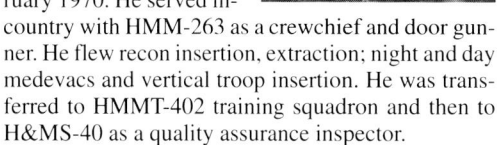

He was discharged on Aug. 20, 1971 at New River as a Sergeant. He was awarded the Combat Aircrew Wings, Distinguished Flying Cross with star, Strike/Flight Air Medal with Numeral 30, Navy Comm. Medal with "V," Navy Achievement with combat "V", CAR, and Meritorious Mast.

Married twice, he has one daughter and one granddaughter. He is disabled due to injuries in Vietnam and afterwards.

RICHARD E. PHILLIPS, "RED," was born on Nov. 20, 1935 in Lancaster, OH. He enlisted in the Navy in 1952 and was trained as a medical corpsman.

He served 26 months in Vietnam and flew more than 900 medical evacuation missions in UH-34D helicopters. During one 24-hour period, he flew a record 35 missions into combat. He was also cited for heroism in saving the lives of five Marines in Quang Nam Province in 1966 when his aircraft was downed by enemy fire. Disregarding his own safety, he remained with the wounded under intense enemy fire, administered first aid and then successfully evacuated them when other aircraft were able to land. He retired in 1974 as a Chief Petty Officer. He earned the Marine Corps Combat Wings and the Aircrew Insignia.

Following retirement Phillips spent the next 19 years helping veterans and their families at the post and district commander level in the VFW. He was also in the Vietnam Veterans, AmVets and American Legion. He has three sons and two grandchildren. He died on Oct. 13, 1993 and is buried in Arlington National Cemetery.

RICHARD L. PHILLIPS, "RICK," Major General, USMC (Ret) is a native of California. He received an engineering degree from California Polytechnic, was commissioned in 1961 and designated a Naval Aviator in January 1963.

General Phillips served in Vietnam during 1966, '67, '70 and '71 both as a helicopter pilot and ground company commander. He was in the first four aircraft H-53 detachment to fly in Vietnam.

General Phillips had commanded three Marine squadrons and MAG-39. During his MAG-39 command, he also commanded both the 11th and 17th MEUs. He was Deputy Commander and Acting Commander of the Naval Space Command and from 1989 to 1991, he was Commanding General First Marine Expeditionary Brigade and dual-hatted as Deputy Commanding General FMFPAC for part of that time.

His decorations include: the Legion of Merit, Meritorious Service Medal, 26 awards of the Air Medal with both Single Mission and Strike Awards, Joint Service Commendation Medal, Navy Commendation with "V," Navy Achievement with "V," Combat Action Ribbon, the Distinquished Service Medal and numerous unit, expeditionary and campaign medals.

He retired in September 1995 after a tour as the Marine Corps Inspector General and his designation as the "Silver Hawk" of the Marine Corps (longest designated aviator on active duty). He currently serves as VP of Wheat International Communications Corp. in Vienna, VA. His son is a Marine Cobra pilot.

JERRY PIATT was born on March 30, 1936. He is a life-long resident of Chelsea, MI. He joined the Marine Corps in April 1953, attending boot camp at San Diego and infantry training at Camp Pendleton. He then spent 14 months in Korea after the war. He entered flight school at Pensacola as a Corporal in May 1961. He received his wings and was commissioned in February 1963.

Piatt served two tours in Vietnam as a Marine helicopter pilot. He flew the UH-34 and CH-46 in Vietnam with HMM-263, 1965; HMM-262, 1966 and HMM-265, 1967.

Piatt retired in June 1971 and joined Air America in Laos from 1972 until the phony truce in 1973. He retired from the Marine Reserves in 1983 as a Major.

Piatt presently is working as a senior research technician for Environmental Research Inst. of Michigan in Ann Arbor. He helped fabricate radar and infrared scanners mostly on government contracts. He also sets up test sites and flies as a non-pilot crew member on a U.S. Navy P-3 and civilian CV-580 aircraft.

GARY E. PIDOCK, "G.E.," was born on Sept. 4, 1942 in Portsmouth, OH. He holds a BS from Ohio Northern University, 1966 and an MBA, Boston College, 1976. He joined Officer Candidate School (OCS) at Quantico, VA and was commissioned a Second

Lieutenant in February 1968. He was designated a Naval Aviator in Pensacola, FL on June 24, 1969.

While serving in-country, he was a pilot with HMM-364 from November 1969 to November 1970. He was also a flight instructor with HMHT-401 and HMT-204 and a pilot with Reserve Squadrons HML-771 and HMA-773.

He completed active duty on Dec. 15, 1972 and was discharged on May 4, 1978 with rank of Captain. He was awarded the Single Mission Air Medal with Bronze Star, 35 Air Medals, Vietnam Service Medal, Vietnam Cross of Gallantry and Presidential Unit Citation.

He has been a special agent with the U.S. Treasury Department since 1977. He belongs to the Federal Criminal Investigators Association and the USMC/Vietnam Helicopters Pilots and Aircrew Reunion. He resides in Columbus OH.

WALTER E. PINKERTON JR., "PINKY," Captain USMCR (Ret) was born on Oct. 24, 1946 in Philadelphia, PA. He graduated from high school in 1964 in Abington, PA. He holds a BS from Philadelphia College of Textiles and Science in 1968 and JD from the University of West Los Angeles in 1979. He was commissioned on Nov. 10, 1968 as a Second Lieutenant from Quantico and designated a Naval Aviator in March 1970 at Pensacola. He served with MAG-26 at New River from April 1970 to September 1970. He qualified in the AH-1G.

Pinkerton joined HML-367 at Marble Mountain, Vietnam in October 1970 and flew more than 400 combat hours.

He was awarded the Purple Heart, two Single Mission Air Medals. Holds a total of 24 Air Medals plus the CAR, NUC, MUC, Vietnam Service and Campaign Ribbons and the Vietnamese Cross of Gallantry. He completed active duty as a Captain with H&MS-24 in November 1973.

As a civilian he was a director of operations for Briles Wing & Helicopter, Santa Monica, CA and flew more than 3,000 flight hours. Subsequently he has been admitted to the State Bar of California in 1979 and has participated in a law practice specializing in civil litigation and personal injury matters, in San Diego, CA. He is married with two children and resides in Ramona, CA.

MARTIN LEE POLESKI, "MAD DOG," was born on Feb. 2, 1947 in Richmond Hts., MO. He enlisted under the 120 day delay plan and reported to San Diego for boot camp in June 1965 and took helicopter mechanic training at Memphis NAS.

He served in RVN from 1966 until 1968. He was based at Ky Ha, Marble Mountain, Dong Ha and LPH *Iwo Jima*. He was crewchief on a UH-34D a/c No. 24, HMM-361. The remaining 18 months of enlistment he served on Search and Rescue duty in Iwakuni, Japan.

He was awarded the Bronze Star, Purple Heart, 23 Air Medals and Combat Aircrew Wings. He was discharged as a Corporal at El Toro MCAS in June 1969.

Poleski was employed by Eastern Airlines in St. Louis, Charlotte, Miami and Sarasota from 1969 until 1988. He is currently an industrial electrician. He is married with two sons and active in politics. He is thankful for having the opportunity to have served under the command of Lieutenant Colonel Tweed and alongside every member of HMM-361. Resides in Sarasota, FL.

LEONARD J. PORZIO, "BROOKLYN," was born on Jan. 8, 1942 in Brooklyn, NY. He enlisted in the Marine Corps on Jan. 28, 1961 and went through boot camp at Parris Island. He achieved the rank of Lance Corporal and was stationed in Key West as an intercept operator. He was accepted and went through the MARCAD program from April 1963 through October 1964 where he received his wings and commission.

He was stationed in Tustin with HMM-361 and shipped out on the *Valley Forge* in May 1965. He arrived in Da Nang in June 1965 and flew daily medevac and re-supply in relatively light action including the first ever night helicopter insert. The war changed dramatically in August during operation "Starlight." That was the start of the escalation in both enemy and U.S. involvement in I Corps. He volunteered to stay in-country when all the experienced pilots were spread throughout the squadrons that recently arrived and went to HMM-263.

He finished his tour with HMM-263 and received two Purple Hearts for minor shrapnel wounds within eight days in December 1965. He received a Distinguished Flying Cross in May 1966 for an emergency medevac mission.

Upon returning to Tustin and training new arrivals in both H-34s and H-46s Porzio went regular for a tour in OV-10s. This required jet transition. He flew A-4s in Yuma and then returned to Camp Pendleton to fly UH-1s until he picked up the new OV-10s from Columbus, OH.

He returned to Vietnam in October 1968 with VMO-6 in Quang Tri. He went past the 1000 mission mark and received three more Distinguished Flying Crosses and a Single Mission Air Medal along with all the other "I was there" medals and ribbons.

Most of his "war stories" would probably be from the "Go-Go-Tent" at Marble Mountain with A.J. Donahue (also received a DFC as his wingman in May 1965) during 1965-66 and the "Swamp" in Quang Tri with Dave Shore (Davie died in a crash of an OV-10 stateside in 1974) and John Pierson the "TRIUMVIRATE" during 1968-69.

He went into the reserves as a Captain in 1970 and until asked to leave for flat-hatting the Colorado River in 1972; an obsession he had with 46s also.

Porzio owned and operated a furniture manufacturing company for eight years then became and still is a mortgage broker in California. He's still quite nuts and still sees several USMC cronies. He has four children (28, 26, 23 and 6, yeah 6!!!) and a lovely wife in Santa Ana, CA. He is an active member of Vietnam Combat Pilots Association and USMC/Vietnam Helicopter Pilots and Aircrew Reunion.

JACK D. PRIEST was born on July 29, 1940 in Lerna, IL. He was enlisted in the Marine Corps on Nov. 18, 1958 and went through boot camp at MCRD San Diego. He was stationed at 29 Palms California with the 1st LAAM Battalion until 1961 when he reported to Memphis for Helicopter Mechanics School. Upon completion of school, he was assigned to HMM-363 at MCAF Santa Ana, CA.

He transferred to H&MS Company in Futema, Okinawa in 1962 and was later assigned to Da Nang, South Vietnam where he served as an H-34 helicopter crew member with HMM-162 and HMM-261 until August of 1963.

Priest arrived back at MCAF Santa Ana in December of 1963 and served again with HMM-363 until he was discharged with the rank of Corporal on June 24, 1964.

He is currently a retired electrician living in Mattoon, IL. He is married with seven children and 10 grandchildren.

JESSE L. PUGH, "JESS," Major, USMC (Ret) was born on July 15, 1940 in Sheldon, IA. He graduated from high school in Cedar Rapids, IA and attended the University of Iowa. He enlisted in the Marines in January 1961 and entered flight school April 1961. In MARCAD Pugh completed primary jet and advanced multi-engine training in November 1962. Pugh's first assignment was to the 2nd Marine Aircraft Wing, Cherry Point, NC in January 1963. In April 1963, he began his helicopter career by receiving $2.58 travel pay and an invitation to helicopter transition at MCAS(H) New River, NC. Upon completion of training in HMM-263, he joined HMM-264.

As a member of HMM-264 he completed one Mediterranean and two Caribbean deployments. During the last Caribbean deployment (1965) he participated in the first night helo-borne combat operation into the Dominican Republic. In March 1966 Pugh joined HMM-161 at Marble Mountain, RVN. As part of HMM-161 (later transferred to Phu Bai) he participated in many operations including "Operation Hastings."

Upon leaving RVN he served as a flight instructor at HT-8 Pensacola, FL. In November 1968, he joined HMH-361, MAG-26 for transition into the H-

53 deploying with 361 to RVN in September 1969. He was assigned to HMH-462 which was later relocated to MCAS Futema, Okinawa.

His later assignments were with HMH-362, MAG-26; OIC MATCU-64, MAG-26; HMM-262, MAG-26; CO H&HS MCAS(H) New River, NC; Embry-Riddle Aeronautical University, Daytona Beach, FL September 1974-December 1975 with BS aeronautics-cum laude; HMH-462, MAG-16, 1st MAW; XO MABS-24, MAG-24 1st Marine Brigade; XO HMH-463.

He retired in June 1981 after flying 13 different aircraft and logging more than 5,000 flight hours. He was awarded 21 Air Medals and other standard awards.

Pugh presently is employed by Allied-Signal as a logistics engineer and has received an MBA from Northrop University. He has been married to the same wonderful wife, Jean, for 32 years. They have two children and four grandchildren.

TIMOTHY H. PYLE, Lance Corporal, USMC was born on Feb. 2, 1949 and served with distinction in HMM-262 in Vietnam. He died on May 2, 1969 at the age of 20 years. He was the son of Mrs. Rachel Pyle of Mobile, AL.

IRA E. RAMSEY, "BLONDIE," Commander, CHC, USN was born on Oct. 16, 1943 in Watervliet, MI. Following graduation in 1962 from high school in Hartford, MI, he enlisted in the Corps in September of the same year. From boot camp in San Diego, CA and Aviation Electronic training in Millington, TN, Ramsey was finally assigned duty with HMM-363 at Santa Ana, CA.

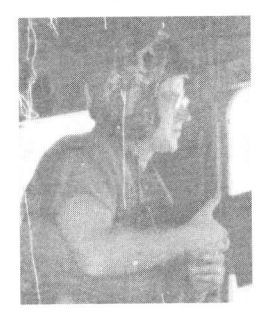

An aviation electronic tech. and helicopter machine gunner, he served with HMM-363 and HMM-362 in the Republic of South Vietnam from July 1965 to November 1966.

Sergeant Ramsey was honorably discharged at NAS Alameda, CA on Jan. 17, 1967. Besides receiving numerous Unit Awards, he has been awarded the Air Medal, the Combat Aircrew Wings with three Gold Stars, the Navy Commendation Medal, the Navy Achievement Medal with Gold Star and the Combat Action Ribbon.

Back in civilian life, he earned college degrees from Western Michigan University, Kalamazoo, Michigan and Baptist Bible College, Springfield, MO. Civilian experiences include: marketing representative with Goodyear Tire and Rubber, deputy sheriff in Kalamazoo, MI and Springfield, MO and six years pastoring Baptist churches.

In 1981 he was commissioned a Lieutenant junior grade in the USN Chaplain Corps. He served as the Staff Command Chaplain at the Navy Supply Corps School, Athens, GA.

ROBERT W. RARICK, "BUFFALO," was born on Aug. 13, 1945 in Coeur d'Alene, ID. He graduated from the University of Idaho with a business degree. He was commissioned a Second Lieutenant OCS Quantico, VA in February 1969; designated a Naval Aviator in Pensacola, FL in May 1970 and graduated as a Cobra AH-1G flight programmer in Savannah, GA in August 1970.

He joined HML-167 in Vietnam in September 1970. and flew 88 combat missions. He returned to New River, NC in July 1971. He joined HMM-263 aboard USS *Guadalcanal* (helo carrier) in the Mediterranean in January 1972 and returned to HML-167 in August 1972. He was designated a maintenance test pilot UH-1N.

Rarick completed his tour of duty in October 1973 as a Captain. He was awarded accompanying ribbons and medals for his service.

Currently living in Hayden Lake, ID, he is married to his beautiful wife with three lovely daughters and two grandchildren. He owns and manages Precision Wood Prod., Inc., Coeur d'Alene, ID.

WILLIAM T. READ, "TED," Lieutenant Colonel USMC (Ret) was born in Kilgore, TX on Sept. 21, 1933. He was commissioned on July 18, 1955, NROTC Unit, University of Texas, Austin, TX. Education: BA in sociology, University of Texas, 1955 and an MS in management, U.S. Naval Postgraduate School, 1970. Military training: the Basic School (TBS March 1955) 1955-56; flight training from 1956-57; Special Weapons Delivery School in 1958; Helicopter Training in 1959; Air Intelligence School in 1959 and Amphibious Warfare School in 1965-66.

Deployments: Lebanon Crisis 1958 on the USS *Antietam* with VMA-324; Med cruise in 1960 on the USS *Donner* with HMR(L)-262, Sub Unit-1; WESTPAC, 1961-62 with HMR(L)-261 (including three months aboard USS *Thetis Bay* in the Philippines, three months ashore in Indonesia, training Presidential Helicopter pilots and two months ashore in Thailand defending against Chinese invasion); Vietnam 1966-67 with HMM-165; and WESTPAC, 1971-72 with MAG-36 Fuji Det (Atsugi) H&MS-36 (Futema).

Major assignments: VMA-324 from 1957-59; HMR(L)-262 from 1959-60; HMR(L)-261 from 1960-62; HMX-1 from 1962-65; AWS, student from 1965-66; HMM-165 from 1966-67; HMMT-302 XO from 1967-68; MAG-36 training officer from 1968-69; USNPGS student from 1969-70; H&MS-36 from 1971-72; Marine Corps Development Center, Project officer from 1972-73; and Naval Air Systems Command CH-53E Deputy Project Manager from 1973-78.

Read retired on July 1, 1978 in Washington, D.C. His awards include Distinguished Flying Cross (two awards), Air Medal (24 awards), Presidential Unit Citation, Marine Corps Expeditionary Medal, National Defense Service Medal, Vietnam Service Medal, RVN Cross of Gallantry with Gold Star (personal award), RVN Cross of Gallantry/Palm (unit award), RVN Campaign Medal and the Presidential Service Badge.

He is currently a cattle rancher in Franklin, TX. He is married to Norma Brock, (Houston, TX on July 29, 1955) and has three children, Steve, Mike and Terri.

DOUGLAS REINIKA, (DOUG) "SUNDANCE," was born on Feb. 22, 1943. He served in-country as a pilot on CH-46D, E, F and also flew AH-1Gs.

He was retired on Sept. 30, 1986 with the rank of Lieutenant Colonel and was awarded the DFC, Purple Heart, Air Medal with 25 awards, CAR, NUC, MUC and PUC. He is married and residing in Bend, OR. He is a retired airport manager.

MICHAEL J.W. REMME Lieutenant Colonel USMCR (Ret) was born on Nov. 24, 1944 in New Orleans. He is a graduate of Ball High, Galveston, TX 1962; Texas A&M 1966; University of Texas Law School 1972 and PLC (A) summer of 1965. He was commissioned on Sept. 9, 1966 and designated a Naval Aviator on Oct. 13, 1967. He qualified PQM UH-1 at VMO-1, New River and FAC/ALO at Little Creek.

RVN service: VMO-6; May-August 1968 (Quang Tri); FAC with 1/7 1st MARDIV from August-November 1968 (Quang Nam); and HML-167 November 1968-June 1969 (Marble Mountain). Most memorable mission: emergency ladder extraction of survivors of recon team 'Paddle Boat' on Jan. 22,1969, 15 miles SW An Hoa (HAC check flight).

Remme was awarded 27 Air Medals w/one Single Mission Air Medal, two PUCs, NUC, MUC, CAR, VCG w/Palm, VS and VC Ribbons.

He was on active duty with HMH-461 on Anorek Express and Teamwork 80 (Northern Norway) from December 1979 to May 1980 and in active reserves from April 1971 to May 1986.

He has been an attorney in general civil practice since April 1973. He is married to Janis and has daughters, Karen and Anne. His hobbies include Texas history, music and soaring.

WILLIAM A. REPKO, "BILL," was born on April 7, 1947 in Bridgeport, CT. He joined the Marine Corps on Feb. 2, 1966 and attended Aviation Helicopter School in Memphis, TN.

He transferred to Vietnam from Dec. 15, 1966 to Jan. 12, 1968 and served with HMM-165 as a qualified crewchief and aerial gunner.

He transferred to New River HMM-261 and HMM-264 and was discharged as a Sergeant on Jan. 30, 1970. Repko was awarded two Purple Hearts, a Letter of Commendation, Combat Aircrew Wings, 16 Air Medals, Boeing Rescue Award, Boeing 1,000 Hrs. Award, Vietnam Service Award, Vietnam Campaign Medal, Vietnam Cross of Gallantry, Presidential Unit Citation and Navy Unit Commendation Medal.

He is single. He attended Conn. Academy of Dental Technology from 1971-1973 and is a certified dental technician and is a partner in a dental laboratory in Norwalk, CT.

JOHNNY DEAN RESTIVO was born on July 16, 1944 in Oakwood, TX, graduating from Bryan High School in 1962. He received an AA degree from Allen Military Academy in 1964; BA in mathematics from Texas A & M University in 1966; MS in aeronautical engineering from the Naval Post Graduate School (NPGS) in 1978; and is a graduate of the Naval Test Pilot School (NTPS), 1972.

He entered the Marine Corps PLC(A) Program on March 10, 1964, was commissioned a Second Lieutenant on March 27, 1967 and designated a Naval Aviator on June 2, 1968. He trained in CH-53As with HMH-461 from June-September 1968-New River.

Restivo was assigned to CH-46As with HMM-265 from October 1968-May 1969 and 3rd MARDIV as DivAirOff/2nd DARVN FAC May-October 1969 in Vietnam. He flew 345 combat missions.

He returned to CH-53As with HMH-363 and HMH-361 in October 1969-May 1971-Santa Ana. He was promoted to Captain in May 1971 and graduated from NTPS and flew test projects on more than 30 different aircraft from July 1971-September 1973-Pax River. He completed TA-4J transition and was assigned to A-6Es with VMA(AW)-224 from October 1973-July 1976 at Cherry Point. Restivo was promoted to Major in September 1977, graduated NPGS in June 1978 and assigned XO/CO of H&MS-15 from September 1978 to August 1979-Iwakuni.

He was Laser Maverick, Sidearm and Hellfire Missiles Program manager with the Air Force and Naval Air Systems Commands from August 1979-April 1987-Dayton/Washington, D.C. He was promoted to Lieutenant Colonel in October 1982.

He retired from active duty on April 30, 1987. Was awarded the Single Mission Air Medal w/Bronze Star, 18 Strike Air Medals, the Combat Action Ribbon, Meritorious Unit and Navy Unit Commendation, Vietnam Meritorious Unit Commendation with Palm, Vietnam Service with four stars and Vietnam Campaign Ribbons and the Vietnam Cross of Gallantry.

He provides technical and management consulting services to weapon systems support contractors in the Washington, DC area. He also flew B-727s for a major airline and remains current in general aviation. He is married, with two sons and one grandson and is a member of the Society of Experimental Test Pilots.

FRANCIS T. REYNOLDS, "FRANK," was born on June 1, 1939 in Malden, MA. He joined the service on Oct. 8, 1962.

While serving in-country in Vietnam he was a CH-37 pilot while stationed at Marble Mountain and Dong Ha.

Reynolds was discharged with the rank of Captain in May 1967 and was awarded 20 Air Medals.

After discharge, he flew OH-6s with the Massachusetts National Guard. He resides in Medford, MA and is married to Mary and has two daughters, Ann and Alice. He is an attorney and is involved in real estate.

EDWARD RICK III, "HULK," was born on Oct. 25, 1941 in Lancaster, PA. He joined the service in July 1964 and was stationed at New River with HMM-264 and at NAS Pensacola.

While serving in-country he was a pilot flying UH-34s for HMM-361.

He was discharged in February 1969 with the rank of Captain. He was awarded the "usual" medals.

After discharge in 1969, he returned to Lancaster, PA and joined Ready Mixed Concrete as owner and operating partner. Ready Mixed Concrete has two concrete plants and 36 Ready Mixed trucks. They produce 100,000 cubic yards of concrete per year for the construction industry in Lancaster County. His wife, Celia, owns and operates a gift shop in Bethany Beach, DE and Brickerville, PA. They have a daughter, Jennifer, and a son, Edward. His leisure time is spent on community functions and playing golf.

ALBERT R. RIDEOUT, "RAY," Lieutenant Colonel USMC (Ret) was born on Jan. 14, 1940 in Whittier, CA. He joined the PLC Reserve on Sept. 1, 1958, MARCAD on Sept. 4, 1960, commissioned and winged in 1962. He was assigned to HMM-364 (UH-34), MCAF (Helo), Tustin.

WESTPAC with 364 in 1963, including a tour in Vietnam, Operation "Shu-fly." He was on an instructional tour (T-28) NAS Whiting (1965-66); became a civilian and reservist for 18 months; went back on active duty in June 1968; and transitioned to UH-1E at Camp Pendleton and joined HML-267 at Marble Mountain. Returned CONUS transitioned to H-53A and joined HMH-363 in 1969. Attended AWS in 1970. Transitioned to fixed-wing and flew A-6s for VMA(AW)-332, MCAS Cherry Point (1971-72); 533, MCAS Iwakuni (1973-74); MATSG 90, Millington, TN (1975-78); CO, VMA(AW)-242, El Toro (1979-80); XO, MCAS Futema (1980-83); and S-1 MCAS Beaufort, where he retired in June 1984.

Rideout was awarded the Meritorious Service, DFC, Air Medal and Combat Action. He was a Lieutenant Colonel at the time of discharge.

He is a corporate pilot for Ball Corp. in Muncie, IN. He has been married 32 to years to Toni. Their first two sons are Army officers (Tanker and Blackhawk pilots), their third son is a Marine officer flight student at NAS Whiting and their daughter is a speech pathology major and cross-country All-American at Calvin College, MI.

DARYL RIERSGARD was born on Sept. 30, 1946 in Minot, ND. He graduated from Minot State University in December 1968 and went immediately into flight school. In January 1969 he was designated a Naval Aviator and went to the West Coast for fleet transition. He was also stationed at Kaneohe Bay.

He was promptly assigned to HMM-364 (Purple Foxes) at Marble Mountain, Da Nang, Vietnam. During October 1970, he was shot down at Hill 270 south, at which time he received the Purple Heart. All of his Vietnam combat time was spent flying the CH-46.

Following Vietnam he completed a 20 year career which included the following: 14 years active fleet flying; Naval War College; duty on five continents; HQMC/Pentagon tour; and Commanding Officer at Quantico MCAF.

After retiring in 1988, he spent a year as an aviation security consultant in Washington, D.C. Subsequently, he moved to Lake Tahoe and presently serves as a deputy for the Washoe County Sheriff's Office. He is happily divorced and the father of three children with custody of his son, Christian.

BILL RING, "THE CHICKEN MAN," being an early morning kind of guy, came out screaming at 0300 hours in the "Windy City" of Chicago. The desire to fly came during his early days so, after two years of college at Wright J.C., the Marine Corps MARCAD program sounded pretty good. He bypassed boot camp, to go directly to Pensacola to fly jets (what else?) in April of 1966, only to find out the Corps needed helicopter pilots more than Ring needed jets. He was designated a Naval Aviator/Second Lieutenant in June 1967. His first assignment was two weeks at NAS Pensacola hospital with mono. His next stop was Southern California and a chance to finally fly jets (jets engines that is) CH-46s.

After a quickie transition into CH-46s, he was off to Vietnam with the "Purple Foxes," HMM-364, and later HMM-165 at Phu Bai and finishing his tour aboard the USS *Tripoli*. Asked about the nickname, "The Chicken Man," "Because I was everywhere!" (I guess you had to either be in Vietnam in 1968 or Chicago listening to WLS Radio to understand.)

He went back to Southern California at the end of 1968 as the Assistant 3rd MAW Aviation Safety Officer, Monterey Safety School and then back to El Toro. This time he had a chance to fly the TA-4 for a year, plus CH-46s, CH-53s, UH-1Es and OV-10s.

Was awarded the Distinguished Flying Cross, Purple Heart, Air Medal (26) plus other assorted ribbons and medals that went along with a tour of Vietnam. In June 1970 he was off active duty and went back to school for a BS degree and two years in the reserves flying CH-46s, in Southern California.

He has held civilian flying jobs as a helicopter pilot flying on tuna boats in an oldie but goodie Bell H-1 and giving rides in a Great Lakes bi-plane. He is an investment broker specializing in "one good deal after another." Founder of the Vietnam Combat Pilot's Association in 1986, he resides in Mission Viejo, CA.

RICK RIVERS, "HAWK," was born on St. Patrick's Day in 1943. He earned his wings on March 10, 1967 at NAS Ellyson and immediately went to HMMT-301 at MCAS(H) Santa Ana and was "blessed" an H2P.

He went on individual orders to MAG-16 (FWD) at Phu Bai in July 1967. HMM-163 was "happy" to have two FNGs as replacements ... T.D. Wilson and "Hawk." They volunteered for night medevacs on the second night to get their feet wet, seems they didn't know enough to be scared. Each received HAC papers at 500 hours and proceeded to baffle the enemy and even some co-pilots and crewchiefs with their aggressive airmanship throughout Northern I Corps for the rest of their tour. Their final "soiree" at Marble Mountain was not only memorable to the squadron but brought the MPs to Da Nang just as their "freedom bird" was taxiing for takeoff. Fortunately some squadron mates restored the "damaged" hooch and

"Hawk" and T.D. arrived at Okinawa without knowledge of their near mandatory RTB to Marble. T.D. finally lost his battle to "Agent Orange" in 1985 and "Hawk" retired from the Pentagon in 1986 as a Lieutenant Colonel.

Rivers is now the equivalent to a wing NATOPS officer in his second career with Federal Express and has a son, Patrick, born on Feb. 20, 1993 who already aspires to be a Marine Aviator ... *Semper Fi!*

SCOTTY ROBERTS, "THE HEAVY EQUIPMENT OPERATOR" born on March 31, 1939 in Pensacola, FL. The son of "Stinky" Roberts, USN, TBF pilot extraordinaire. Enlisted in the USMCR in 1957. He went to Parris Island, Camp Lejeune and earned a BS at Florida Southern College. Commissioned in the USMC in May 1963, went to basic school and was designated a Naval Aviator in February 1965.

He served in Vietnam from 1965-1966 with HMM-263, 363 and 364. He was an instructor pilot VT-1, NAAS Saufley Field from 1966-1969. He resigned his regular commission and completed service with 19 Air Medals and two Letters of Reprimand.

Roberts worked for Continental Airlines from 1969-1985, then worked for Pride Air, People Express, and Air Cal. He is presently an MD-80 Captain for American Airlines, ORD. He has two sons and had two wives. He purchased a Navy T-28, Buno 138340 in September 1991. It was "almost" as expensive to maintain as a wife. Presently, he is single and residing in Laguna Beach, CA.

ROBERT W. ROBINSON, "ROBBY" born on Aug. 22, 1944 in Bismarck, ND. Graduated from North Dakota State University in 1966. He received his commission on Dec. 8, 1967 and was designated a Naval Aviator on April 8, 1969. He served in HML-367 flying UH-1E and AH-1G models at Phu Bai and Marble Mountain, RVN from August 1969 to August 1970.

Awards: Single Mission Air Medal, 57 Air Medals, Army Commendation Medal, Army Achievement Medal, Vietnam Cross of Gallantry, Vietnam Service/Campaign Ribbons and the Army Broken Wing Award.

He joined the North Dakota Army National Guard in 1974 and has been employed with them full-time since 1977. He is currently a supervisory maintenance test pilot and flies the C-12, UH-1 and OH-58. He has about 5,500 hours flight time and holds the rank of Lieutenant Colonel.

He is married with two children. He is a member of the AMVETS, American Legion, VFW, El Zagal Shriners and Masonic Lodge.

ERIC L. ROBISON, "ROBBY" born on Oct. 10, 1948 in Logan, UT. Graduated from North Central High School in Spokane, WA in 1966. Enlisted in the Marines on Dec. 22, 1966 on the 120 Day Delay Program and reported to boot camp at San Diego, CA on April 19, 1967. After AIT, he reported to Memphis, TN for Aviation Hydraulics School and was assigned to H&MS-26 at MCAF New River in January 1968.

He was assigned to HMM-364 (Purple Foxes) on Oct. 8, 1968 at Phu Bai. The squadron was transferred to Marble Mountain Air Facility on Dec. 10, 1968. He flew as a gunner and worked as a hydraulic technician on CH-46 helicopters.

Robison was awarded Combat Aircrew Wings with three stars, National Defense Service, Combat Action Ribbon, 12 Air Medals, Good Conduct Medal, Vietnam Combat Medal, Vietnamese Cross of Gallantry w/Palm and Vietnam Service Medal. He returned to the U.S. on June 15, 1970, got married and reported to HMMT-401 at MCAF New River where he was assigned duty as barracks police sergeant. He was discharged from active duty as a sergeant on April 16, 1971. He drove an over-the-road truck until 1984 when he started working in the oil fields.

Robison is a Production Operator I with Amoco Production Company in Evanston, WY. He is married (but currently separated) with nine children (seven girls and two boys) and two grandsons.

THOMAS J. ROEHRIG, "DUCKFOOT," was born on June 1, 1946 in Detroit, MI. He enlisted in the USMC on Dec. 31, 1963 and went through boot camp San Diego. He took infantry training at Camp Pendleton and was assigned to HMM-362 MCAF Santa Ana, CA. Later he was transferred to HMM-363. He participated in flood relief operations in Eureka, CA during Christmas, 1964.

Roehrig left for Vietnam in June 1965 and embarked with all of MAG-36 aboard the USS *Princeton* from Long Beach, CA. He arrived in Chu Lai, Vietnam about 10 days later. During his 15 months in Vietnam, he participated in medevac troop assaults and various missions as a door gunner with HMM-363.

Awarded Combat Aircrew Wings and five Air Medals. He received a combat promotion to Corporal and was discharged March 30, 1967 as a Sergeant.

He is a franchise agent in Florida for a national pizza chain. He is married with three children and soon to be four grandchildren.

DEAN A. ROGERS, "ROG," was born on Pearl Harbor Day, 1944 in Colorado Springs, CO. Enlisted with the Marine Corps in September 1963. He went to boot camp at MCRD San Diego and received avionics training in Memphis, compliments of the Navy, before joining MAMS-37 at El Toro Marine Base, CA.

He joined HMM-165 after a two-week crash course on CH-46 avionics systems and arrived in Vietnam in September 1966. He served with HMM-265 and HMM-164 before returning to the real world.

"Among my close calls and misadventures in Vietnam, I would count emergency night extractions as the most frightening. I was fortunate enough to have been involved in only a handful of these operations. One especially predominates in my memory of those missions. The night was unbelievable black and it seemed like it took forever to locate the Army Special Forces Team. With no LZ in sight it seemed like another eternity to find a hill we could back into. Among the routine questions we asked; 'When was your last contact with Charlie?' After a short pause the response came back, 'Just a minute, I'll let you talk to him.' We hovered, waiting for the Green Berets to reach the ramp, and listened as the small arms fire got closer and closer. At the moment when the bullets started coming through the chopper the team ran on board and to my complete and utter amazement kicked, beat and shot out the chopper porthole window to return fire. We lifted off, hosing down the entire area with lead and returned to base without incident. The crewchief and I spent the following day locating and replacing the broken windows, asserting that the Army should pick up their own, even though their chopper pilots, with only six months of training, were unqualified to fly on instruments."

He was awarded Combat Aircrew Wings, a Purple Heart, Air Medals, Presidential Unit Commendation with Bronze Star, Vietnamese Cross of Gallantry and the Vietnam Civil Action Citation. He was discharged as a Sergeant in late August 1967.

Rogers earned a bachelor's degree in electronic engineering technology on the GI Bill and is an electrical contractor in the Chicago area, living happily with his wife and stepson, three cats and a dog.

TOM ROGERS, "ROGERS NO. 1," was born on June 30, 1948 at Everett, MA and graduated from Everett High School in 1967. He enlisted in the Marine Corps in November 1967 and attended boot camp at Parris Island (Plt. 2079). He reported to Memphis and New River before joining HML-167 at Marble Mountain during March 1969.

His first three missions resulted in crashes or hard landings (two engine failures and a "loss of turns" taking off from Hill 55 with crewchief Melkivitch who grabbed him on the ground and kept him from walking into a rotor blade. For a short while, flight crews suspected he was carrying bad luck and some avoided him like the plague.

He was the crewchief who flew under the telephone wires in Da Nang with a pilot recently featured in a Marine helicopter periodical. He won't reveal names as he was sworn to secrecy. The chase bird chickened out.

"My favorite story involves crewchief Russ Stakes. I arrived at Marble six months prior to Stakes so when he arrived I was an old salt. I remember sending him on wild goose chases all over Marble. Finally he smartened up and told me to take a long walk off a short pier. That was just before Christmas 1969. A short time later, a radio disc jockey from Cleveland (Stakes' home town) arrived on the flight line with a tape recorder. He was recording Christmas messages from Marines who came from the Cleveland area which were to be played on the radio Christmas week. I immediately ran back to the hootch area and woke Stakes up (he just finished a night medevac shift). I told him about the disc jockey and he told me to get lost-he wasn't falling for any more of my BS. Christmas came and went then came a letter from his Mom asking why he wasn't on the radio with the rest of the Marines from Marble Mountain. Stakes still talks to me but he still calls me bad names."

He was discharged on Sept. 3, 1971 with the rank of Corporal and is presently a police Sergeant in Acton, MA.

ERLING O. ROLFSON JR. Col USAR, was born on Nov. 2, 1945 in Bismarck, ND. He graduated from Concordia College, Moorhead, MN in 1967 with a BA. He entered the USMCR with a PLC(A) commission. He received his wings in October 1968 in Pensacola and was assigned to MCAS New River in CH-46s.

He spent four months in the Caribbean at Vieques and flew off the USS *Boxer*. He was in Vietnam in June 1967 at Marble Mountain with HMM-263.

Rolfson returned to MCAS New River and was released in July 1971. He was awarded 38 Air Medals and one DFC.

He returned to New Rockford, ND and entered the insurance business. He joined the North Dakota Army NG in November 1971 and flew UH-1s and OH-6As until 1980. He was transferred to the USAR in 1984 and was a battalion commander of 439 Engineer Battalion for three years. He was promoted to Colonel in August 1988 and is presently assigned to 5555 USAR Det. He served on AD in Honduras from December 1988 to May 1989 with TF-164.

Married to Dee and has a son 25, a daughter, 24 and owns a Cardinal 177 and has more than 3,000 hours of flight time.

GEORGE A. ROSS, Lieutenant Colonel USMC (Ret), born May 4, 1939 in Newark, NJ and joined the USMC in February, 1963, commissioned a Second Lieutenant in May, 1963, assigned as Platoon CMDR 6th Marines, November, 1963-May, 1965. He was designated Naval Aviator June 14, 1966, served with HMM-162 and HMM-161, New River, NC, joined HMM-163 RVN in June, 1967 to July, 1968 flying the UH-34D, HMM-164 flying the CH-46 A/D with 1,000+ combat missions, 52 Air Medals with Star, Purple Heart, CAR, PUC, MUC, RVN COG, RVN CA, RVN MUC. He was the jet instructor at NAS Meridian; flew OV-10As with VMO-2, and VMO-6, and with H&MS-24. He was CO of HMA-369 at Camp Pendleton flying the AH-1J.

He retired in February, 1984. He flew as a production test pilot and chief pilot for McDonnell Douglas Helicopter Co., 1984-91, and now flies for Southwest Airlines. He married Nan Seligman Ross, a great lady from Pensacola, FL, and they have three children.

RICHARD LEE RUSSELL, "RUSS," was born on Jan. 20, 1934 in Geneva, NE. He joined the service on Dec. 10, 1952. He was stationed in the States in MCAF Santa Ana, CA; Quantico, VA and New River Air Station, Jacksonville, NC. He was also stationed with HMMT-407 and MAG-40.

He served in-country in Korea on HRS-3s, then on UH-34s and CH-53s in Vietnam from 1965 to 1966 and 1968 to 1969. He was involved in Operation Hastings 1965-66, Tet Offensive in 1968 and the resupply of Khe Sanh in 1968-69. He is also the original builder and designer of Bearing Staking Kit used on CH-53 Sea Stallions. He was stationed at Ascom City, Korea, Oppama, Japan and served in Okinawa, the Philippines, Chu Lai and Ky Ha, Vietnam.

He was discharged on June 30, 1972 with the rank of Gunnery Sergeant. He was awarded the Bronze Star with combat "V" and seven Air Medals with Aircrew Wings.

Today he is retired from ITT Cannon. He had been with them for 14 and one half years since 1989. He worked in several casinos in Laughlin, NV. He has been married for 38 years and has four children and eight grandchildren. He resides in Bullhead City, AZ.

ERNEST P. SACHS, known to fellow Marines as "Gunny," grew up in Vermont and enlisted in the Marine Corps MARCAD program on July 14, 1964 after two years of college. He trained in Pensacola and was designated as a Naval Aviator at Ellyson Field on March 29, 1966, receiving his wings from his initial flight instructor, then Captain Gregory Aloysius McAdams. He served as a squadron pilot with the Ugly Angels of HMM-362 from July 1966 to October 1967.

He received numerous Air Medals, unit awards, participation decorations and nothing particularly heroic. Following his return to CONUS, he instructed in H-34s in HT 8 until his release from active duty.

In late 1969, Sachs became disillusioned with the conduct of the Vietnam War. In the most painful decision of his life, he joined with other combat vets to form Vietnam Veterans Against the War, protesting the prolonged conduct of war without definition.

He has since received an AB from Harvard and a JD from Vermont Law School. One of the most fascinating puzzles of mankind is that such a hurtful pursuit as war nevertheless brings out the finest qualities of mankind—cooperation, self-sacrifice, devotion to a cause, common effort and discipline. "There is nobody I admire more than the folks I served with in the Marine Corps."

JAN SALLAZ, "SAL," born Sept. 29, 1945 in Phoenix, AZ and enlisted in the USMC on Aug. 26, 1963. His military locations include Parris Island, Camp Lejeune, Cherry Point, NC, Ky Ha and Marble Mountain, Vietnam. He was assigned to HMM-361 and never comprehended why an engineer equipment (mechanic-diesel) was needed in a UH-34 helicopter squadron, but he was sure it beat an Engineer Battalion. His memorable experiences include sneaking his favorite "34" pilot (Louse) into the NCO club at Da Nang and "every night medevac he ever flew" ... his 21st birthday at Cam-Duc, the afternoon he saw that bullets don't bounce off 34s.

His awards include: National Defense Medal, Vietnam Service and Campaign Medals, Marine Corps Air Crew Wings, seven Air Medals, Good Conduct, PUC, NUC, Vietnam Civil Action Award. He was discharged in August, 1967 at the rank of Sergeant.

He resides in Warren, OH and is married with two children and two dogs.

MARTIN E. SALTER "GENE" was born on Dec. 13, 1927 in Warren, AR. He enlisted as a Private in the USMC in June 1945 and was commissioned Second Lieutenant in June 1953.

While serving in-country he flew UH-34s while stationed at Marble Mountain with HMM-363's Red Lions.

He retired in August 1969 with the rank of Major, USMC. He was awarded the Marine Corps Good Conduct Medal, the Silver Star and the Purple Heart.

Today he is retired and living in Saraland, AL with his wife, Dot. They have three sons: Gene Salter, Corporal USMC, Richard Salter, Sergeant USAF and Maj. David D. Salter, USMC (RET). *(Salter was the pilot of the UH-34D shown on the cover of this history).*

WALTER L. SANDERS, "BUDDY," Col USMCR (Ret) joined the Marine Corps Reserve in April 1954 in Norfolk, VA. He was commissioned a Second Lieutenant in Quantico 1962 and designated a Naval Aviator in New Iberia, LA on Oct. 25, 1963. He went through helicopter transition in 1964 at Santa Ana, CA and flew with HMM-361 in the Republic of Vietnam from 1965-66.

He was released from active duty as a Captain on Oct. 15, 1966. He was awarded Meritorious Service Medal (2), Distinguished Flying Cross, Purple Heart, Air Medals and Republic of Vietnam Campaign Medals.

Sanders continued to fly with HMM-765 (H-34) and HML-765 (UH-1) in Atlanta. He transitioned to the "Cobra" with HMA-773 and served as commanding officer from 1977-1980. He transferred to 4th FSSG until 1982 when he was selected as the first Reserve Commanding Officer of MASD, Andrews AFB, 1982-1984. He became the Operations Officer with the FMFLANT RAU and served as the RAU Commanding Officer from 1987-1989. He was Officer in Charge, Quantico MOB Station from 1990-1992.

Sanders retired on July 4, 1992 after more than 38 years. He is now a retired Eastern Airlines Captain practicing law in Georgia.

PETER G. SAWCZYN, "PETE," was born on July 19, 1945 in Clifton, NJ. He graduated from Montclair State in 1967. He was commissioned a Second Lieutenant at Quantico in May 1968 and was assigned to Army Helicopter School, Ft. Wolters, TX and Hunter Army Airfield, Savannah, GA. He earned

93

Army Aviator Wings in March 1969 and Naval Aviator Wings at New River, NC in June 1969.

He qualified in the CH-46 and was assigned to Prov. MAG-39, Quang Tri in August 1969 with HMM-262. His squadron relocated to Phu Bai and finally to Marble Mountain with MAG-16. He left Vietnam in July 1969 as a division leader with approximately 500 combat hours.

Sawczyn was reassigned to New River with HMM-261 as avionics officer and MABS-26 as operations officer. He was released from active duty December 1972 as a Captain. He was awarded 42 Air Medals, NUC, Vietnamese Cross of Gallantry, Navy Commendation with "V" and various Campaign Ribbons. After his release from active duty he was employed by IBM in Dayton, OH. Currently, he lives in Knoxville, TN and works in sales and marketing. Has two children, Peter and Deborah.

MARTIN J. SCANLON, "MARTY," was born on July 13, 1942 in Jersey City, NJ. He enlisted in the Marine Corps on March 15, 1960 and went through boot camp at Parris Island. After boot camp he went to Camp Geiger for infantry training and then reported to Memphis for training as an aviation electronics technician. His next assignment was HMR(L)-262 at MCAF New River.

He participated in the Cuban Missile Crisis aboard the USS *Thetis Bay* (LPH-6). He later served in Southeast Asia in 1963 with HMM-262 at Da Nang, 1968-69 with VMO-6 at Quang Tri and 1972-73 with VMGR-152 Nam Phong, Thailand He later earned Naval Aircrew Wings an H-46 crewchief, with HMM-162 at MCAS New River.

Scanlon retired in March 1980. He earned Combat Aircrew Wings and six Air Medals plus the CAR, two NUCs, two MUCs, Vietnam Service and Campaign Ribbons, the Vietnam Cross of Gallantry and five Good Conduct Medals.

He holds an AS in mid-management from State Technical Institute at Memphis and a BS in professional aeronautics from Embry-Riddle Aeronautical University as well as an Airframe and Powerplant license.

Scanlon and his wife, Marguerite, son, M.J. and daughter, Marguerite reside in Memphis, TN. Currently, he is a manager of Service Planning and Systems Support at Federal Express Corporation.

DENNIS E. SCHREINER, "DENNY," Master Sergeant, USMC (Ret) was born on June 19, 1936 in Unadilla, NE. He worked in the oil fields until joining the Corps in September 1956. He went to boot camp at San Diego, 1st 75 AAA Gun Battalion 29 Palms. In November 1958 he had embassy duty in Nicosia, Cyprus. He retrained in helicopters and in March 1962 he completed ADH "A" School at Memphis. He joined HMM-363 at MCAF Santa Ana. In November 1963 he completed ADJ "B" School and in November 1964 he was with HMM-164.

In March 1966 he arrived at Marble Mountain, RVN with CH-46As and in January 1967 he was with HMM-265. From 1967-71 he was an instructor with NAMTRADET Santa Ana. In 1971-1972 he was in WESTPAC with HMM-165 and MARTD Whidbey Island in October 1972.

Schreiner retired in September 1978. His decorations include: Combat Aircrew Wings, nine Air Medals, Navy Commendation with "V," CAR, PUC, NUC, seven Good Conducts, ND, Vietnam Service, Philippine PUC, Gallantry Cross and Civic Action Unit Awards and Vietnam Campaign.

He has two daughters and lives in Oak Harbor, WA with his wife of 32 years. He works as a helicopter mechanic for USN/SAR. He fishes and is active in the FRA and VFW.

BRUCE A. SCHWANDA, Colonel USMC (Ret) was commissioned in 1965 after graduation from The Citadel. He earned his wings in October 1966 and was deployed to Phu Bai, RVN in December with VMO-3, flying UH-1Es. After being medevac'd in September 1967, he transitioned to A-4s, served with VMA-214 at El Toro and returned to RVN, where he was assigned to VMA-311. He served subsequent tours as an instructor with VAT-203; MOI at The Citadel; as a maintenance officer, executive officer and operations officer with HML-268; and HOC officer aboard USS *Sampan* LA-2.

After attending the Marine Corps Command & Staff College 1979-1980, he served on the college staff from 1980-1982. He commanded HML(A)-369 from Nov. 22, 1982 until Feb. 15, 1985, after which time he was assigned as executive officer, MAG-39. Promoted to Colonel in July 1986, he was reassigned to HQMC, where he served as head, Team CIG Division. In 1988 he was reassigned to an accompanied tour on Okinawa, where he served as 1st MAW G-1 and Commanding Officer MCAS Futema.

He retired in 1993 from his duty station, NTSC, Orlando, FL where he is currently a consultant and in the financial services industry. Col Schwanda accumulated 4,500 flight hours including 842 in Vietnam. He has been awarded the Legion of Merit, two Distinguished Service Medals, two Distinguished Flying Crosses, the Purple Heart and 41 Strike/Flight Air Medals.

He is married to the former Anne Swicegood of Larchmont, NY. They reside in Orlando, FL with their two daughters, Hillary, 21, a senior at the University of Central Florida and Shannon, 19, a sophomore at Florida State University.

H. RAE SCOTT, "SCOTTY," Captain USMC was born on June 1, 1938 in Rockford, IL. He holds a BS degree from Iowa State University (1960) and a master's degree from Memphis State (1978). He was commissioned in May 1960 and served as an infantry officer in the 5th and 9th Marines. He was designated a Naval Aviator on March 31, 1964, and was assigned to HMH-462 at MCAF Santa Ana. He went to the Sikorsky factory with the first group of Marine pilots to fly the CH-53. He went to Vietnam with HMH-463 from 1967-68.

Scott was awarded the usual many Air medals, Navy Achievement Medal, PUC, MUC, NUC, CAR, Armed Forces Expeditionary (Cuba), U.S. and Vietnamese Campaign and two Vietnamese Crosses of Gallantry. He completed active duty in July 1970.

He joined the federal law enforcement community as an inspector with the IRS. He was promoted to the senior executive service as an assistant inspector general at the U.S. Dept. of Transportation. He retired from federal service in January 1994 after 21 years of civilian service.

WILLIAM F. SCOTT, "FOX," Col USMC was born on Nov. 18, 1945 in Norfolk, VA. He holds a BA from Transylvania College, Lexington, KY and an MBA from National University. He was commissioned in February 1967 and after flight school joined HMT-301 and 363 before going to Vietnam in January 1970. After a short stint in HMH-463 joined HML-167 (seems as they were out of pilots).

Passing though VMO-1, he joined HMA-269 as a plank owner. After holding each job in maintenance and a Med cruise with HMM-162 and the STD group tour as the S-4, he was then off to Quantico for AWS and TBS instructor duty (he thought he had avoided it). In 1978, a tour with the Army at CCAD allowed him to become qualified in all Bell helicopter series aircraft. Next, a two year tour as the operations officer of the *Peleliu* (LHA-5) occupied his time. Enough salt water, he joined HMA-169 as the AMO then on to HMT-301 as the XO. No longer being able to avoid the Washington tour, he went to NAVAIR as the AH/UH class desk. In an attempt to learn the system, he attended ICAF. He escaped to the fleet to be the CO of MALS-26 and XO of MAG-26. Well all good deals come to an end so he went back to DC and a tour in the Pentagon as the keeper of the POM for aviation in CNO.

His special moment was seeing the happiness and glee in the eyes of a seven man recon team (Elkskin) that they had pulled out of a SH_T Sandwich (surrounded by 750 VC) when we got back to the recon pad.

He accumulated 1,000 hours of flight time and 42 Air Medals.

Currently he is assigned as the Amphibious Warfare Branch Head on the N88 staff of CNO, the non-flying tour has caused him to return to his other love, Harleys.

JAMES L. SHANAHAN, "JIM," Colonel USMC (Ret) was born on Sept. 24, 1930 in Vicksburg, MS. He graduated from high school in Vicksburg in 1948, received a BA from the University of Oklahoma in 1952, and an MA from Auburn in 1971. He was commissioned a Second Lieutenant in the USMC on June 2, 1952. He was designated a Naval Aviator on Oct. 14, 1955; a helo pilot on March 22, 1957; went to USMC Basic School 15th SBC in 1952; USMC Command and Staff College in 1965; and USAF Air War College in 1971.

Other duty assignments included: Korea as Rifle Platoon leader A/1/7 from March-December 1953; and TAO VMO-6 from January-April 1954. After flight training he was assigned to VMF-122, HMH-461 and HMH-162. Vietnam tours: XO, MABS-15 from June-September 1966; XO, HMM-263 from October 1966-June 1967; ACofS G-3 9th Marine Amphibious Brigade during Easter Invasion from February to November 1972.

After Vietnam tours: director, Strategy Applications Branch, J-5 OJCS; Commanding Officer MAG-26; and Commanding Officer MCAS New River, NC.

Shanahan retired on Feb. 28, 1979. His decorations include: RVN Cross of Gallantry, NCMs, MSM, AMs, DFC and LOMs.

After retirement he represented USPA&IRA, a financial planning firm with concern focused exclusively on financial independence for military families for eight years.

He is married to Anne and they have three children and three grandchildren. His favorite military flying was night medevacs. He now tools around VFR in a 1947 C-140 out of Northwest Regional Airport near Roanoke, TX.

WALTER H. SHAVER, "EASY WALT" was born Jan. 27, 1931 in Great Falls, MT. He was commissioned on June 5, 1953 at the University of Idaho as a 2nd Lt. in the NROTC.

He served in RVN from November 1967 to November 1968 as CO HMM-362 with the UH-34D. He was awarded the following awards: Legion of Merit, Combat "V", DFC, Single Mission Air Medal, 50 Air Medals, Vietnam Cross of Gallantry.

He retired from the military in August 1974. He is married to Thelma and is self-employed as a consultant in computers and accounting.

BERLE JOHN SIGMAN III, was born Oct. 24, 1948 in Charlotte, NC. He was graduated from Belmont Senior High School in 1966 and attended N.C. State University.

Enlisted in USMC on Feb. 7, 1968. Graduated from MCRD Parris Island, SC. Promoted to Corporal upon completion of Basic Helicopter School at Memphis in February 1969. Transferred to 1st MAW, RVN July 1969. Served with HMM-362, H&MS-16 and HMH-463 as crewchief/gunner until July 1970. He was awarded Combat Aircrew Wings, six Air Medals, VN Gallantry Cross, Civic Action Colors, VN Service Medal and Campaign Medal. Other duty stations include MARTD Dallas, TX; and HMX-1, Quantico, VA. More than 2000 hours as crewchief in UH-34D, CH-46D, UH-1E, CH-53A/D, RH-53D and VH-3D.

Personal awards include Navy-Marine Corps Medal, Defense Meritorious Service Medal, Navy Achievement Medal, Marine Corps Expeditionary Medal, and Presidential Service Badge.

Promoted to Master Gunnery Sergeant in September 1990. Currently DAPML for Presidential Helicopters at NAVAIRSYSCOM. Married, three children. Hobby is stockcar racing.

GARY F. SIMONS was born on Sept. 2, 1939 in Reading, PA. He earned a BS in mechanical engineering at Yale University and was commissioned in the Marine Corps on June 6, 1962 through the NROTC program. At first he yearned to be in artillery, but applied for aviation while at Basic School in Quantico and arrived at Pensacola in January 1963.

He flew H-34s and served with HMM-264 at MCAS New River. He participated in NATO exercise Steelpike in 1964 and a Med. cruise in 1965. In 1966 he flew with HMM-363 at Ky Ha, RVN and aboard LPH-5 and LPH-2 in the area of Dong Ha and Quang Tri. He also served at 9th MAB HQ at Camp Hansen.

Simons was released from active duty as a Captain in June 1967 and flew with HMM-772 at NAS Willow Grove until 1970. In civilian life he worked with IBM Corp. as a systems engineer from 1968 through 1992. He is now a consultant in software development. He is married and lives in Marietta, GA.

ARTHUR MARC SLAGLE was born on Dec. 13, 1948 in Dayton, OH. Graduating from De Witt Clinton High School, Bronx, NY, he enlisted in the Marine Corps in July 1966. Following Parris Island and Camp Lejeune, Slagle reported to NAS Memphis where he joined his twin brother Louis already going through aviation training. In June 1967 he went to Marine Corps Air Facility, Tustin, CA to join HMM-364 "Purple Foxes" a CH-46D Squadron preparing to leave for Vietnam.

From November 1967 to December 1968, Slagle served with HMM-364 and HMM-164 at Phu Bai and Marble Mountain. As an aerial door gunner, he flew in support of the Marines in the battles at Hue and Khe Sanh. In December 1968 he joined VMF(AT)-101 a newly formed F-4 Phantom Squadron at El Toro. In July 1970 he transferred to the VMFA-531 "Grey Ghosts."

Discharged in September 1970, Slagle attended Walla Walla College in Washington where he earned his BA degree. In 1979 he earned an MA degree and was commissioned an ensign in the Naval Reserve. From 1979-1984 Slagle served with the Marines in Garden City, NY and South Bend, IN.

Slagle is a highly decorated Marine/Chaplain with 26 decorations including the Purple Heart, Air Medals and Combat Aircrew Wings.

Following post-graduate studies in Michigan, Slagle accepted a superceding appointment into the Chaplain Corps. He reported to San Diego in October 1984 as the chaplain onboard the guided missile cruiser USS Halsey CG-23. Following this Slagle served as battalion chaplain for 3D BN 7th Marines and 1st Combat Engineer Battalion. In December 1989 he reported to Bethseda Naval Hospital as a Staff Chaplain. In July 1992, he attended post-graduate school at The Catholic University of America in Washington, D.C. He is currently assigned as the Group Chaplain Marine Air Control Group-38, MCAS El Toro.

Slagle is married with two sons and a daughter. He is a chaplain for the 1st MAW Vietnam Service Assoc. and the USMC/Vietnam Helicopter Pilots and Aircrew Reunion.

LOUIS EDWARD SLAGLE was born on Dec. 13, 1948 in Dayton, OH. Graduating from Aviation High School, NY with an airframe and powerplant license and private pilots license MEL, he enlisted in the U.S. Marine Corps in May 1966. Following boot camp at Parris Island, SC and ITR at Camp Lejeune, NC, Slagle reported to NAS Memphis for aviation training.

He joined VMA-324 in Beaufort, SC and participated in a Med. Cruise. In September 1968, he departed CONUS for Okinawa and Vietnam. Serving as crewchief with H&MS-11, HMM-164 and HMM-364, he participated in operations from Da Nang, Dong Ha, An Hoa, Phu Bai, Marble Mountain and Quang Tri. Returning to CONUS in early 1970, Slagle served as admin. chief with VMO-2 El Toro, CA.

Slagle received his BS degree in 1971 from Northrop Institute of Technology, Inglewood, CA. In 1972 he returned to Beaufort, SC serving tours with VMFA-312 and VMFA-451, F-4B Phantoms.

Slagle's decorations include the N&MC Medal, Purple Heart (two), Air Medals, Combat Action Ribbon, Vietnamese Cross of Gallantry and 18 other awards, Combat Aircrew Wings and Gold N&MC Jump Wings.

Upon retirement Slagle joined the NYC Transit Police Dept. Having served three years undercover, he was promoted to detective and was medically retired after 10 years of service.

He has three children, son, Lance Corporal David Slagle (21) stationed at 2d FSSG, Camp Lejeune, NC; daughter, Diana (13) and daughter Michelle (12). Louis serves as the South Regional Coordinator board of directors, 1st MAW-Vietnam Service and a member of USMC/Vietnam Helicopters Pilots and Aircrew Reunion. Gunnery Sergeant Slagle's twin brother is Lieutenant Commander Arthur M. Slagle, chaplain for Marine Air Control Group-3 MCAS El Toro.

WALTER W. SMITH was born on May 21, 1941 in Mt. Hermon, LA. He attended LSU in Baton Rouge, LA and joined the service on June 6, 1963. He was a Lieutenant in the Marine Corps Reserves.

While in-country he served with the HMM-362 and was involved in Double Eagle and Double Eagle II.

He was discharged on June 1, 1993 with the rank of Colonel. He was awarded the DFC, Purple Heart, 24 Air Medals, MUC, NUC-OMCR, Vietnam Campaign and Vietnam Service medals.

Today he is married to Dianne Wine and they have three children. He was an administrator with the state police and has retired to Baton Rouge, LA.

WILLIAM J. SMITH, "BILL," Colonel USMC (Ret) was born on May 17, 1937 in San Francisco, CA and commissioned a Second Lieutenant on June 3, 1959 upon graduation from the University of San Francisco. He was designated a Naval Aviator on Jan. 13, 1961, he was originally an F-8 pilot with VMF-312 and later a KC-130 pilot with VMGR-152.

He then transitioned to CH-46s, and was with HMM-164 in-country in 1969. Subsequently, he commanded HMM-463 in 1973 and HMM-462 in 1976-1977.

A 1981 graduate of the Air War College, he was the CO, MACG-48 from 1983-86. He retired on Sept. 30, 1986. His decorations include the Legion of Merit, two Meritorious Service Medals and 20 Air Medals.

Currently, he is a DHC-8 Captain for Mesaba, a Northwest Airlink, and domiciled in Detroit, MI. He and his wife, Yumiko, reside in Ypsilanti, MI. Their daughter, Jennifer, is a senior at the University of Michigan.

MARC THOMA SOHM was born on April 18, 1947 in Oshkosh, WI. He joined the Marines in March of 1966.

He served two tours of duty in Vietnam with HMM-362, "Ugly Angels." Flying night medevac missions was the most meaningful experience of his life, then and now.

As a helicopter gunner, he earned his Combat Aircrew Wings and two Air Medals. He was discharged in California in 1970 and bounced around the aircraft industry as a structural mechanic and radiographer.

After modifying aircraft in Edmonton, Canada, for the Shah of Iran, he went on to work as a quality control inspector at the Tom's River Nuclear Plant in New Jersey. He returned to California to try college; drive taxis; tend bar in a biker bar in Pacific Beach; and sell seafood. Eventually, he returned to Wisconsin to become a youth counselor, then a state and federal correctional officer. He is happily married with three children.

RONALD E. SORENSEN, was born on Jan. 27, 1944 in Athens, GA. He graduated from high school in 1962 in Lincoln, NE and the University of Nebraska in January 1967. He was commissioned on June 2, 1967 as a Second Lieutenant from Quantico. He was designated a Naval Aviator on June 27, 1968 in Pensacola, he served with HML-561 at MCAS Santa Ana from July 1968 to November 1968. He qualified in UH-34s.

He joined HMM-363 in Vietnam in December 1968, was assigned to HMM-362 in February 1969 and joined HML-367 in August 1969.

Sorensen was awarded three Single Mission Air Medals and a total of 37 Air Medals plus the CAR, MUC, Vietnam Service Medal and Campaign Ribbons and the Vietnamese Cross of Gallantry. He completed active duty as a Captain with HML-267 at Camp Pendleton in September 1971.

He served with the Nebraska National Guard as UH-1 pilot from 1972 to 1975. Sorensen has served as the director of the Nebraska Department of Labor and as a consultant with the Gallup organization. He is currently the deputy director for Operations of the Nebraska Department of Social Services.

He is married to Margie Rine and has two daughters, Taylor and Hayley. He is a member of the board of directors of the Nebraska Veteran's Leadership Program and is a member of the American Legion.

TERRENCE J. SPEICHINGER, "SPIKE," was born on Oct. 17, 1946 in East St. Louis, IL. He joined the service on Nov. 8, 1965 and was stationed in the States at New River, NC and Santa Ana, CA.

While in-country he was a crewchief and gunner on UH-34Ds and was stationed at Marble Mountain, Vietnam.

He was discharged in November 1971 with the rank of Sergeant and was awarded 22 Air Medals and Combat Aircrew Wings.

From December 1969-1975 worked with Missouri Young Marines in St. Louis, and has been an auto and heavy truck mechanic until hired on at McDonnell Douglas in St. Louis. Worked on many older aircraft with friends at local airports. He recently did most of the rewiring of a 1942 PBY-5A at Spirit of St. Louis Airport with a group of friends.

Speichinger is married and has four children and three grandchildren. He resides in St. Peters, MO.

DAVID L. SPENCER, "DAVE," was born Nov. 22, 1948 in Richwood, WV. He enlisted in the Marine Corps on May 19, 1967, went to boot camp at Parris Island, ITR at Camp Geiger and basic aviation maintenance training at NAS Memphis. He was next assigned to HMH-461 at MCAF New River.

He was transferred to HMH-463 in Vietnam where he worked in the Hydraulic Shop and flew a gunner.

While serving with HMH-463 he received a meritorious combat promotion to Corporal and Sergeant and was awarded the Distinguished Flying Cross, Purple Heart, Air Medal, Navy Commendation Medal with "V" GCN, CAR, Combat Aircrew Wings, Vietnamese Cross of Gallantry and other service ribbons. He was discharged on June 26, 1971.

Spencer joined the WVARNG on April 25, 1972 and was commissioned in the Corps of Engineers. He is still serving with the WVARNG and holds the rank of Lieutenant Colonel.

In civilian life he is self-employed and is married with one daughter.

GARY D. SPEVACEK, "SPEVER," born March 6, 1948 in Iowa City, IA and enlisted in the USMC in February, 1966, with recruit training in California. He was sent to Camp Lejeune for Truck Mechanics School. He was sent to helo school in July, 1967, choosing to go to New River and discovering his mistake too late.

He volunteered for WESTPAC arriving in Vietnam in November, 1967, assigned to HMM-265. His most memorable experiences were the flying personnel of 265, still the best friends he ever had, and landing in the mountains fully loaded, having the pilot pull too much collective, falling like a rock through the trees, saying, "Not much power, crewchief," his knees shaking too much to answer. His awards include 32 Air Medals, Combat Air Crew Wings, two Navy Unit Citations, and Bronze Star with "V." He was honorably discharged in June, 1969. He is married to Patti and has four children.

EARLY W. SPIARS, "EARLY," Lieutenant Colonel USMC (Ret), born Dec. 24, 1934 in Greenville, MS. He graduated from high school at Rolling Fork, MS in 1952 and has an MBA from National University in San Diego, CA. He was commissioned a Second Lieutenant and designated a Naval Aviator in June, 1956 at Corpus Christi, TX. He served with VMA-211 and as a flight instructor at Whiting Field. Transitioned to helicopters in 1959 and served with HMM-361 and 362 at Santa Ana and HMM-362 in WESTPAC and in Vietnam (Operation Shu Fly) in 1962. Served with HMX-1, VMO-5 (Pendleton) and VMO-2 at Marble Mtn. in 1968 and 1969. He was awarded two single mission Air Medals and 14 Air Medals. After serving as XO and CO of Det. MAG-16, he retired from I MAF at Camp Pendleton in May, 1980.

He is currently the manager of support services for Texas Municipal Power Agency, near Bryan, TX. He is married to the former Jocelyn Rohner and has six sons, one daughter, and nine grandchildren.

DAVID M. SPOTTSWOOD, "SPOTTY," Major, USMCR was born on Sept. 20, 1944 in Tampa, FL. He graduated from Boca Ciega High School in St. Petersburg, FL in 1962. He received a BA from Florida Southern College in 1967. He was commissioned in October 1968 as a Second Lieutenant from Quantico, went through Army Flight School in Fort Wolters, TX and Savannah, GA, and was designated a Naval Aviator in September 1969 while stationed with HMH-461 in MCAS New River.

He joined HMH-463 in Vietnam in June 1970, returned to MCAS New River in March 1971 and was assigned to HMH-362. He began jet transitioning in March 1973 at NAS Meridian, MS. He was selected for the Harrier program in 1974 and reported to VMA-231, MCAS Cherry Point. He left the Marine Corps in July 1977 and became a pilot for Delta Air Lines in Atlanta. In 1983 he joined the Reserve Cobra Squadron at NAS Atlanta.

Spottswood died on March 11, 1984 in the crash of a Cobra helicopter while on reserve duty. He is survived by his widow, Nancy Spottswood Vice and three daughters.

GORDON W. SQUIRES Major USMCR CWO4 USAR (Ret) was born on Jan. 21, 1939 in Detroit, MI. He attended Wayne State University and joined the Marine Corps in May 1962 as a MARCAD. He completed basic jet training, was commissioned and received his helo wings in April 1964. He was assigned to HMM-363, 362, 164 in Santa Ana. He went to Comm. O School, Quantico in August 1965.

Assignments included: Comm. O MACS-6, pilot HMM-161, Okinawa from May 1966 to December 1966; pilot HMM-163, Phu Bai; Camp Commandant, Air Base Commander, S-4 MAG-16 FWD (Dong Ha); Khe Sanh from December 1966 to May 1967.

He was released in May 1967 and joined the USMC Reserve in January 1968. He flew the SH-34J and the OV-10A. He was awarded an Army Commendation, Air Medals, etc. He spent two tours as ALO 1/24 Marines. He joined MI ARNG in August 1977 as CWO-2 and flew UH-1H, UH-1C, UH-1M with 1/238th CAU and 338 ATTK BN. He was aviation safety officer with the 46th AVN BN (STARC) flying OH-58A.

Squires earned a BS, MS, AEP (FAA) from Western Michigan University, Kalamazoo, MI. He taught four and a half years in AEP Program at WMU. Currently he is an instrument procedures specialist/flight inspection co-pilot, FAA, Battle Creek FIFO, MI. He is married with one daughter. He retired from MIARNG in May 31, 1991 with 29 years service.

LYN STAUTZENBACH, "STATZ," was born on Feb. 12, 1946 in Toledo, OH. He joined the Marines in June 1965 and attended boot camp at San Diego and aviation technical schools in NAS Memphis, TN.

He served two tours of duty in Vietnam. He was crewchief on UH-34D YN-2 for HMM-361, Tweed's Tigers. He was discharged in June 1969 with the rank of Corporal.

Today he is an aircraft mechanic and resides in Toledo, OH.

DONALD L. STIEGMAN Colonel, USMC (Ret) enlisted in the Marine Corps on March 12, 1954, took boot camp at MCRD Parris Island and served tours with 3rd Tank BN Japan, 2nd MAW Cherry Point and 3rd RECON BN MCAS Kaneohe.

In 1960 he was commissioned a Second Lieutenant. He was designated a Naval Aviator in August 1962 and served with the 1st, 2nd, 3rd and 4th Marine Aircraft Wings and saw action in Vietnam with HMM-363 and VMGR-152.

Independent duty included: OIC Marine KC-130 support to the Blue Angels and on the staff of COMSIXTHFLT in the Mediterranean. Command tours included Commanding Officer VMGR-352, 3rd MAW 1979-81 and Commanding Officer MWSG-47, 4th MAW 1982-85.

Colonel Stiegman's awards include the Legion of Merit, Distinguished Flying Cross, 31 Air Medals, Marine Good Conduct Medal and numerous campaign and unit commendations.

Colonel Stiegman left active duty on Feb. 1, 1989 and resides in Orlando, FL with his wife, Carolyn. They have two children, Christy and Matthew.

JIM STIGER, "STIGS," born May 8, 1945 in Seattle, WA and joined the USMC in June, 1967 at Oregon State University. He was assigned as a HAC, Maintenance Test Pilot. His military locations include Quantico, Pensacola, Santa Ana, Futema, Atsugi, Marble Mountain, and Kaneohe.

His most memorable experience includes being shot at during the night while on a test flight with no co-pilot or guns on board. He also has fond memories of his hoochmates. He remembers the smells: the showers at Marble Mountain, the medevac bunker, the "head," the Ready Room, the cockpit, the DDT sprayer at the "O" club. He returned to Kaneohe, HI, where he enjoyed some of the best flying. He was honorably discharged from USMC active duty in July, 1974 with the rank of Captain, USMCR in September, 1983, with the rank of Major, and the USAR in May, 1993 with the rank of CWO-3. He currently resides in Redmond, WA and investigates aircraft accidents.

WILLIAM W. STILWAGEN, "DINGER" Corporal, was born in Oceanside, NY. Enlisted in U.S. Marine Corps and attended the Parris Island boot camp.

Arrived in Vietnam in August 1969. Assigned to 12th Marines, 3rd Marine Division as field radio operator along the DMZ. Served at Rockpile, Cam Lo, and Dong Ha. When 3rd MarDiv was redeployed to Okinawa, he volunteered to stay in-country and was transferred to the 1st Marine Air Wing at Phu Bai, and subsequently to Marble Mountain where he served as helicopter door gunner with "Purple Fox" squadron HMM-364. Wounded May 4, 1970 while medevacing ROKs (Korean Marines) engaged in firefight with the Viet Cong southeast of An Hoa.

Decorations: Air Medal, Bronze Star, Air Medal with numeral 10, Purple Heart, Combat Action Ribbon, National Defense Service Medal, Vietnam Service Medal with four stars, RVN Cross of Gallantry, RVN Meritorious Unit Citation, RVN Campaign, Combat Aircrew Wings, Rifle Expert, and NYS Conspicuous Service Cross.

Has two beautiful children: Malia Jade and Braden Joshua. Owns Quality Marine Service & Commerical Boat Works in New York. A founder of the Vietnam Veterans of Suffolk, Inc., a combat veterans organization dedicated to helping children in need.

FRED D. STITH was born on Oct. 16, 1946 in Jefferson City, MO. He enlisted in the Marine Corps on Sept. 13, 1964 and went through boot camp in San Diego, CA. After boot camp and infantry training, he attended Hydraulic Mechanic School at Memphis, TN. He was assigned to VMF-334 in El Toro, CA and after three months was transferred to HMM-164 at LTA Santa Ana, CA.

He went to Vietnam with HMM-164, aboard USS *Princeton*. HMM-164 was the first CH-46A squadron to go to Vietnam. He served with HMM-164 from February 1966 to March 1967.

He was assigned to HMH-461, a CH-53A squadron in New River, NC until discharge on Sept. 13, 1968 as a Corporal. He was awarded the standard Air Medals, citations and awards.

He went to Airframes and Power Plants School in Kansas City, MO and received his A and P License in 1970. From there he attended civil engineers tech school until 1972. From 1972 to 1990 ran a mobile home sales lot in Lake Ozark, MO. He is retired and is in the rental and mobile home park business, and raises registered Yorkie and Maltese pups as a hobby.

LARRY D. STOLLAR, Lance Corporal, USMC, was born on Jan. 28, 1947 to Mrs. Mary E. Perry in Marietta, OH. He served with distinction with H&MS-36 in Vietnam, where he died on May 2, 1969 at the age of 22.

DONALD B. STONEKING was born on April 28, 1943 in Keokuk, IA and graduated from high school in 1961 at Warsaw, IL. He holds a BA from Iowa Wesleyan College, 1966. He was commissioned in March 1967 as a Second Lieutenant at Quantico and designated a Naval Aviator in August 1968 at

Pensacola. He qualified in the CH-46D in HMMT-302 at MCAS Santa Ana in December 1968.

He joined HMM-263 in January 1969 at Da Nang and completed his Vietnam tour with HMM-265 at Phu Bai and afloat. He served two months with HMM-165 as part of 1st MAW (Rear) both in Okinawa and afloat.

Stateside, he served with HMM-163 as QA officer and Post Maintenance Inspection Pilot on the CH-46F.

His active duty was completed with HMH-361 as CH-53A HAC in November 1971. He was awarded the Distinguished Flying Cross, 26 Air Medals, NUC, MUC, CAR, National Defense Service Medal, Vietnam Service and Campaign Medals and RVN Cross of Gallantry.

In 1972 Stoneking served short-term reserve duty in East and West Coast USMC Reserve Squadrons. From 1986-1989 he served as A/C on UH-1M with the National Guard. His civilian aviation career includes: international aviation in 1975 in Indonesia as a BH-212 aviator doing jungle "heli-rig"; offshore flying in BH-212 from Abu Dhabi and Dubai during 1976 and 1977; North Sea offshore flying in 1978 and 1979 in SA-330J from the Shetland Islands; offshore flying from Houma, LA in 1979 until 1981; flying SK-76A and BH-212 and serving as deputy base manager; 1981-1984 additional international offshore flying in Egypt as chief pilot and flying BH-205A, BH-206 and BH-212; 1985-1989 domestic corporate aviator flying SK-76A for Merrill Lynch and Squibb; and 1989 until present with Weatherstone Air, Inc. flying SK-76B. He has been an active aviator since 1967 with 10,000 hours of flight time.

Stoneking has been married 32 years to the former Joyce Hempen and they have one daughter, Renee, who is in college at East Stroudsburg University.

RICHARD STRICKLAND "DICK" was born on March 31, 1943 in Phoenix, AZ. He graduated from high school in 1961 in Calexico, CA. He attended New Mexico Military Institute in Roswell, NM and graduated with a BS from San Diego State in 1966. He joined the MARCAD program in 1964 and was commissioned a Second Lieutenant upon graduation on June 19, 1966. He was designated a Naval Aviator on Oct. 27, 1967, Pensacola and served with HMH-462 at Tustin, CA from December 1967 to February 1968.

He qualified in the H-53 and joined HMH-463 in Vietnam in March 1968. He served 3rd BN 1st Marines as forward air controller from July 1968 to November 1968 and then returned to HMH-463 to complete his tour. In May 1969 he was stationed at MCAS New River and served in HMH-461 and HMH-365 aboard the LPH *Guam*. While aboard the *Guam*, he flew humanitarian missions into the Peruvian Andes after the 1970 earthquake that destroyed towns and villages.

Strickland completed active duty as Captain in August 1970.

He began law school at California Western University, San Diego, CA in 1970, graduated and was admitted to California Bar in 1974. He has practiced law in El Centro, CA since that time.

MARION F. STURKEY, "STURK," was born on Nov. 9, 1941 in Plum Branch, SC. He enlisted in the USMC on April 3, 1961. He received the Leatherneck Magazine Marksmanship Award at Parris Island on June 29, 1961. As a Corporal in 1963, he was selected for the MARCAD program. He graduated from Pre-Flight Class 43-63 as a Cadet Regimental Commander on March 19, 1964, was commissioned as Second Lieutenant on Nov. 1, 1964 and was designated a Naval Aviator No. T-9219 on March 2, 1965.

He was assigned to HMM-265 the first USMC H-46 squadron. He was a helicopter pilot in Vietnam in 1966-1967.

Sturkey's awards included: the Purple Heart, a Single Mission Air Medal, Vietnamese Cross of Gallantry, PUC and VSM. He set the world endurance record (9.1 hours) for the CH-46D/Vertol-107 aircraft at MCAF New River on March 1, 1968. He was discharged as a Captain on March 2, 1968 after seven years of service.

He flew helicopters commercially in Louisiana and Texas in 1968-1969. He returned to native South Carolina in 1970 and was employed as a corporate security manager by a national communications company. In 1993, he completed a 372 page manuscript, *Bonnie-Sue: A Marine Corps Helicopter Squadron in Vietnam*. He lives in Plum Branch, SC with his adult daughter, Christie Sturkey.

THOMAS J. SULLIVAN, born March 22, 1947 in Nyack, NY and joined the USMC in September, 1965 designated as "7562 Pilot" and was assigned to HMM-161 and HMM-364. His most memorable experiences were Boot Camp, Flight School and Vietnam which were a gas! His awards include the Silver Star, three DFCs, Purple Heart, 49 Air Medals. He was honorably discharged on June 23, 1970 with the rank of First Lieutenant.

He and his wife, Nanci and dog, "Phred," reside in Alameda, CA. He is a friend of "Frenchy."

STEPHEN A. SWAIM, "SWAMI," Captain USMC was born on Oct. 4, 1946 in Aurora, IL; grew up in Denver, CO; graduated from Aurora Central High School in 1964; and attended Western State College, Gunnison, CO, where he graduated in 1969 with a BA in Psyche-Soc. He "swore in" in April 1969 and flew F/W charter and aerial photo out of Gunnison until reporting to OCS Quantico in September. He was commissioned a Second Lieutenant in December 1969 and began Army Rotary Wing Flight School in January 1970 at Ft. Wolters, TX. He was awarded Army "pewter" wings in August 1970 and reported to HMH-301 Santa Ana, CA to learn the CH-53C and received naval aviator wings January 1971.

He spent his Vietnam tour with HMM-165 aboard the USS *Tripoli* (LPH-10) 1971-72 and returned to Santa Ana to HMH-361 until separation in June 1974.

Swaim was discharged in June 1974 with the rank of Captain. He was awarded all Vietnam Service Awards and Air Medals.

He returned to Colorado to work on F/W Inst. Multi-Eng., ATP, F/E Turbo Jet, Lear-type, etc. He lived in Georgetown, CO, teaching skiing in the winter and crop-dusting in the summer. In April 1977 he joined the Aurora Fire Department and presently works as a firefighter/paramedic.

With no USMCR aviation in the Denver area and previous Army flight training, he joined the Colorado Army National Guard in 1979, flying Huey-M model and Cobra gunships. In February 1991 he joined a Rapid Deployment Air Assault Unit at the High Altitude Training Site, Eagle, CO and is currently flying H-model Hueys.

RICHARD H. SWOSZOWSKI, "SKI," born May 4, 1931 in Baltimore, MD, joined the U.S. Army in April, 1950-53 and the USMC January, 1954-71 being assigned avionics/gunner in the UH-34D and CH-53A. He served at Parris Island, SC; VMF (AW)-542 MAG-15 MCAS El Toro, CA; VMF-324 MAG-31 MCAS Miami, FL; VMA-212 MAG-13 MCAS Oahu, HI, Iwakuni, Japan, NAS Cubi Point, P.I., USS Coral Sea CVA-*19*, USS *Michel* (TAP-114); HMM-265 MAG-26 MCAF New River, NC (UH-34D) USS *Boxer* LPH-3 Vieques, P.R.; USS *Okinawa*; USS *Raleigh LPD-1* (MED Cruise); H&MS MAG-16, 1st MAW Okinawa and Sub Unit-1 MABS-16 MCAS Da Nang, RVN; HMH-361, MAG-36, 1st MAW Da Nang, Tam Ky, RVN. He retired with full ceremony on January 31, 1971 with the rank of Staff Sergeant.

He resides in Orlando, FL, and is married with one daughter, one deceased son and two stepchildren and five grandchildren. He has worked for the U.S. Postal Service in Boston, MA, and Orlando, FL for 22 years.

JOSEPH A. SYSLO JR., Chief Warrant Officer-3 TXARNG was born on Aug. 30, 1946 in Washington, D.C. and graduated from high school in 1964 in Philadelphia, PA. He enlisted in Baltimore and went through boot camp at Parris Island in 1964. He applied for the MARCAD program and was selected in 1965. He was commissioned as a Second Lieutenant and designated a Naval Aviator in October 1967 from Pensacola.

After a short tour with HMM-264 at New River, he joined HMM-362 in Vietnam. He flew more than 1,300 combat hours in the UH-34D.

Syslo was awarded all the same "atta-boys" as everybody else. Upon return to CONUS, he transitioned to the A-4 and served with VMA-214. He left active duty in 1967 and attended ERAU in Daytona Beach. After graduation in 1974, he served in the Rhodesian Air Force as a helicopter pilot until 1977 and joined the TXARNG in 1987, where he is currently serving as a UH-60L aviator and unit aviation safety officer.

As a civilian, he worked as a commercial helicopter pilot around the world until joining AMSI as an aviation consultant in 1981. In 1984 he joined Bell Helicopter Textron as an accident investigator representing Bell on major domestic and international military and civil accidents. He is currently married with two sons still at home. He is a member of the USMC/

Vietnam Helicopter Pilots & Aircrew Reunion, VHPA, American Legion, NGUS and NGAT.

ROBERT M. TALENT, "4-STORY," was born on July 3, 1943 in Astoria, OR. He joined the Marine Corps PLC program in March 1963. He graduated from Washington State University and was commissioned a Second Lieutenant in June 1965. He was designated a Naval Aviator in March 1967 and flew with HMM-163 "Ridgerunners" from August 1967 to September 1968.

He was awarded 41 Air Medals and a Purple Heart.

He was a flight instructor, VT-2 in Pensacola from September 1968 to March 1970 and flew H-46s and UH-1Ns in reserve squadron HMM-770 from 1972 to 1981. He participated in the MOB and MTU program until 1992.

Talent retired as Colonel with 26 plus years and 3500 flight hours. He works for the Governmental Department of Public Works, coaches high school football and participates in an occasional triathalon.

CLAUDE E. TAVERNIER JR., "TAV," was born on Oct. 23, 1947 in Trona, CA. He graduated from Burroughs High School in Ridgecrest, CA in June 1965.

He enlisted in the Corps on July 20, 1965 and was assigned to H&MS-36, MAG-36, Ky Ha, Vietnam on June 6, 1966. As a gunner he flew in UH-34s with HMM-361 "Tweed's Tigers," and HMM-364, among others and UH-1Es with VMO-6. He was transferred to HMM-165 metal shop in March 1967 and flew as a gunner in CH-46s. He rotated on July 7, 1967 to MCAF, Santa Ana. He is an expert with 38s, 45s and M-14s, as well as M-60s and 50 caliber machine guns. He qualified as a crewchief on the CH-46A, D, and F.

His awards include: Combat Aircrew Wings, AM's, MUC, NUC, PUC, CAR, Vietnam Service and Campaign Ribbons, Vietnamese Cross of Gallantry, Sea Service Deployment, National Defense, Letters of Appreciation, Meritorious Masts and GMCs. Tavernier retired on Nov. 8, 1985.

He was a member of HMM-161, (1978), the first complete CH-46 squadron to deploy from MAG-16 to MAG-36 for unit rotation. He served tours in 1st, 2nd and 3rd MAWs. He has lost some dear friends, as have others, but also made some for life.

After retirement, he obtained an AA degree in business administration and a BS in health care administration.

A.J. TAYLOR, "BUD," enlisted in the Marine Corps on July 21, 1967 and went through boot camp at Parris Island. After boot camp, he went to M CO., 3rd BN, 8th Marines as a shooter in the M-16 rifle test. He reported to Memphis in February 1968 for training as a hydraulic mechanic. From there he went to HML-567 at Camp Pendleton and trained in OV-10s and UH-1Es.

After California, he transferred in October 1968 to HMM-362 "Ugly Angels" in Vietnam and served as a mechanic and door gunner. In August 1968 he transferred to VMO-6 at Quang Tri to finish his tour and earn the Navy Achievement Medal with Combat V.

Staying on active duty, he retired in August 1991 after 24 years of active duty as a Master Gunnery Sergeant. He earned Marine Corps Combat Aircrew Wings, 11 Air Medals and other campaign and service medals.

Presently, he is working as a consultant and an A&P mechanic.

LARRY S. TAYLOR, "KIM-CHI," Major General, USMCR, born March 28, 1941 in New York, NY, commissioned a USMC Second Lieutenant upon graduation from Georgia Tech in 1962 being assigned to flight school at NAS Pensacola, FL, designated a Naval Aviator. He served with HMM-264 and 263 in 1964-65 as a squadron pilot. He was transferred to the 3rd Bn., 8th Marines as Forward Air Controller/Liaison Officer in 1965, completed active duty in December, 1966 as the CO of the Air/Naval Gunfire Platoon, 2nd Marine Division. He served in Laos and Vietnam as a pilot for Air America flying H-34s in 1967-68.

In August, 1968 he served in the 4th Marine Division as CO of Brigade Air/Naval Gunfire Platoon, 3rd ANGLICO, West Palm Beach, FL. In May, 1984, he served as Assistant Chief of Staff for Readiness/Safety, 4th Marine Aircraft Wing. From November, 1984-October, 1986, he served as Assistant Chief of Staff G-2 (Intelligence), 4th Marine Aircraft Wing and since 1986 as G-3 (Operations), FMF, Atlantic, until becoming Chief of Staff, 2nd Marine Expeditionary Brigade in 1987. He was promoted to Brigadier General in 1990 with duty as Assistant Wing Commander, 4th Marine Aircraft Wing and in 1992 he became the Commanding General, 2nd Marine Expeditionary Brigade at Camp Lejeune, NC. He advanced to Major General on May 1, 1993 and assumed his current mobilization assignment on July 2, 1993. He has 19,000 hours of flight time, resides in Atlanta, GA and is an Airbus A-320 Captain for Northwest Airlines.

TOM THURBER, "THUMPER," was born in Klamath Falls, OR on Dec. 13, 1942. He joined the service in October 1965.

While serving in-country, he was a HAC on UH-34s and served with the HMM-362. He was stationed at Ky Ha, Phu Bai and was aboard the USS *Iwo Jima* and the USS *Princeton*.

He was awarded 35 Air Medals, NDF, VSM, VCM, PUC, CAR, NUC and MUC. He was discharged on July 8, 1975 with the rank of Major.

Today he is a Captain for Alaska Airlines-SEA. He is married with two sons, ages 19 and 24, and resides in Kent, WA.

THOMAS K. TIERNEY, "TOM," was born on May 15, 1941 in Brooklyn, NY and grew up on Long Island. He went to Chaminade High School and after graduation from Boston College and commissioning (PLC) as a Second Lieutenant in June 1963, he reported to Basic School at Quantico. He went on to Pensacola in January 1964 receiving his wings in June 1965.

He went to MCAF New River until March 1966 then to Ky Ha, BLT, Ky Ha with 363 and 362, and CONUS in April 1967. He was a driver for H-34s. He was involved with the Kaman "K" Mission of Mercy.

He was awarded Air Medals, NUC, MUC, Vietnam Service and Campaign Ribbons. He was OIC of SAR MCAS Beaufort, SC until discharged in June 1968 with the rank of Captain.

Tierney has been in the financial services industry since then in Florida. He sees Navy and MC jets daily from there in JAX, still a welcome sight. He reunites with seven or eight flight school and VN buddies every two or three years in New Hampshire, and has three fellow "Dog drivers" there to continue the lying! Currently, he is empty nesting-four grown children with wife, Kathy, whom he met during flight school.

ROBERT E. TRIGALET, FIRST LIEUTENANT, USMCR was born on June 9, 1944. He served with distinction in Vietnam with HMM-262 and died on May 2, 1969. He was the husband of Marcia L. Trigalet and the father of Brian Andrew Trigalet. He was 24 years of age.

MICHAEL W. TRIPP, born on March 25, 1947 in Fall River, MA. Graduated from LaSalle Academy, Providence, RI in June 1964 and the Plus School of Business in March 1971. Also graduated from Bryant College, North Smithfield, RI in May 1975 and the Rhode Island State Board of Accountancy. Enlisted in the USMC on April 28, 1965 at Hartford, CT. Attended boot camp at Parris Island, ITR at Camp Lejeune and aviation training at MAD, NATTC, Memphis.

He served as a designated crewchief/door gunner on UH-34D helicopters for 21 months in RVN with HMM-363.

"On May 14, 1966 I was crewchief on lead

medevac bird (YZ-77) out of Dong Ha, responding to 1-9 Delta approximately 10 kilometers south of Con Thien. On the way out of the zone, with two medevacs aboard, the tail rotor drive shaft was sliced by a .50 caliber round. Luckily, we crashed back into the zone. The entire aircrew and medevacs survived the crash. We then spent the next day and half waiting for a relief column to reinforce the company. After a rather "interesting" night, I got caught on top of the bird in the middle of a fire fight while in the process of removing two of the rotor blades in order to clear the landing zone for another attempt at a medevac. When the relief column reached us, we pulled back to a church which was just outside the village. We spent the next night in the church seeking shelter from the mortar attacks of the NVA (the mortars were so close you could hear the rounds leaving the tubes). The next morning, medevac birds from HMM-363 were able to land in the rice paddies just east of the church. After assuring that all WIAs were loaded and after a rather heated "discussion" with the FAO (who was now the senior officer of 1-9 and wanted us to walk back to Con Thien with the remnants of the company) the aircrew boarded on the last of the medevac birds for the ride back to Dong Ha.

At some point during our stay in the church, a combat photographer (Frank Johnston) took a picture of me in my borrowed tin-pot next to a statue of an angel and a crucifix. The picture won an award for one of the top-ten pictures of the year depicting the war in Vietnam. After a number of years, I was able to obtain a print from UPI which I have on the wall in my office as a constant reminder of my days as a "grunt." (UPI Wire Photo-SGP1556338 (FILE) UNIPIX.)

His awards include: Combat Aircrew Wings with three service stars, 21 Strike/Flight Air Medals, Navy Commendation Medal with combat "V", Purple Heart, Presidential Unit Citation, Navy Unit Citation, Good Conduct Medal, Vietnam Service Medal with three stars, Vietnam Campaign Medal with device and National Defense Medal. He was separated from active duty on March 1, 1969 at NAS South Weymouth, MA as a Corporal.

He is a member of the Rhode Island Society of Certified Public Accountants, American Institute of Certified Public Accountants, A New Leaf, board of directors (A "sheltered business" under the auspices of the State of Rhode Island, Department of MHRN), Bryant College Alumni Association, LaSalle Academy Alumni Association, Appalachian Mountain Club, Barrington Yacht Club, Narragansett Bay Yachting Association, U.S. Sailing Association, treasurer of Manny Moniz Memorial Hockey League, Republican Town Committee; past chairman of LaSalle Academy Annual Fund Phonathon, past treasurer of Barrington Yacht Club, Bryant College Alumni Association and Hampden Meadows Parents Teachers Association.

He was been a finance manager trainee for ITT Aetna Finance Co., computer programmer for B.A. Ballou & Co., Inc., administrative assistant for Manasett Corporation, professional staff for Peat, Marwick, Mitchell & Co., CPA's, a partner in Turosz, Maccarone, Keenan & Tripp, CPA's, firm administrator for Licht & Semonoff, Attorneys-at-Law and is now self-employed in East Providence, RI.

He has been married since May 24, 1966 to Ella Middlebrooks and they have two daughters, Sandra, 27 and Wendy, 24.

GORDON E. TUBESING, "TUBE," Major, USMC (Ret), was born in Seattle, WA on Feb. 23, 1942. He graduated from high school in Kent, WA in 1960 and holds a BA in economics from Central Washington, University, a BS in aviation mgmt. from Southern Illinois University and an MBA from National University, San Diego, CA. He was commissioned a Second Lieutenant at Quantico, VA on May 27, 1966 and designated a Naval Aviator on Sept. 1, 1967.

Served in Vietnam as a CH-46A pilot with HMM-265 from January 1968 to March 1969; and served with HMM-261 at New River from April 1969 to March 1971 before transitioning to the OV-10 at VMO-1 from March 1971 to June 1973. Participated in Operation Frequent Wind, the evacuation of Saigon as a member of the original HMM-165 (January 1975-May-1975) and as a member of the Marine Corps C9B Squadron at Cherry Point NC from November 1975 to July 1979. He flew more than 900 combat hours.

He completed his career as the director of Safety and Standardization for Marine Air Bases-West El Toro, CA March 1983-April 1986, and left the Marine Corps with more than 7000 hours of flying time. He was awarded 36 Air Medals.

"Tube" is currently a Captain with American Airlines flying the Boeing-727 on int'l. routes out of Miami, FL. He resides in Sequim, WA and has a daughter in San Diego, CA and a son in Boulder, CO.

GEORGE TWARDZIK, was born on Dec. 31, 1938 in Norway, MI. He joined the service in April 1958 and served until June 1961 and re-upped in May 1965 and served until May 1969. He was inducted at Milwaukee, WI.

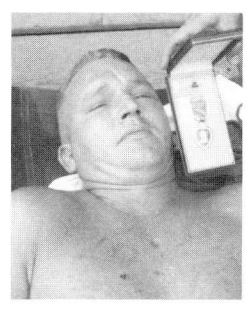

He was a gunner and crewchief on a UH-34s and CH-53s. He was stationed at MCAF New River, NC, MCAF Okinawa, Iwakuni, Japan, Search and Rescue, Phu Bai and Da Nang, Vietnam. He received the Combat Aircrew Medal, 10 Air Medals, Navy Commendation Medal with Combat V, Purple Heart, PUC, Good Conduct Medal with one star, NDS Medal, Vietnam Service Medal with three stars, and the Vietnam Campaign Medal with device. He was discharged in June 1961 and May 1969 with the rank of Sergeant. He is on 100 percent V.A. disability.

He is married and has two children and two grandchildren.

MCDONALD D.S. TWEED, Colonel USMC (Ret), was born on Sept. 27, 1920 near Marshall, NC. He enlisted in the Naval Aviation Cadet Program on July 4, 1942. He went through Civilian Pilot Training at Chapel Hill, NC and Hendersonville, NC. Pre-Flight training at the University of North Carolina and "E" Base at Bunker Hill, IN. He received his "Navy Wings of Gold" and was commissioned a Second Lieutenant on Feb. 29, 1944.

He received operational training was as an attack bomber pilot in B-25s at MCAS Edenton, NC and MCAS Cherry Point, NC. He was assigned a B-25 Aircraft with a crew of five Marine airmen, and flew the aircraft from Cherry Point, NC to Emirau, located in the Solomon Islands in the South Pacific where he joined Marine Bombing Sdn.-443. While based in Emirau, he made bombing attacks on the Japanese strong-holds located in and around Rabaul. In August 1945 along with Sdn. 443, he was transferred to Malabang, Mindanao, in the Philippine Islands where special training for low level bombing attacks on Japan were rehearsed.

Tweed returned to the States and was placed on inactive duty in the Marine Corps Reserve. He went to work for the Standard Oil company of New Jersey in Charlotte, NC, however he remained in the Inactive Reserve and applied for regular commission status in the regular Marine Corps. In January 1947, Tweed received a regular commission and returned to active duty for Amphibious Warfare School at Quantico, VA. After graduation he was transferred to MCAS Cherry Point where he joined VMR-252 and became Transport Plane Commander. He was transferred to VMR-352 at MCAS, El Toro, CA where he qualified as an R5D Plane Commander. Tweed went to Japan in May 1951 where he flew combat supplies to Korea with VMR-152's forward echelon. For flying emergency ammunition and medical supplies to a Marine Regiment on the main line of resistance in North Korea he was awarded the Distinguished Flying Cross. After returning to the States, Tweed was assigned to VMF(N)542 where he was personnel officer. He was adjutant and trained as a night fighter pilot. After serving two years in VMF(N)542, he was transferred to the Naval Air Training Command at Pensacola, FL where he served as a flight instructor, taught aerology and engineering to aviation cadets and was later assigned officer in charge of the Naval Aviation Cadets Regiment. Tweed went through helicopter training at Pensacola in the summer of 1957 and was transferred to Oppama, Japan where he served as Commanding Officer of H&MS-16. In September 1958, he returned to the States and became Operations Officer of HMR(M)-461 at New River, NC where he participated in the American Space Shuttle Missions for "Project Mercury" at Cape Canaveral, FL as officer in charge of the Helicopter Rescue Detachment to rescue the astronauts in case of an emergency abort immediately after lift-off. In 1962, Tweed graduated from the University of Maryland followed by a three year tour at FMFLANT where he was assistant G-3 plans officer and prepared plans for Marine Amphibious landings in General War and the Cuban Missile Crisis. He graduated from the Armed Forces Staff College at Norfolk, VA in February 1965, and was sent to duty overseas in Vietnam for his third combat tour. In Vietnam he served as Commanding Officer of Marine Air Base Sdn.-16 where his unit completed the Helicopter Landing Facility at Ky Ha, Republic of Vietnam. Tweed then served as Commanding Officer of Marine Helicopter Sdn. 361, a combat squadron known as "Tweed's Tigers" where under his leadership the squadron established a Marine combat record high for a 24 aircraft squadron flying 10,774 combat hours, 31,959 combat sorties while maintaining an average aircraft availability of 18 aircraft in commission (AC ready for flight) per day during six months of sustained combat operations.

Tweed returned to the States in the spring of 1967 and served as Assist Chief of Staff G-1 FMFLANT until August 1969. He then graduated from the Naval War College at Newport, RI in 1970 and was selected to be Commandant of the Naval ROTC Unit at Vanderbilt University in Nashville.

Tweed retired from the Marine Corps in July 1974 and accepted the position as chief of police and director of security and safety department at Vanderbilt University. He now resides in Tennessee with his wife, the former Mary Mullis of Charlotte, NC. They have two children.

Tweed is a graduate of Mars Hill College and obtained a BS degree from the University of Maryland and an MS degree in international affairs from George Washington University.

During his 32 years as a Marine Aviator, Tweed flew more than 7,000 "accident-free flight hours and flew more than 750 combat missions in three wars.

Besides campaign and commendation medals his awards include the Legion of Merit, with combat "V," four Distinguished Flying Crosses, 27 Air Medals, the Navy Meritorious Service Medal, the Vietnamese Cross of Gallantry with Silver Star, the Presidential Unit Citation and the Helicopter Rescue Award.

JOHN H. UPTHEGROVE, "U-GROVE," was born on Feb. 26, 1943 in Detroit, MI. He graduated from high school in 1961 at Port Austin, MI. Holds a BS degree from Central Michigan University 1967. He was commissioned as a Second Lieutenant at MCB Quantico in March 1968 and designated an Army Navy Aviator in November 1968. He was designated a Navy Aviator on Feb. 18, 1969 while assigned to HML-267 at Camp Pendleton and transitioned into the UH-1E. He joined HML-367 at Phu Bai, Vietnam in April 1969 and transferred to HML-167 Marble Mountain in December 1969. He flew 900 plus combat hours.

Upthegrove was awarded two Distinguished Flying Cross Medals, three Single Mission Air Medals and holds 39 Air Medals. He transferred to New River Air Station, served with H&MS-26, HMA-269, and completed active duty with MAG-29. He was discharged in March 1974 with the rank of Captain.

He was hired by Bell Helicopter International in April 1974 for work in Isfahan, Iran. He held positions as a maintenance test pilot, flight instructor, flight commander and branch chief. He departed Iran on Feb. 20, 1979 at the personal invitation of Ayatollah Khomeini.

Upthegrove was hired by Port Austin Level and Tool in April 1979 to present and has been promoted to VP sales. He is an active member of American Legion Post 499 and is married with two sons.

F. WILLIAM VALENTINO, "STRAWBERRY STUDHORSE" was born on Nov. 1, 1945 in Massapequa, NY. He received a BA from Long Island University in 1968 and an MPA from the University of West Florida in 1975. He was commissioned in August 1968, completed Basic School in December 1969 and Army Flight Training in September 1970. He transitioned to CH-46s at Santa Ana and went to WESTPAC in May 1970. He served briefly with HMM-165 on Okinawa, then split a tour in Vietnam flying 500 combat hours with HMM-263 and serving as ALO/FAC with First Recon Bn.

He was awarded the Bronze Star with "V," two Single Mission Air Medals, 15 Air Medals, Navy Achievement Medal with "V," two NUCs, MUC, Vietnamese Cross of Gallantry and Vietnamese Civil Action Award. He completed active duty with the rank of Captain, Standardization Instructor Pilot, HT-8 Pensacola 1973.

He married Kay Seligman from Pensacola. He joined the National Guard and was commander of an attack helicopter squadron.

As a civilian he is the president of the New York State Energy Authority and serves on a number of university and governmental boards. He is also the president of the National Association of State Energy Research Organizations. He lives in Delmar, NY with his wife Kay, son Matthew and daughter Jane.

ROCCO VALLUZZI, "ROCK," was born on Sept. 7, 1938 in Brooklyn, NY. He graduated from Saint Francis College with a BS in chemistry. He was commissioned on Sept. 21, 1960 as a Second Lieutenant and designated a Naval Aviator on Oct. 26, 1962. He served with HMM-263 at MCAF New River from October 1962 to May 1964. He was a flight instructor with VT-2/VT-3 from 1964-66.

He joined HMM-263 (UH-34) in Vietnam, participated in the following operations: Prairie, Colorado, Mississippi, Mcon, Lincoln, Cleveland, Tuscaloosa, Independence and Stone; he was part of the Special Landing Force, participating in Beaver Cage, Beau Charger and Bear Bite from May 1966-June 1967; joined HMH-461 (CH-53) New River and was promoted to Major in 1967-68; joined HMH-463 (CH-53) in Vietnam: Golden Forest, Pickens Forest, Victory Dragon, Tailwind, Dewey Canyon II and Lam Son 719; joined 1st MarDiv, MCB Camp Pendleton, Div. Air Officer joined VMO-2(OV-10) MAG-16, MCB Camp Pendleton; joined the USS *New Orleans* (LPH-11) as the HDC officer promoted Lieutenant Colonel in August 1977; CO H&MS-39 in May 1980, CO VMO-2 in January 1981; promoted to Colonel in July 1982; CDO CINCPAC, Hawaii; joined MCAS Cherry Point, (T-39) Base Inspector, Personal Services Officer from May 1984 to May 1986; and joined MCB S.D. Butler, Okinawa, Japan as the A/C of Staff, Manpower from June 1986 to June 1988.

Valluzzi retired in September 1988. He flew more than 2,000 combat hours and was awarded the Silver Star, Legion of Merit, the DFC 4 GS, the Purple Heart one GS, the AM/85, NCM, CAR and DMS Medal.

He lives in Vista, CA with wife Katherine. He is involved in the citizens' group, Vista Visions 2000, was selected to serve on the Vista Community Block Grant Committee and elected to the Chamber of Commerce board of directors. He joined the local Lions Club and was elected twice as secretary then president and also as the Zone Chairman of the district under its first female governor. He manages and teaches at a computer training facility in Vista and ran for Vista city council in 1994.

JOHN VAN NORTWICK, Lieutenant Colonel USMC (Ret) was born in Geneva, IL on Nov. 11, 1933. He earned a BA from Chapman in 1969 and an MS from USC in 1973. He enlisted in September 1954; NAVCAD from June 1955-December 1956; was commissioned a Second Lieutenant in the USMCR; and designated a Naval Aviator in December 1956. His squadrons included: HMR(L)-263, 262, 163, 264 from 1956-1961; and Presidential service with HMX-1 from 1961-64. Three Vietnam tours: HMM-363 in 1965-66; HMM-263 (XO), H&MS-16 (Group S-3A) in 1969-1970; HMH-463 (CO) in 1972-73. Vietnam flight time: 860 hours in UH-34D, CH-46D, UH-1E, AH-1G and CH-53D.

Memorable experiences: Being shot in the center of his overlapped flak jacket, the round penetrated layer one, dented layer two and bounced away. Second, being CO of HMH-463 during ENDSWEEP and getting daily "frags" directly from Henry Kissinger at the Paris peace talks.

Decorations: Legion of Merit, 32 Air Medals, two Purple Hearts and others. He retired in August 1976.

From 1980-1989 Van Nortwick worked for SAIC in El Paso and San Diego as a training coordinator, facility development coordinator and a project manager. He was project manager of design, engineering and relocation of NBC's WMAQ-TV in Chicago. In 1989, he retired again after 35 years on the road and is now a manager of two real estate offices that his wife owns in El Paso. He has been married for 35 years to Sonja. They have two sons, two daughters-in-law and five granddaughters.

DAVID JULIAN VEAZEY, "VEESY" born Dec. 29, 1944 in Washington, DC, completed the PLC course at Quantico in 1964. He joined the MARCAD program at Pensacola in 1966 and was commissioned a Second Lieutenant and designated a Naval Aviator on Sept. 19, 1967. He served with HMM-264, "Black Knights," at New River. He joined HMM-163, "Ridgerunners," in Quang Tri in March, 1968.

He volunteered to fly with VMO-6, "Tomcats," flying with Lieutenants Ralph Aye and Pat Partridge. He was transferred to HMM-363, "Red Lions," at Phu Bai. He flew off the USS Princeton, then to H&MS-16 at Marble Mountain, then to HMM-362, "Ugly Angels," flying off the USS *Iwo Jima*. His awards include Single Mission Air Medal, 41 Air Medals, CAR, PUC, and two MUCs, Vietnam Cross of Gallantry, Vietnam Service & Campaign Ribbons. He completed active duty as a First Lieutenant with VMO-1 in October, 1970 with 1,655 flight hours.

DARCY VERNIER was born on March 29, 1944 in Palo Alto, CA. He joined the service on June 2, 1967 and was commissioned as a Second Lieutenant. He was stationed in Pensacola and New River.

He was also in HMM-261 on a Carib cruise in March 1968. While in-ountry he was a HAC in CH-46D/Fs for HMM-263. He was discharged on Jan. 1, 1972

with the rank of Captain. He was awarded three DFCs, 50 Air Medals, a Navy Commendation Medal and the Combat Action Ribbon.

Today he resides in Nairobi, Kenya and has one daughter and one son. He flies relief missions for the UN in the Sudan, Rwanda and Angola.

JIMMY VILLARREAL born Aug. 21, 1946 in Beeville, TX, and joined the USMC on Dec. 30, 1965, being assigned as a crewchief/gunner with HML-267. His military locations include Memphis, TN; New River, NC; and Camp Pendleton. His awards include Combat Air Crew Wings, 31 Air Medals, Vietnam Cross of Gallantry, Combat Action Ribbon, PUC, three NUCs, MUC, Vietnam Service & Campaign Ribbons with Silver Star device.

His most memorable experience was an incoming attack in Khe Sanh when they fled into an underground operations bunker. An explosion demolished the shitter with a fellow crewman, Jessie, being hit by a flying "turd."

He was honorably discharged on Dec. 12, 1969 with the rank of Sergeant. In 1985 he joined the Naval Reserve. He is married to Luci. They have two daughters, one son, one granddaughter, and one grandson and he has a private practice in San Antonio, TX.

FRANK VISCONTI was born on Oct. 16, 1934 in Syracuse, NY. As a baptized Catholic, he attended the Cathedral of the Immaculate Conception and graduated from the Christian Brothers Academy High School where he was lettered for excelling in football. He went on to Syracuse University where he was teammates with Jim Brown (Cleveland Browns). He married on Nov. 24, 1955 and he finished with a degree in physical education from Ithaca College. Their daughter Geri was born on Jan. 16, 1957 and Frank on Oct. 10, 1960. After a year of teaching P.E. and coaching football Frank joined the USMC and was commissioned a Second Lieutenant in December 1961. He went on to basic school and then on to the flight program. His pre-graduation gift was his second son, Tony, on Jan. 25, 1964 and on March 4, 1964 he was awarded his wings.

His first duty was with HMM-261, MAG-26, 2DMAW FMFLANT MCAF New River, Jacksonville, NC. Visconti went on to Da Nang, Vietnam in June 1965 and in September to HMM-362, MAG-36, 1st MAW, Chu Lai, Vietnam. He went MIA on Nov. 22, 1965. The most compelling information to date, is the March 8, 1994 Joint Task Force Full-Accounting Report. Intelligence reports from 1967 and 1968, indicate three of the four men on Frank's chopper could have made it to shore. Tom Douglas, Frank's crewchief was a known POW as of June 1967.

ALBERT J. VITI JR. Master Gunnery Sergeant USMCR (Ret) was born on Feb. 10, 1950 in Brooklyn, NY. He entered the service on Oct. 24, 1967. He was in the USMC as a UH-1E crewchief serving in VMO-1 in 1968, HML-167 in 1969 and VMO-6 in 1970. He was in Search and Rescue in 1971 in Beaufort; HMH-772 in 1972; HML-771 from 1972 to 1986; DC/S Air HQMC 1988; and ASL Branch Maintenance Chief 1989 to retirement. He was also an aviation maintenance chief for MALS-49 in Newburgh, NY.

His most memorable experience was in San Diego. It was his last flight in a UH-1 from Boston to San Diego. "Five days of the lowest and slowest most beautiful flying time I have ever experienced-a happy time."

He was retired on Feb. 28, 1994 with the rank of Master Gunnery Sergeant. He was awarded Combat Aircrew Wings and Air Medals.

He is married to Lynne and has daughters, Nicole and Celine and a son, Al.

DONALD L. WAGONER, "WAGS," was born on Sept. 28, 1947 in Ft. Wayne, IN. He enlisted in the Marines in June 1965 and went through boot camp at San Diego, CA. He reported to Memphis for helicopter training.

He transferred to Vietnam in July 1966 and served as a crewchief on the UH-34D with HMM-361 at Ky Ha; Marble Mountain, Dong Ha and aboard the USS *Iwo Jima*.

He was awarded one Single Mission Air Medal, 15 Air Medals and the Purple Heart. He returned to NAS Los Alamitos in March 1968 and was stationed there one year and discharged in April 1969 with the rank of Corporal.

Wagoner was married in California and lived there one year. He returned to Ft. Wayne, IN in April 1970 and has worked for the Gas Company since then. He has two daughters, ages 19 and 23. His hobbies include tennis and bicycle touring.

JAMES BRIAN WALLACE born March 31, 1938 in Boston, MA and served as a USMC Aviator from 1961-1966 with one combat tour serving with HMM-363, MASS-2 and HMM-163 in Vietnam. His service stations include Naval Flight Training School, Pensacola, FL, 1961-63. He was honorably discharged with the rank of Captain. He joined the USCG retiring on Aug. 1, 1982, with the rank of Lieutenant Commander. He had 22 years of service time, five years with the USMC and 17 years with USCG flying as a search and rescue aviator from, 1967-82. He has a total flight time of 5,500 hours with 211 combat missions, UH-34, 1965-66. His awards include the Distinguished Flying Cross (USCG), Mariner's Medal, City of Gloucester, MA, nine Air Medals, and eight to ten other military awards.

He resides in East Sandwich, MA, is divorced with three sons ages 23, 26, and 28. He is a member of the Reserve Officer's Association.

HUEY C. WALSH received his wings on April 2, 1965 and reported to HMM-265 New River. In the summer of 1965 he deployed on the USS *Okinawa* for a "fun filled" three months Carib cruise, with the H-46A, followed by many "fun filled" months in sunny Vietnam!

Following his Vietnam tour, he instructed in Pensacola, his hometown, for two and a half years and got out of the Corps on Jan. 1, 1970.

He became a pilot for Delta on Jan. 5, 1970 and flew until June 1992, when he took early retirement. At that time he was a Captain/flight instructor on the B-757 and B-767.

He remained in the reserves until 1977 and then later got his 20 good years by completing correspondence courses.

Walsh has been married to Judy since 1965 and his oldest son is a former Marine and now an airline pilot. His youngest son, (PLC Air) was commissioned on April 2, 1994 by "Major Dad." Two sons, both Marines and both pilots! Makes the ole man proud!

JOHN A WARGO was born on June 14, 1945 in Ambridge, PA. He joined the service in September 1963. He was stationed in the States in Santa Ana and Cherry Point.

While serving in-country he was stationed at Marble Mountain, Vietnam. He was a gunner on the CH-46A.

He was discharged in September 1967 with the rank of Sergeant.

Today he is married to Cindy and has a daughter, Jennie, and a son, Brian. They reside in Freedom, PA.

TOM WARNING, "FREE," a resident of the Detroit area, enlisted after the "Gulf of Tonkin" incident. Following training in San Diego, he completed helicopter training at Memphis and was assigned I & I duty at Willow Grove, PA.

He went "in-country" December 1966 and was assigned to HMM-361 "Tweed's Tigers" and from November 1967 to August 1968 was assigned to HMM-362 "Ugly Angels."

He was awarded the "Winged S" from Sikorsky Aircraft, two Single Mission Air Medals and a combat promotion to Sergeant.

He joined the Detroit Police Dept. in June 1969 and retired in September 1975 with a duty disability. He graduated with an MA from Wayne State University in 1981. In 1986 he incorporated his own company (Printing Broker), today he serves as President of 22nd Century, Inc. Two years ago, he was finally remarried, after being unmarried for 18 years. He has two grown sons who he is extremely proud of. He is very active with a number of charitable organizations that provide him a sense of accomplishment. For fun, he belongs to a flying club (Piper Archer II & Dakota) and has "tailwheel" acrobatic experience in a Stearman.

ZIN DAVIS WATFORD was born on May 3, 1937 at Graceville, FL. He graduated from Campbellton High School, Campbellton, FL in May 1956. He joined the United States Marine Corps in June 1956 and went to boot camp at Parris Island, SC-Service No. 1612276. His first tour of duty was with motor transport maintenance in Okinawa. On his return to the States, he was stationed at Parris Island Motor Transport Section.

He re-enlisted in 1960 and went to Aviation Machinists School at Memphis, TN. After finishing, he was stationed at New River, NC. His squadron, HMM-162, was sent to Okinawa where they embarked on the USS *Valley Forge*. Colonel R. Leu was CO. They went to the Philippines and Hong Kong, then were called to go to Cuba during the Cuban missile crisis. When it was called off, they were sent to Udorn, Thailand for a month or so, then back on the Valley Forge and around to Da Nang, Vietnam, arriving around July 1962. While serving in RVN, all HMM-162 crewmen received their Combat Aircrew Wings. They were the first crew members to receive their wings since World War II. They left Vietnam to return to the U.S. in May 1963.

He returned to Vietnam in August 1966 to serve with HMM-361's Tigers which were at Ky Ha. Then they rotated to Okinawa, then returned to Vietnam in Jan. 1967 to a new base, Marble Mountain. HMM-361 set new records for flying hours and sorties.

Watford returned to the U.S. in October 1966. In 1967 he went to Photo Interpretation School in Denver, CO and he finished out his time in the USMC as a photo imagery interpreter. He retired from the Marines in June 1976.

While in the USMC, Watford received the Air Medal Ninth Award W/GS, Pres. Unit Citation, Navy Commendation Medal, National Defense Medal, M.C. Expeditionary Medal, Armed Forces Exp. Medal, Good Conduct Medal W/GS, Vietnam Cross of Gallantry, Vietnam Service Medal and Aircrew Combat Wings with three stars.

Since 1978, he has been working for Rockwell International, Rocketdyne Division as an industrial security officer.

DENNIS D. WEATHERS, "DRIPPY," was born on July 25, 1946 in Miami, FL. He enlisted in the Marine Corps on Nov. 23, 1965 at Parris Island. He was assigned to VMA(AW)-533, MAG-14, Cherry Point on Sept. 5, 1966, after attending Aviation Ordnance School at Jacksonville NAS. He also attended Conventional Weapons and Nuclear Weapons Loading Schools at Cherry Point; A6A Armament School at Norfolk NAS; and Walleye Missile School at Cecil Field NAS.

He was deployed to Chu Lai on April 1, 1967 with VMA (AW)-533, MAG-12. He extended his tour of duty and joined VMO-2, Marble Mountain on April 30, 1968 as door gunner/ordnance man, UH-1E. In May 1968, he was assigned to HML-167 during formation of the squadron when VMO-2 began to receive the OV-10 aircraft. He was the first ordnance man/door gunner to be assigned to HML-167 during the transition. He served with HML-167 until Nov. 23, 1968 when he transferred Stateside to VMA-331, MAG-31, Beaufort.

During his Vietnam service he was awarded the Bronze Star w/Combat "V"; Air Medal with 79 stars; CAR; Combat Aircrew Wings with three stars; PUC; NUC; Vietnamese Service with six stars; Vietnamese Campaign Medal and Letter of Appreciation. He was discharged as a Sergeant on Sept. 5, 1969

He attended and graduated from Florida State University in 1973. He is employed at the Osceola County Sheriff's Office, Kissimmee, FL as a Sergeant, CID.

J.B. WEGENER II, "WHEELS," had been in the Corps from 1960 to 1964, was working at Gordon's Jewelers in Port Arthur, TX and re-enlisted in 1967 after reading about our men getting killed in Vietnam. He left a good job and girlfriend to do what he thought was right for his country, as many others had done. He came from a long line of Americans who had fought for this country such as his great-great-grandfather, Acting Colonel William James Wendell Donnel who fought the "bluebellies" out of Murfreesboro, TN. Other members of the family fought for the North. Both sides of his family go back as early as the 1600s in Kentucky, Tennessee and Texas.

His father, Technical Sergeant Joseph B. Wegener, was killed in a plane crash in Venezuela in 1948 while flying on a mapping mission for the U.S. Air Force, but not before he had served as a rated mechanic and pilot in the U.S. Marines as one of their first pilots. After serving six years with the Marines and flying off of the USS *Saratoga*, he transferred to the U.S. Army Air Corps and stayed with them until they evolved into the U.S. Air Force. His brother, J.B.'s Uncle Fred, served at Guadalcanal. Wegener, felt that because of all of his family history, that it was his duty to re-up and fight in Nam.

Vietnam anecdote: Wegener arrived in-country Feb. 3, 1968 during "Tet." He was sent to Headquarters Wing pad at Da Nang where he worked in supply. He was the squadron scrounge after a while and learned where to find what others could not. He earned the nickname, "Wheels," from his NCOIC GySgt Pelt after finding parts for his OIC's weapons carrier that no one else could find. After putting his OIC back on the road, the term, "Wheels," as in wheeling and dealing for parts, stuck as his nickname. Soon after arriving at the Wing pad compound at Da Nang across Highway 1 from an infamous little village known as "Dog Patch," Wegener became bored with nothing but a rocket or two passing over each night and wheeled his way north to Phu Bai where quite a bit of action was going on. One Bosses' Night at the EM Club Wegener introduced himself to the Squadron CO Lieutenant Colonel "Red Dog" Keller, a man who had fought with General "Chesty Puller" himself, explained his family's history and why he re-up'ed, etc. Lieutenant Colonel Keller told him to come to his office and he would start the transfer. Wegener said that he would start through the chain of command the next day and would see him soon, and said, "Thanks a lot, sir." His CO looked him straight in the face and said, "Don't request mast, just come see me when you're ready to go."

That could get a man killed in a combat area such as Da Nang, walking though a squadron office without permission, but that's what the man said, so that's what Wegener did. He explained everything to his NCOIC Gunnery Sergeant Pelt and their OIC, Captain Tobin. They explained that if he was mistaken about what was said, he would end up in the brig and there would be nothing that they could do. After a couple of days of thinking it over, Wegener walked down to the squadron office, opened the door and started walking past the Duty NCO and all the other Sergeants and officers while hollering out at them, as they were standing up and commanding him to stop. "The CO said for me to come see him without going through the chain of command!" When he got to the back of the Quonset hut he made a right turn, two left turns and stared into the face of the Sergeant Major and his pistol. Once again, and very loudly Wegener repeated what he had said before. Lieutenant Colonel Keller called the Sergeant Major in, gave him papers to give to Wegener. He was soon headed for Phu Bai and HMM-165, where he flew as a machine gunner. He was shot down Oct. 10, 1968 in Happy Valley where he earned the Purple Heart and the Silver Star for saving the lives of several men from 1st Mar. Div. Recon Battalion.

Promoted to Lance Corporal, Wegener flew many supply missions to find parts that were in demand for his squadron when they were based at Phu Bai, Marble Mountain and aboard the LPH-10 USS *Tripoli* i.e., "The Famous Fighting Ship," which was decommissioned in 1995. Because of wounds suffered in Vietnam, Wegener was discharged in 1970 with the rank of Sergeant.

Today he is retired and enjoys fishing and hunting. He resides in Ash Fork, AZ.

ROBERT CHARLES WELCH, Lieutenant Colonel, USMCR, (Ret) was born May 14, 1946 in Miami, FL and on August 5, 1969 he joined the USMC. He graduated from OCS in 1970, commissioned a 2nd Lieutenant. He then attended and graduated the Basic School. He attended the Aviation Supply Officer Course in Camp Lejeune, NC and was assigned as Aviation Supply Officer to VMA-331 at Beaufort, SC. His next tour was in Vietnam with a composite helicopter squadron (HMM-165) participating in combat operations. He returned to serve at MARTD Atlanta, GA. He joined HMA-773 1973-76 and was transferred to HQ 4th MAW DET A, New Orleans, LA in November, 1976 as Wing Supply Officer and MORDT Evaluator. He was a member of the selected Marine Corps Reserve serving as Wing Supply Officer, 4th Marine Aircraft Wing, Detachment A, New Orleans, LA, until April 15, 1990.

He was awarded the Vietnam Service Medal, Republic of Vietnam Campaign Medal, and the Navy Unit Citation, the National Defense Medal, the Armed Forces Reserve Medal, four Organized Marine Corps Reserve Medals and the Philippine Presidential Unit Citation. He retired from the military on Jan. 1, 1993. He has three children and two grandchildren and is an engineer at South Central Bell.

GREGORY DEAN WELLS was born on Dec. 30, 1946 in Robinson, IL. He enlisted in the Marine Corps Sept. 29, 1965. He attended boot camp in San Diego, CA and helo machine school in Memphis.

After graduation he was sent to Chu Lai, Vietnam attached to HMM-363. He became crewchief of YZ-65. While in-country, HMM-363 did a tour in Chu Lai, on the USS *Kitty Hawk*, Dong Ha and Da Nang, and then transferred to New River, NC where, before being discharged, he spent time in both HMM-262 and HMM-365.

He was discharged with the rank of Sergeant on Sept. 26, 1969. His awards include: Combat Aircrew Wings w/three stars, National Defense Vietnam Campaign, Letter of Commendation, Winged "S," Air Medals with 22 stars, Vietnam service with two stars, Purple Heart with two stars and Presidential Unit Award.

After returning home, he started farming with his family: parents, Leroy and Anna Dee; wife, Sharon; daughter Heidi; son, Chad; daughter-in-law, Deann; grandson Clayton and one grandchild.

ROBERT F. WEMHEUER, "BULL," Colonel USMC (Ret) was a native of Illinois. He graduated from Western Illinois University and was commissioned a Second Lieutenant in 1965. Upon his designation as a Naval Aviator in 1966, he was assigned to the 3rd Marine Aircraft Wing for duty with Marine Helicopter Squadron HMH-463. In May, HMH-463 deployed to Vietnam where he served as the squadron's Ordnance Officer. Upon returning from overseas he attended the Aircraft Maintenance Officer School in Memphis, TN and then became a flight instructor in HMHT-301 at MCAS Tustin, CA.

In 1970, Wemheuer was again assigned to HMH-463 already in Vietnam, where he participated in Lam Son 719 and served as logistics and embarkation officer until the unit's relocation to Kaneohe, HI. While assigned to HMH-463 in Hawaii, he participated in Operation End Sweep which involved helicopter mine sweeping in Haiphong Harbor and other locations in North Vietnam.

In 1975, after graduating from Pepperdine University with a master of arts degree in education, he was assigned to HMH-363 as the Aircraft Maintenance Officer until 1977. He then served as Executive Officer and later Commanding Officer of Marine Air Base Squadron-16. From 1978 until 1981, he was assigned to the Community Planning and Liaison Office at MCAS El Toro, CA. In 1981, he graduated from Pepperdine University with a doctorate degree in institutional management and was transferred to MAG-36 at Futema, Okinawa where he served as Commanding Officer of Headquarters and Maintenance Squadron 36.

Returning to the United States in 1982, Wemheuer was assigned to Headquarters Marine Corps in Washington, D.C. where he served as Head, Facility Planning and Programming Section, Facility Branch. Facility and Service Division, Installation and Logistics Department until 1984. He was reassigned as the Branch Head of the Land Resources Management and Environmental Branch, a newly created encroachment control organization in the Facilities and Service Division, where he served until 1986.

In 1987, he assumed command of the Marine Corps Air Station Tustin, CA, where he served until his retirement from the Marine Corps in October 1989. Wemheuer's personal decorations include: three Legion of Merit Medals, two Distinguished Flying Cross Medals, two Single Mission Air Medals and Air Medals with Numerals 39, the Navy Commendation Medal, the Navy Achievement Medal with Combat "V" and the Combat Action Ribbon.

In February 1990, Dr. Robert Wemheuer became a project manager for the Fluor Daniel Inc. He is presently assigned to the Fluor Daniel Advance Technology Operating Company in Las Vegas, NV working on the Civilian Radioactive Waste Management System Project for the U.S. Department of Energy.

R. LARRY WERT, "STOGGIE," Captain USMCR (Ret) was born on Nov. 6, 1943 in Bellefonte, PA. He graduated from high school in 1960 in State College, PA. Holds a BS from Penn State University 1964. He was commissioned on June 14, 1964 as a Second Lieutenant and was designated a Naval Aviator on Dec. 10, 1966 at Pensacola. He was assigned HMH-463 Santa Ana, CA in January 1967.

He was deployed with HMH-463 to Vietnam on May 1, 1967 where he served from May 1967 through February 1968 with more than 500 combat hours.

He served with MAG-16 from February 1968 through May 1968. He completed active duty as a Captain with HMH-461 in December 1969. Wert was awarded 10 Air Medals, NUC, MUC, Navy Commendation Medal, Vietnam Service and Campaign Ribbons.

As a civilian, he worked in the human resources field. He currently is Labor Relations Manager for Beech Aerospace Services, Inc. Madison MS. He has a daughter and a son. He is currently an active member of the USMC/Vietnam Helicopter Pilots and Aircrew Reunion, American Arbitration Association, Mississippi Mun. Association, Society Human Resources Management and the Mississippi High School Athletic Association.

EUGENE WESOLOWSKI JR., "GENE," OR "SKI" was born on May 17, 1948 in Warren, OH. He graduated high school in 1966, joined the Corps in October 1967 and left PI in 1968 for Aviation Structural Mech. training at Memphis. He and his wife, Linda, honeymooned at the Millington Trailer Park in April 1968. Soon after, they headed to Cherry Point, where their son Chris (presently a second enlistment Sergeant, USMC) was born in April 1970.

He went on to WESTPAC in September 1970 and to H&MS-16, Marble Mountain, RVN. He worked as a Sergeant, structural mech. at H&MS-16, then as a gunner with HMM-263 "Peach Bush." He earned six Air Medals, NUC, MUC, RVN MUC and RVN COG. Was discharged in 1971. He returned to the USMC in 1973 as a PFC. In 1976 he was promoted to Staff Sergeant and in 1978 left the Corps with a medical discharge.

He presently works for General Motors in Lordstown, OH. Gene and his wife Linda, also have a daughter, Erin and reside in Warren, OH. Gene restores, shows and drag races old Chevys. He is a member of the USMC/Vietnam Helicopter Pilots and Aircrew Association, VFW and life member of the DAV.

BOB WHALEY, "SEAWORTHY WHISKEY," Lieutenant Colonel USMC (Ret) was born on Jan. 7, 1935 in Missoula, MT. He graduated from Carroll College, Helena, MT on May 15, 1958 after three summers of smoke jumping out of the Missoula Base. He entered the 23rd OCS as an Air Officer candidate in September 1958.

Commissioned a Second Lieutenant and entered flight training in Pensacola in January 1959, receiving his wings on June 8, 1960. He joined HMM-362 (UH-34Ds) which participated in Operation Shu-Fly as the first Marine Squadron to enter combat in Vietnam at Soc Trang in April 1962. He served a second tour in VMO-6 (UH-1Es) in 1965-66 at Ky Ha. His last tour in-country was again with VMO-6 at Quang Tri, Prov. MAG-39, flying OV-10A Broncos and UH-1Es.

Whaley flew more than 800 missions and was awarded the Distinguished Flying Cross (2), Bronze Star, 40 Air Medals, Single Mission Air Medal, Purple Heart, Vietnamese Cross of Gallantry, Joint Services Meritorious Commendation Medal and various theater and unit awards.

He retired in October 1979 with more than 5,000 accident free flight hours and, since September 1980, has been in the brokerage business where he currently serves as branch manager of the Missoula Office of AG Edwards. He is a life member of the American Legion, DAV and TROA. His wife, LaWana, is an RN who recently retired from the American Red Cross. They have two dogs, Kootenai and Cheyenne.

GERALD WHITE, "WEASEL," was born on Aug. 5, 1945 in Jacksonville, IL where he joined the service in 1967. He attended flight school at Pensacola in 1968 and then served with HMM-162.

While serving in-country he was a pilot for HMM-364 flying CH-46s. He was at 2nd Anglico in Cuba. He was awarded the single mission Air Medal and 30 Air Medals and was discharged in 1972. He resided in Murrayville, IL and reared two daughters. He is deceased.

JAMES A. WHITE JR., "SNEAKY WHITE," was born on Sept. 20, 1937 in London, England. He completed more than five years in the USMC as a helicopter crewchief, and was the first enlisted Marine to win the DFC since WWII, before transferring into the Army. He served as 11 Bravo before Army Flight School. Other military schools include Cold Weather School (Norway), Aviation Electronic Technical, Rotary Wing Mechanical Course, Escape and Evasion (both in the Philippines and Pickle Meadows, CA),

Jungle Expert and Aviation Maintenance Officer Course.

He served in Guantanamo Bay, Cuba, Dominican Republic and multi-tours in Southeast Asia, as well as other overseas assignments. He was awarded Navy/Marine Corps Combat Aircrew Wings with three Gold Stars; CIB; Army Aviator Badge; numerous parachute wings, Silver Star (with two OLC); DFC (with Gold Star, four OLC); Bronze Star (with V, four OLC), Air Medal (with three Vs, numerals '133'); Purple Heart (with two OLC); Joint Service Commendation (with V); Army Commendation (with V, three OLC); Navy Commendation (with V, GS); Navy Achievement Medal with "V"; Army Good Conduct; the Marine Corps Good Conduct; Vietnam Cross of Gallantry with six Gold Stars, one BS; VN Air Cross of Gallantry (w/Silver Wing); VN Honor Medal (w/GS); Laotian Order of Merit (w/Rosette and Palms) and numerous unit awards.

His combat units include HMM-263; HMM-262 "Bill's Bastards"; 1st BN 6th INF/198th LIB; 108th FAG "Sneaky Whites"; C4th 77th ARA "Griffins"; C/2nd/17th CAV "Condors"; and 22 months with Air America. All Vietnam service was in northern I Corps and his call sign "Sneaky White" was well known.

White is now living, semi-retired in Northern California. He has four children (two who are still in Vietnam) and is a grandfather. He spends his time involved in several veterans organizations: Vietnam Helicopter Pilots Assoc., Vietnam Combat Pilots Assoc., Vietnam Helicopter Crew Members Assoc., Special Operations Assoc., Disabled American Veterans and Veterans About Face.

WILLIAM D. WHITEHURST, "WHITEY," was born on March 7, 1949 in Miami, FL. He joined the service on Dec. 7, 1967. He was stationed at MCAF New River with HMH-461.

He was a gunner on the CH-53 and participated in Operation Lam Son 719. He was stationed at Marble Mountain with HMH-463 during 1970 and 1971.

He was discharged on Sept. 7, 1971 with the rank of Sergeant. He was awarded 11 Air Medals, Combat Aircrew Wings, and a Commendation.

Today he resides in Lakewood, CO and is a certified public accountant.

FRANK G. WICKERSHAM III, "WHISKEY THREE," Colonel, USMC was born on April 14, 1943 in Daytona Beach, FL. He is a graduate of Georgia Military Academy 1961, the University of Florida (1965) and the Catholic University of America (1987). He has a BA degree in Education, a MA degree in International Affairs and Ph.D. (ABD) in World Politics and American Government.

Wickersham was commissioned a Second Lieutenant with the 40th OCS in May 1966. He was designated a Naval Aviator in June 1967 at Pensacola, FL. He flew UH-34s with HMM-363 from November 1967 until January 1969. With this squadron he participated in numerous major combat operations in RVN, NVN and Laos. This included Operation Mighty Yankee in which he flew from the USS Dubuque and assisted in the diplomatically sensitive repatriation of 14 North Vietnamese prisoners of war to the North Vietnamese city of Vinh.

Extending his initial combat tour, he flew the UH-1E and AH-1G gunships with VMO-2 from January 1969 to January 1970. Completing this 27 month combat assignment, he was posted to VMO-1, MCAS New River, NC. With this squadron he flew UH-1E, AH-1J and OV-10A aircraft. Additionally he was the first Marine to attend the U.S. Army AH-1G Flight Instructors Course at Hunter AAF, Savannah, GA. Returning from this course, he volunteered for assignment to the Marine Corps AH-1J Combat Evaluation Team. With this unit he participated in operational and ordnance testing of the AH-1J at NATC Patuxent River, MD. In January 1971 he returned with the AH-1J DET to RVN for his second combat tour.

During this tour he flew AH-1J and AH-1G gunships while assigned to HML-367. He participated in Operation Lam Son 719 in Laos. In May 1971 his AH-1J unit was withdrawn from Vietnam and reassigned as Sub Unit-One, H&MS-36, MCAS Futema, Okinawa. He returned to CONUS in January 1972.

Upon completion of this second overseas tour, then Captain Wickersham had accumulated more than 2,100 combat flight hours, a highly experienced combat aviator in the Marine Corps. He subsequently completed a series of tours to include: Officer Selection Officer duty; Army airborne training at Ft. Benning; Amphibious Warfare School; AMO/OpsO with VMO-2; sea duty on board USS *Mount Whitney* as the Air Officer, 4th MEB; Marine Corps Command and Staff College; military advisor to the OV-10 Foreign Military Sales Team, Rabat, Morocco; and Commanding Officer, MCAF Camp Pendleton, CA.

In June 1981 then Major Wickersham assumed command of H&MS-39, MCAF Camp Pendleton, CA. Promoted to Lieutenant Colonel, he relinquished command of H&MS-39 and assumed command of VMO-2 in March 1982. He led his squadron through numerous special projects to include OV-10A/D launch operations from the USS *Nassau* and USS *Belleau Wood*. Under his command, VMO-2 achieved the milestone of 28,000 mishap-free flight hours and earned a FMFPAC safety award. Lieutenant Colonel Wickersham was cited for having personally accumulated more than 5,000 mishap-free hours.

Colonel Wickersham relinquished command of VMO-2 in April 1982. He was frocked to Colonel in August 1986 and promoted in March 1988. From 1982 to 1993 he served in a wide variety of staff and field assignments to include: AC/S, G-3 (training), 3rd MAW; Deputy Air Combat Element commander, Berbera, Somalia; Director, Joint Strategic Plans Division, HQMC; special advisor to CMC on U.S.-USSR bilateral negotiations; AC/S, G-1, 2nd MAW; JFACC Project officer with the Commander, U.S. Second Fleet; Marine Corps advisor to Commander, Naval Forces Central Command (Desert Shield/Storm) and Fleet Marine Officer, U.S. Seventh Fleet, USS *Blue Ridge,* Yokosuka, Japan.

Colonel Wickersham retired from active duty Sept. 1, 1993. His personal decorations include the Legion of Merit with Gold Star, the Distinguished Flying Cross with three Gold Stars, the Purple Heart, the Meritorious Service Medal, the Air Medal with four Gold Stars and the numeral 85, the Navy Commendation Medal the combat "V," the Navy Achievement Medal with combat "V," and the Combat Action Ribbon. He has more than 5500 flight hours, 60 parachute jumps and four years of sea duty.

Colonel Wickersham is married to the former Nancy Taylor of St. Petersburg, FL and they have a daughter, Taylor Lee. Colonel Wickersham and his family live in Falls Church, VA. He is the president of Falcon Group Worldwide, an international executive consultancy. Additionally, he is an officer/member of the board of the directors with the northern Virginia Chapters of the Retired Officers Association, the Military Order of the Purple Heart and the Youth For Understanding International Exchange Program. He is listed in the Who's Who Worldwide Registry.

BENJAMIN L. WILLIAMS, "GENTLE BEN," (the hookers friend) a.k.a. "Ben-Joe-Ditch" and "Wino," Major (yes, he's untrainable) USMC (Ret) was found in Atlanta, GA on April 26, 1944. He sniveled and bribed his way through San Jose State and Webster Universities pilfering a bachelor's and master's degree along the way. Commissioned in April 1968 at Quantico and designated a Naval Aviator in May 1969, "Gentle" reported to MMAF Da Nang in June 1969 and became the founding father and first secretary of Alcoholics Unanimous. Concurrently joining the "Purple Foxes" of HMM-364 he flew 863 missions (mostly in Vietnam) striking fear and trepidation into the hearts and minds of both the "friendlies" and "Charlie" alike. During his blackouts he earned three DFCs, one Single Mission Air Medal and 43 Strike Flight Air Medals, in addition to all the usual "been there-done that" multi-colored stuff. During his inadvertent 20 years, five months, two days, 23 hours and 60 minute career, "Gentle" flew choppers five years, the AV-8A/C 12 years, and the C-12 two years, circumnavigating any tours in DC. Retiring in June 1988 he was immediately snapped up (for his superior piloting skills) by American Airlines.

"Gentle" flies the 767 out of LAX, has three known children (two of which he claims) and resides in Laguna Beach where he just finished a manuscript for his first novel. He is a Patron of the Pirates Alley Faulkner Society, is barred from the local VFW and is currently tolerated by the USMC/VHP&ACR.

PETER D. WILLIAMS, "PD," was born on Oct. 24, 1945 in Oakland, CA. He lived in Walnut Creek, CA, attended Las Lomas High School and Diablo Valley College until joining the Marine Corps in March 1965. He went to NAS Memphis from August 1965 to February 1966 and joined HMM-262 in New River, NC in February 1966.

He took a Caribbean cruise in February 1966 on the LPH-7 and went with HMM-262 (via MAC C-130) to Vietnam in December 1966. He flew as a gunner for five months. (C.A. Williams got the credit for most of my flying) He received two Air Medals. He served in the "float phase" of his tour aboard the USS *Tripoli* in August 1967 until the Boeing Modification Program started in Okinawa. He was an airframe quality control inspector for the Mod Program and completed active duty with HMMT-302 at MCAF Santa Ana as NCOIC of the Airframe Shop. (I can eddycurrent your blades!) He was released from active service in March 1969 as a Sergeant.

As a civilian he has been a pilot since 1966, a flight instructor since 1970 and a corporate pilot since 1973. He has flown more than 11,000 hours in recips, turboprops and jets, from Cessna 150s around the Arizona desert to Gulfstream III's to India. He is currently flying a Cessna Citation for the ALLTEL Corporation based at the Akron/Canton Airport in Canton, OH.

He is married to a special education teacher with two sons living in Canton.

THOMAS E. WILLIAMS, First Lieutenant, USMCR was born on Dec. 26, 1943. He was the husband of Janis L. Williams. He served with distinction with HMM-262 in Vietnam and died May 2, 1969. He was 25 years of age.

THOMAS L. WILLIAMS joined the Marines in October 1965 four months after high school along with four other high school friends. They signed up for an aviation guarantee, so after boot camp in San Diego and ITR they were off to Aviation School in Memphis, TN. After school they were given 20 days leave then off to Vietnam.

Ten and half months after joining he was in Vietnam assigned to H&MS-16 Marble Mountain. He spent his tour there and extended for six months. He was put into VMO-2 where he was a mechanic and gunner on UH-1E gunships. After 19 months in Vietnam he got orders to New River Air Facility, North Carolina where he spent the next one and half years in VMO-1 as a crewchief. He was on a Caribbean cruise just before getting discharged with the rank of Corporal.

"The only memorable story I think of now was one of a young Texan I went through boot camp with. His name is David A. Padilla. We went through boot camp, ITR and were sent to Memphis, TN for aviation training together. Dave didn't finish his schooling and was transferred to the infantry. I thought I had seen him for the last time but I was wrong. After extending for six months in RVN we were sent home for 30 days leave. On the way home all Marines stop at Okinawa and get sent back from there. When I was in the barracks I ran into Dave Padilla who had himself extended for six months and was going home on leave. We went home and came back and met in Okinawa again. Dave was stationed at the DMZ as a radio operator, had already earned three Purple Hearts and other ribbons and medals. On his second go around he was killed but is still listed as a POW-MIA. He was a hell-of-a Marine and a good friend. All of us know of someone like him and we will never forget them."

Currently he is a heating and cooling contractor in Detroit, MI. He has completed a five year apprenticeship in Pipe Fitters Local Union 636 and has been doing service work ever since. He has one son and one daughter who keep him busy.

FREDERICK J. WILSON III, "RICK," was born on Jan. 4, 1942 in South Kingstown, RI. He graduated from the University of Wisconsin 1964 with a BA and in 1965 with an MA, and from the University of Florida in 1967 with an MS. He was commissioned through NROTC Second Lieutenant in 1964. He was designated a Naval Aviator in May 1967.

He flew more than 1,000 hours in Vietnam with HMM-163 and HMM-164 primarily in CH-46s.

Wilson was awarded the Silver Star, Distinguished Flying Cross, Bronze Star, Navy Commendation Medal, 46 Air Medals and two Single Mission Air Medals. He completed active duty in 1970 as a Captain, Public Affairs Officer, MCAS New River.

Currently, he is president and publisher of Wilson Newspapers in Wakefield, RI. He has been president of Rhode Island Press Association and New England Press Association, as well as several other civic organizations. He lives with wife, Nancy, and five children in Wakefield, RI.

J.J. WILSON was born on March 27, 1943 in Rock Springs, WY. He graduated from high school in Coalgate, OK in 1961. He received a bachelor of science in chemistry from East Central University in 1966. He was commissioned a Second Lieutenant, 40th OCS Quantico, VA. He transferred to NAS Pensacola, was designated a Naval Aviator on Aug. 25, 1967 and transferred to HMM-162, MCAF New River.

In February 1968, he transferred to HMM-265 at Marble Mountain Air Facility. During Vietnam, he earned 42 Air Medals, Vietnamese Service Medal, Vietnam Campaign Medal and was promoted to Captain. In March 1969, he returned to MCAF New River and trained CH-46 pilots. In August 1970 he completed active duty.

From September 1970 to May 1975 he was chief chemist, Vickers Petroleum Ardmore, OK. In 1975 he began work as a scientist for GSA, Fort Worth, TX. Currently, Wilson is a zone administrator for Defense Logistics Agency, Defense National Stockpile Zone, Fort Worth. He is married with a daughter and a son.

MIKE WOLTER grew up in the farming community of East Chain, MN. He graduated from high school in 1961 and from Wisconsin State University-Stevens Point in 1966. Marine recruiters got him on campus. He completed the 41st OCS at Quantico with Jim Campbell, George Duerre and Byron Gigler, all chopper pilots and Vietnam vets. Naval aviator wings were pinned on Dec. 8, 1967 at Pensacola. He trained in the CH-46 at Santa Ana.

Wolter served in Vietnam with HMM-165 from June 1968 to July 1969. He flew with the White Knights out of Phu Bai until September 1 and went with the squadron when it boarded the USS *Tripoli* (LPH-10). He off-loaded at Marble Mountain on December 5. They had pilot nicknames of Chickenman, HotRod, York, Scarfer, Dinky, Tank and a whole gaggle of Horsenipples. Lady Ace was the call sign and they considered themselves the load hackers! Chip, their maintenance officer, kept them flying along with crewchiefs like Bacon, Brank, Breen, Hoffman, McMillion and Petritis to name a few. Number one sound was the gunner opening up with the .50 machine gun. Number one feeling was the first sip of the "salty dog" at the O Club. The number one sight was of waving nurses on the USS *Sanctuary* and USS *Repose*. Memories include mushy Brussels sprouts and cookers full of greasy bacon and of course, the burning crappers! The saddest memories are of grunts, the Marines they transported while flying medevac.

He was discharged in December 1970 with the rank of Captain. He was awarded 59 Air Medals and Navy Commendation Medal.

Wolter has been employed with the California Department of Fish and Game since 1972 where he started as a warden in the Mojave Desert. He is now a patrol Lieutenant in Alturas.

LARRY L. WOODRUFF, "WOODY," was born on Jan. 12, 1944 in Ft. Worth, TX. He attended Arlington State College until reporting to Pensacola in June 1965 as a MARCAD assigned to class 19-65. He was commissioned a Second Lieutenant and designated a Naval Aviator in September 1966.

After a Caribbean cruise with HMM-162 New River, NC he reported to 1st MAW in June 1967. He was initially assigned to HMM-263 at Ky Ha then MASS-2 serving as the helicopter director of the Phu Bai DASC during the 1968 Tet offensive. He was reassigned to 3rd MAW in August 1968 and served as the ASO of HMM-361, HMM-265 and HMM-161.

After a short in tour in CH-46s, he augmented to the regular Marine Corps and transitioned to KC-130s in 1971. Between 1971 and 1986 he served in a variety of billets in 1st, 2nd and 3rd MAW. He has held positions as AMO, ASO, NATOPS, OpsO, XO, CO and college student. His last tour was the Operations Directorate of the OJCS.

He retired from active duty as a Lieutenant Colonel in 1986 and became an airline pilot. He resides in southern California with his wife, a real sweetheart from Kentucky, and his teenage son.

ALEX WRIGHT, "BUZZARD," was born on Dec. 13, 1944 in Union City, TN. He graduated from UCHS in May 1962. He received a BS in transportation from the University of Tennessee in March 1967 and began OCS, Quantico, VA the next day. He was commissioned a Second Lieutenant in June 1967 and designated a Naval Aviator at Pensacola, FL in July 1968. He served with HMM-162 and HMM-261 at MCAS, New River, Jacksonville, NC.

Wright joined HMM-164 in Vietnam in March 1969. He was awarded 13 Air Medals, Vietnam Service, Vietnam Campaign and Vietnam Cross of Gallantry. After his Vietnam tour, he served as an instructor with HMMT-402, MCAS, New River, Jacksonville, NC. He completed active duty as a Captain in October 1971 with more than 1300 flight hours.

He is assistant vice-president with Marsh & McLennan, Nashville, TN. He has one daughter and one son plus a grandson, Andy.

ROBERT T. YANCHIS, "BOB" Gunnery Sergeant USMC (Ret) was born on Feb. 2, 1940 in Brooklyn, NY and enlisted in the USNR in 1957. Upon graduating from Aviation Trades High School in 1958, he transferred to the USMC. He was assigned to New River HMM-262 in 1958.

He volunteered to form HMM-162 in 1963 and then was assigned with the unit to Da Nang, Vietnam until 1965. After a six month tour in California, he was assigned to H&MS, in the T-58 and 1820 engine shop at Ky Ha until 1967 and in 1969 returned to HMM-262 in Quang Tri, Vietnam until 1970. During his tours in-country, he was assigned as crewchief, gunner, engine shop mechanic, section leader, line chief and maintenance control and was awarded Aircrew Combat Wings, Air Medals and various medals pertaining to Vietnam. He was assigned in 1970 to NAS New Orleans until transferred in 1974 to Okinawa. In 1975 he transferred to New River and retired on Dec. 31, 1977.

Upon retirement, he worked for Carson Helos in Pennsylvania until relocating to the Daytona, FL area and achieving an A&P license from Embry-Riddle University in 1980. He temporarily worked for Petroleum Helo in Lafayette, LA until starting work with current employer in 1981, McDonnell Douglas at the Kennedy Space Center in Florida in quality assurance. He married in 1965 and has a son and daughter. He is currently a member of the Vietnam Vets of America and Vietnam Vets of North Florida and the VFW.

ROBERT ANDREW YASKOVIC, "YAZZ," Lieutenant Colonel USMC (Ret) was born on April 16, 1943 in Yonkers, NY. He attended Texas A&M University and holds a BA in business administration and a BS in animal husbandry. He was commissioned a Second Lieutenant on March 12, 1969 and designated a Naval Aviator Number T-12962 on May 1, 1970.

He served with HML-367 in Vietnam, Cambodia and Laos during 1970 and holds a Bronze Star with combat "V," two Single Mission Air Medals, 27 Strike/Flight Air Medals, RVN Service and Campaign Ribbons, NUC RVN MUC, RVN AF MUC, Navy Commendation and Navy Achievement Medals.

Returning stateside to HML-269, he was the first pilot to achieve 1,000 flight hours in the Bell AH-1J Sea Cobra helicopter. His most rewarding tour was a four year stint as an instructor at NABTC Pensacola where he won accord as an outstanding flight instructor (VT-6 Instructor of the Year, Lions Club recognition). A tour with 2nd Anglico allowed Yaskovic to win his military parachutist "jump" wings. Memorable was his tour as XO of 2nd Maintenance BN, 2nd FSSG where his mechanical interest and aptitude served him well. Numerous billets over the years were generally as Aviation Safety and Operations Officer and as PMIP (Post Maint. Insp. Pilot) and NATOPS Instructor in every model aircraft qualified. Land and carrier operations took him to the Mediterranean, England, Caribbean and several tours in the Western Pacific.

Retiring on June 1, 1991 with 23 years of service, his love for flying resulted in 7,032 flight hours in the T-34, TH-1L, T-28, AH-1G/J, UH-1, OV-10A/D, UC-12B and Sabreliner aircraft.

Yaskovic is currently farming his 100 acre farm in Ontario, Canada. He misses the troops and comradeship. With no regrets he says he'd do it all over again in "A New York Minute."

MICHAEL YOUNG, "MIKE," was born in Chicago on June 12, 1944, graduated from St. Rita High School in 1962 and enlisted in the USMC in September 1962. He went through boot camp at MCRD San Diego, ITR at Pendleton and aviation schools at Memphis. He was assigned to HMM-263 at New River, transferred to HMM-262 and went on two Mediterranean and two Caribbean cruises. He was transferred in June 1965 to HMM-161 at Hue Phu Bai, served there until November and transferred to HMM-263 at Da Nang and served there until July 1966. He flew as crewchief with both squadrons.

He was discharged from New River in October of 1966, he received Combat Aircrew Wings and Air Medals.

"The funniest part of my Vietnam experience was taking a load of allied troops into a zone where we were taking light small arms fire. They did not feel like fighting that day so they wouldn't get out of the plane. The gunner was pushing them towards me and I was throwing them out. As they fell out, one tried to get back in. As he came up the step I put my foot on his chest and kicked him out. I accidentally keyed up my mike and said, "Get out of here, you son of bitch!" The pilot said, "Who me?" and jerked the plane 30 feet up into the air as I was still throwing out more troops. I said, "No, not you. Put it back down," and continued throwing them out. "That was my contribution to the Allied fighting fraternity."

Happily married for 20 years, he and his wife have a daughter, and he drives for Yellow Freight System in Chicago.

ROBERTO G. ZAPATA, "Z," was born on June 9, 1941 in Brownsville, TX. He graduated from Brownsville High School in May 1961 and joined the Marines on June 1, 1961. He attended boot camp in San Diego. While serving in the States he was stationed at MCAS Tustin, MCAS El Toro, NAS Dallas, TX and MCAS Hawaii.

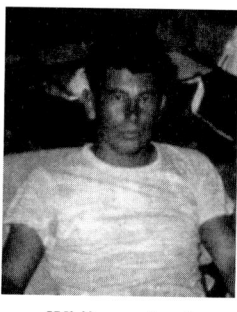

While serving in-country he was an aerial gunner on UH-34s and CH-46s. He was stationed in Vietnam from 1966-67 and in Okinawa, Japan.

Special memory: "All day re-supply mission for some infantry unit. We flew all that day late into the evening. Everyone was pretty tired when we returned to Ky Ha, I asked the crewchief if I could leave my flight helmet in No. 18, I would get it in the a.m. I never saw my helmet again cause No. 18 went down at sea on its way to the hospital ship with medevacs. No. 18 was our first HMM-165 helicopter loss in-country with a very good friend and fellow Marine, Stan Leroy Corfield, Staff Sergeant USMC aboard. He was our first casualty on May 1, 1967. May God bless him and his family."

Zapata was discharged on June 1, 1991 with 30 years service with the rank of Master Gunnery Sergeant. He was awarded the USMC Combat Aircrew Wings, four Air Medals, the Meritorious Service Medal and the Navy Commendation Medal.

He is married with three daughters and two sons, three granddaughters and two grandsons. He is retired and plays golf.

GARY RICK ZILLA, "ZEKE," was born and educated in Darby, MT. He graduated from the University of Montana in 1968. He completed Army flight school in September 1969 and transitioned to CH-46s at Santa Ana. He went to WESTPAC in May 1970. He served briefly with HMM-165 on Okinawa, and then flew about 800 combat hours with HMM-263 in Vietnam.

He was awarded one Single Mission Air Medal, 31 Air Medals, NUC, MUC and Vietnamese Cross of Gallantry. He flew CH-46s in New River prior to his discharge as a Captain in 1973.

He returned to Montana and became a school teacher. He was killed in a National Guard helicopter crash.

CLOYD HENRY ZIMMERMAN, "MONGOOSE," born in May 1945 in Harrisburg, PA. He graduated from Milton Hershey School in June 1963. He enlisted in the Marine Corps in August 1963. After boot camp at Parris Island, he went to Heavy Equipment Mechanics School; then to Engineer Maintenance Company, Camp Lejeune.

He was sent to Chu Lai, Vietnam in August 1966 and then to Dong Ha, Vietnam in December 1966. He served with HMM-363 as a door gunner on UH-34s.

During his tour, he was awarded the Purple Heart, Air Medals, Combat Aircrew Wings, National Defense Medal, Vietnam Service Medal, Vietnam Campaign Medal, Vietnam Cross of Gallantry, Navy Unit Commendation and Presidential Unit Citation. He was released from active duty as a Corporal at El Toro, CA in August 1967 and discharged in August 1969.

He has received diplomas from International Correspondence School-Diesel, General Motors School-Advanced Diesel, Chrysler Air-Temp, Perfect Circle, National Institute of Automotive Service Excellence and holds a Pennsylvania State Inspection License. He worked for 17 years for a Texaco garage and then became owner/operator of Zimmerman's Automotive. He is currently an active member of the USMC/Vietnam Helicopter Pilots and Aircrew Reunion, Vietnam Veterans of America, Veterans of the Vietnam War the VFW and the DVA.

He has been married 27 years, has four daughters and six grandchildren.

TOM ZUPPKE MOS 7564/ aka "HOOPER" OR "DASH-1" was born on July 11, 1944 in Stevensville, MI. He graduated from Kansas State University (BA)

and earned a BS from UNH. He entered the delayed entry program to OCS. Life after Quantico: Flt 18, VT-2, HT-8. Fleet Service: HMH-461, 462, 463,HMM-261, 264, 265. USS *Guadalcanal, Inchon, Guam, Tripoli.* H&MS-13, 26, Camp SD Butler (staff slots).

"One of my most gratifying experiences was Easter 1974 when we (CH-53D crew) landed on a northern beach of Hilton Head and "winched" out a 4WD International truck. No medals, but many really great smiles."

He was ordered home in July 1980 and discharged in July 1981. Presently he is in MNARNG (finally a REMF and loving it). Previously in CTARNG and USAR. TT at 13,000, NVD qualified but not current (darn!). Checked out in 17 helo models and 20 stiff wings. He quit flying commercially in 1991 and military in 1993. He is presently building and remodeling homes. He recently passed a MNDOT check ride in a 50 pass "BlueBird" (diesel school bus). Yes, after all these years, he's still hauling reluctant kids into LZ's that aren't to their liking.

He has two lovely and wonderful ex-wives, three children (confirmed) scattered about Minnesota, Ohio and Connecticut. He holds membership in the USMC/Vietnam Pilots and Aircrew Assoc., American Legion, and NAGUS.

"When I grow up I would like to be a fishing guide/scout for whoever is in control at the time. Am willing to wear my hair short (some of it is permanently that way) and occupy inadequate housing."

... and here's to the one percent who can't get their bio in on time ...

JAMES L. BOLTON, TRAPPER, CAPT. PERFECT, Lieutenant Colonel USMC (Ret), was born March 2, 1933 in Oregon, OH. He is a 1951 graduate of Clay High School, Oregon, OH and a 1956 graduate of Mount Union College, in Alliance, OH.

He was commissioned a Second Lieutenant in June 1955 and was designated a Naval Aviator on February 1959, HAC HMM-262, Flight Instructor HT-8, HAC HMM-161, C. O. and HAC HMH-462. Received DFC, 36 Air Medals, and the Navy Achievement Medal for flying action with HMM-161 and HMH-462.

He received distinguished recognition as USMC Aviator of the Year in 1975 for flying action as CO HMH-462 during the Evacuation of the Embassies at Saigon, RVN, (Operation Frequent Wind) and Phnom Penh, Cambodia (Operation Eagle Pull) He retired in June 1978.

VMO-2 UH-1E gunships escort a HMM-163 UH-34D medevac helicopter over the Marble Mountains south of Da Nang, South Vietnam, November 1967. (Courtesy of A.M. Leahy)

LIFE

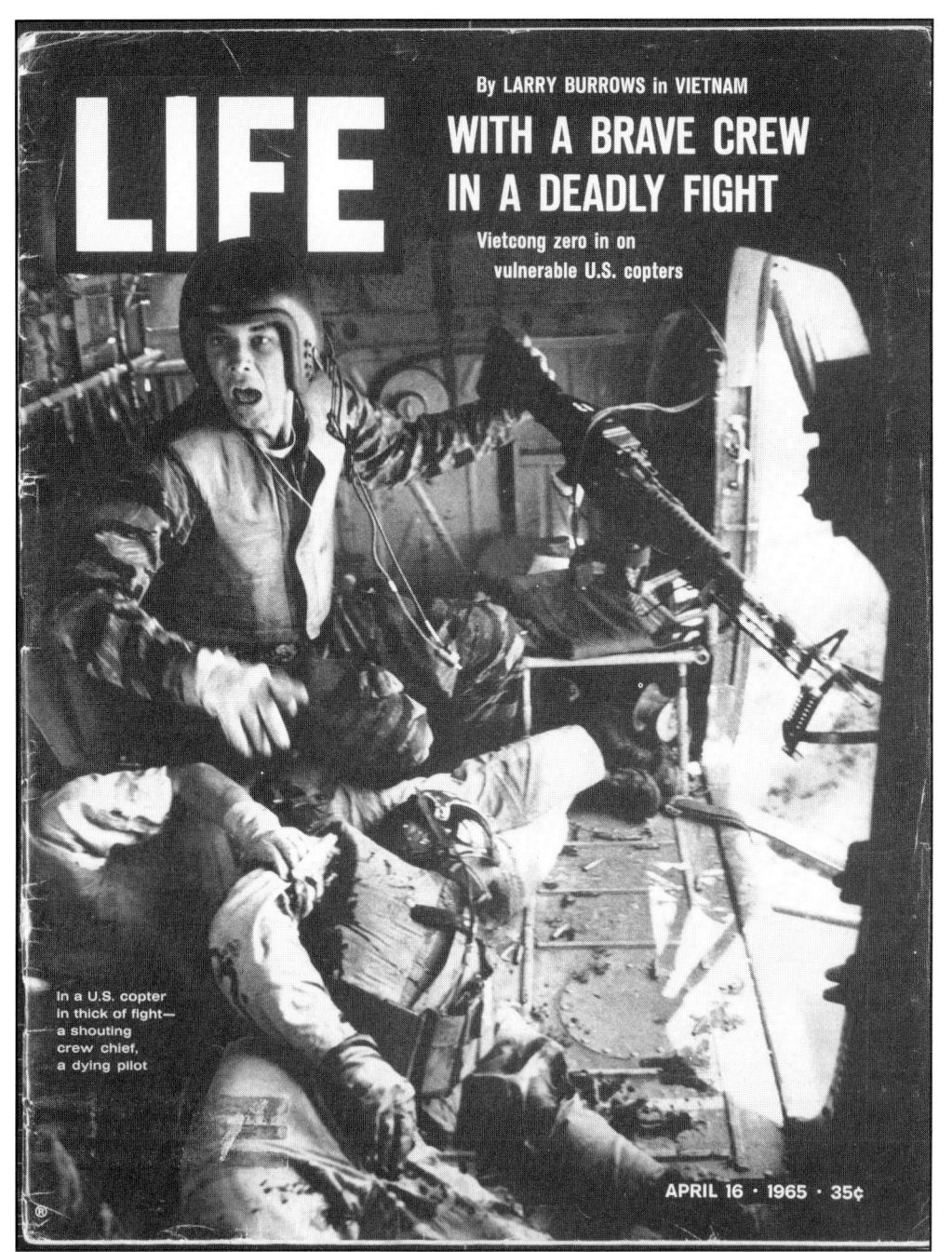

By LARRY BURROWS in VIETNAM

WITH A BRAVE CREW IN A DEADLY FIGHT

Vietcong zero in on vulnerable U.S. copters

In a U.S. copter—in thick of fight—a shouting crew chief, a dying pilot

APRIL 16 · 1965 · 35¢

Private First Class Wayne Hoilien, Gunner

Lance Corporal James C. Farley, Crewchief

LIFE
Vol. 58, No. 15   April 16, 1965

Life *Magazine, April 16, 1965. A doleful account of an HMM-163 mission during early helicopter action in Vietnam.*

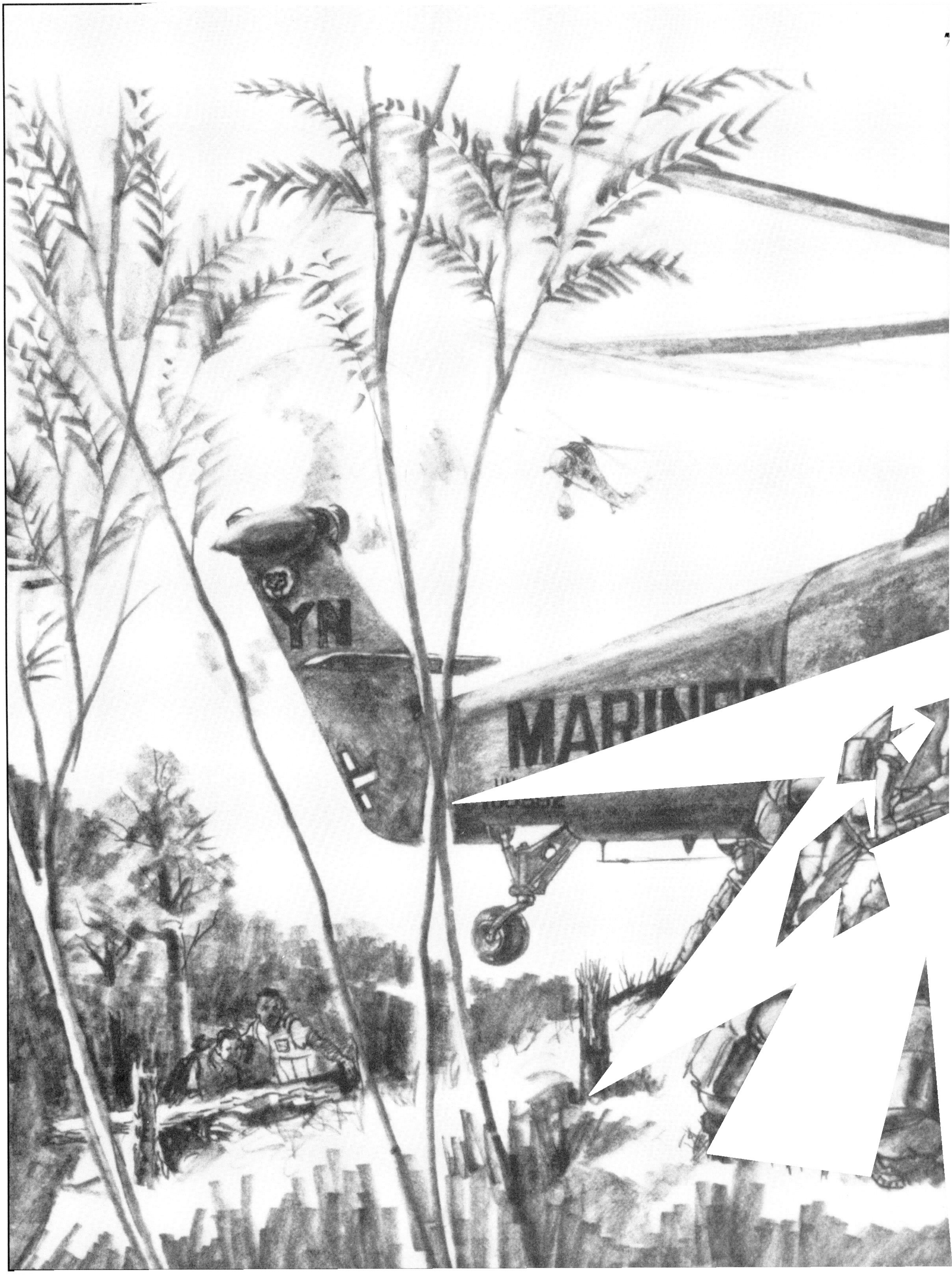
YN
MARIN

Medina Medevac October 1967. (Courtesy of A.M. Leahy)

Roster

NAME	IN COUNTRY SQDN
— A —	
ABERN, JOHN F.	HMM-362
ABERNATHY, MIKE	HMM-162/361
ABRAHAM, DAN	HMM-165/263/H&MS
ACCOMANDO, FRANK ("ACCY")	HMM-263
ACREBACK, JAMES R. ("JIM")	VMO-6
ADAIR, RON ("FAT")	HMM-161/364
ADAMS, ANDREW	HMM-362
ADAMS, JAMES E. ("JIM")	HMM-364
ADAMS, JOHN P.	HMM-164
ADAMS, LARRY	HMM-165
ADAMS, LARRY G.	HMM-361/161/VMO-6
ADAMS, NICK	
ADAMS, RICHARD J.	HMM-161
ADAMS, VERNON F.	HMM-361
ADZIMA, STEVE	HMM-265
AGNEW, ROBERT W. ("AGGIE")	HMM-162/164/364
AGUIRRE, MARIANO ("CHIEF")	HMM-365/165/H&MS
AINSWORTH, DONALD ("DON SAN")	HMM-163/VMO-2
AKERLEY, JOHN	HMM-262
ALBER, JOHN	HMM-265
ALBIN, GARY R.	HMM-161
ALDRICH, STEVEN	
ALDRIDGE, JAMES R.	HMM-165
ALDWORTH, JIM	HMM-362
ALEXANDER, BRUCE	HMM-263
ALEXANDER, JOHN	HMM-262
ALEXANDER, TYLER ("DASC DICK")	VMO-6
ALFORD, PAUL D. JR.	HMM-161
ALLANSON, WILLIAM A.	
ALLEGA, FRED	HMM-165
ALLEN, AL	HMM-262
ALLEN, ANSEL	HMM-163
ALLEN, BILL	HMH-461
ALLEN, BOB	HMM-161
ALLEN, DAVID	HMM-161
ALLEN, DWIGHT R. JR.	HMM-261/163/VMO-2
ALLEN, EDDIE ("ODD JOB")	HMM-161/363
ALLEN, GERALD R.	HMH-463
ALLEN, JIM	HMM-163/VMO-2
ALLEN, RICHARD ("RICK")	VMO-6/HML-167
ALLIGOOD, E.L. ("LEE")	HMM-361
ALLISON, DONALD ("AL")	HMM-364
ALLISON, JOHN	HMM-161
ALLMAN, LEONARD	HMM-163
ALLSHOUSE, RICHARD G. ("ABBIE")	HMM-162/364
ALMY, DICK	VMO-6
ALTAZAN, KENNETH A.	HMM-364
ALTHOFF, DAVE	HMM-262
ALVARADO, FRANK J. ("BIG AL")	HMM-262/365
AMESQUITA, DAVID	VMO-6
AMEY, DAVE	HMM-363
AMISH, PETE	HMM-161
AMOS, BGEN. G.R. ("GRANNY")	HMH-463/HMM-163
AMTOWER, MICHAEL	HMH-463
ANDERSON, DENNIS H. ("DENNY")	HMM-362/165/263
ANDERSON, EARL	1st MAW
ANDERSON, ED	HMM-163/VMO-2
ANDERSON, GEORGE	HMM-161
ANDERSON, HOWARD	HMM-164
ANDERSON, JIM ("ANDY")	HMM-164
ANDERSON, MARC	HMM-163/263
ANDERSON, MGEN. NORMAN	C.G. 1st MAW
ANDERSON, RALPH C.	HMM-261
ANDERSON, RICK	
ANDERSON, WAYNE	
ANDERTON, R.H. ("DOUGHBOY")	HMM-163
ANDHURST, JIM	
ANDRES, GARY	
ANDREW, TOM	HML-167
ANDREWS, JAMES ("CRASH KIMO")	VMO-6
ANDREWS, JAMES F. ("SHADOW")	HMM-161/265
ANGLE, PETER F.	HMM-162
ANGLIN, QUINTON G. ("GUY")	HMM-161/262
ANTONE, JAMES E.	VMO-6
APKER, NEIL	HMR-163 [Korea]
APP, JOHN ("NASTY")	HMM-161
APPEL, R.J.	
APPLEGATE, DOUGLAS S. ("APPLE")	HMM-263
ARCHER, BRUCE	
ARICK, BGEN. JOHN	VMO-6/HML-367
ARMSTRONG, DAN	HMM-361/161
ARMSTRONG, GREG	HMM-362
ARMSTRONG, MGEN. V.A. ("VIC")	MAG-36/MWSG-17
ARMSTRONG, MIKE ("CAPT LATREC")	VMO-6
ARNOLD, JOHN W.	HMM-262
ARNOLD, STAN ("ARNIE")	HMM-26
ARTHUR, DAVID	HMM-161
ASHBAUGH, TIM M. ("MR. POOP")	VMO-2
ASHWORTH, TOM ("SWEAT HOG")	HMM-265
ASKEY, DAVID H. ("SKINNY LT.")	HMM-162
ASKINS, WILLIAM	
ASKMAN, JAMES R.	HMM-165
ASSURAS, JIM ("GREEK")	HMM-263
ASTLE, JOHN ("ACE")	HMM-164
ATAD, BASIL	HMM-365/161
ATKINS, DON C.	HMM-364/363
ATKINS, RONALD W.	VMO-6
ATKINSON, JIM ("BOURBON")	HMM-361
ATTEBERRY, GEORGE W. ("TUNA")	HMM-163/262
ATTELL, STEPHEN ("ABE")	HMM-362/363
ATYEO, WILLIAM E.	HMM-164
AUCELLA, JOHN ("J.R.")	HMM-361
AUST, RICHARD D.	HMM-263/MAG-16
AUTREY, W.G. JR. ("GILL")	HMH-463
AVERY, FREDERICK S. III ("SCOTT")	HML-367/VMO-3
AVERY, GENE	HMM-363
AYE, RALPH	HMM-163/363/VMO-6
AYOTTE, JERRY D. ("EVIL TWIN")	HMM-161
AZBILL, LOWELL	HMM-164

NAME	IN COUNTRY SQDN
— B —	
BABBIT, CARL	HMM-265
BABCOCK, BILL ("BABBERS")	HMM-262
BABCOCK, MIKE	VMO-2
BABITZ, DONALD M. ("DON")	HMM-165/H&MS-36
BABOS, ROBERT L. ("BOB")	HMM-361
BACCITICH, DAVE ("BASS")	VMO-2
BACHERT, M.W.	HMM-161
BACHMAN, DENNIS A.	MAG-16
BACHMAN, VICTOR A.	MAG-16
BACKLUND, STEVE	HMM-363
BACON, LARRY ("ZIP")	HMM-165
BAGBY, MIKE ("BAGS")	HMM-165
BAILEY, BOB	VMO-2
BAILEY, GARY L. ("BEETLE")	HMM-164/165
BAILEY, GENE E. ("BEETLE")	HMM-362/HML-367
BAILEY, GLEN ("CHIEF")	HMM-161
BAINS, FRANK ("UMATILLA")	HMM-361
BAIZ, MARIN G. ("BUZZ")	HMM-361/362/363
BAKER, CHARLES R. ("CHUCK")	HMM-161/263
BAKER, DAVE	
BAKER, GARY ("TURKEY BUZZARD")	HMM-364
BAKER, GENE	HMH-463
BAKER, JERRY	
BAKER, LARRY	HMM-263
BAKER, OWEN ("O.C.")	HMM-364
BAKER, SID	HML-367
BALCH, ROBERT M. ("BOB")	HMM-361/463/H&MS
BALDERRAMA, VINCE	HMM-263
BALDWIN, DAVID	HMM-364
BALDWIN, RICK	HMM-161
BALLATO, NAT ("GUINEA GARDENER")	HML-167
BALLENTINE, DAVID	VMO-6
BALSEY, JOHN	HMM-262
BALTEZORE, AL	HMM-165
BANCELLS, LORENZO R. ("BANO")	HMM-163/263
BANCROFT, DICK	HMM-365/262/H&MS
BANDA, ROBERTO E.	H&MS-16
BANGERT, JOSEPH V.	VMO-6
BANKS, ROLAND	
BARBA, BILL	HMM-262
BARBEN, D.L. ("DOOR SLAMMER")	HMM-364/VMO-2
BARBER, JOHN D. ("J.D.")	HMM-263
BARBES, ALDEN ("AL")	HMM-364/163
BARBIER, W.C. ("BILL")	HMM-364
BARBOUR, AL	VMO-2
BARBOZA, ANTHONY C.	HMM-164
BARCLAY, BOYD	VMO-3
BARCLAY, LAWRENCE A.	
BARDEN, ARNE	VMO-2/H&MS-16
BARENT, BRENT	HMM-263
BARGANSKI, CHARLES D.	HMM-165
BARKER, AL	
BARKSDALE, JAMES E. ("RUGS")	1st MAW
BARNES, DON ("D.B.")	H&MS/HMM-164/361
BARNES, LEW JR.	HMM-362
BARNES, WILLIAM G. ("BILL")	HMM-263
BARON, PETER T. ("PETE")	HMM-364/263
BAROUSSE, JIM ("BARF")	HMM-263
BARR, AL	VMO-2
BARR, DAVE	HML-167
BARR, JIM ("ONE-SHOT")	HMM-161/363/167
BARRACLOUGH, HAL	HMM-365
BARRETT, DENNY ("MA")	HMM-165
BARRETT, JOHN	HMM-263
BARRETT, ROBERT A. ("LUCKY BOB")	HMM-164/265
BARROW, TOM	HMM-263/265
BARRY, JACK	HMM-362/363
BARSKY, CONNIE A. JR. ("IGOR")	HMM-262
BARTEL, HUBERT M. ("BART")	
BARTEL, LARRY	HMM-163/362
BARTH, FRANK	HMM-161/263
BARTLETT, MIKE	HML-367
BARTLETT, TOM	HMM-164
BARTLETT, WILLARD W.	MAG-16
BARTLEY, MIKE ("BLACK BART")	VMO-6
BARTON, DICK ("BLACK BART")	HMM-262
BARTS, DENNIS	HMM-361
BASTIEN, FRED	
BASTIEN, JAMES J.	VMO-6
BATES, AUSTIN ("ABE")	HMM-163
BATES, CHARLES ("CHUCK")	HMM-161/263/H&MS
BATES, JIM	VMO-2
BATHA, FRANK	
BATTEY, WILLIAM A. ("BILL")	HMH-463
BAUER, JAMES F. ("JIM")	HMM-165
BAUERNFEIND, ED ("FAST EDDIE")	HMM-362
BAUMLER, RAYMOND ("BUTCH")	HMM-164
BEABOUT, DONALD	
BEAGHAN, JIM	HML-167
BEAL, DENNIS E.	HMM-263/161
BEAL, JOHN E.	VMO-6
BEAMON, SAM	HMM-262/164
BEARD, BILL	1st MAW
BEARDALL, BILL	HMH-463
BEARS, LARRY E.	HMM-362
BEAVER, DON	HMM-164
BEAVER, JACK	HMM-261
BEAVERS, FRED W.	HMM-363
BECHTOL, DANIEL G. ("BUNGER")	HMM-165
BECK, GABRIEL B. ("GABE")	HMM-364
BECK, RICHARD E. ("THE NOSE")	HMM-164/264/369
BECKERICH, PHIL ("THE HUS-MAN")	HMM-165
BECKMAN, DENNIS ("DENNY")	HMM-162/362/H&MS
BEEBE, BILL ("GREASE GUN")	HMM-364
BEEBE, CHARLES B. ("CHUCK")	HMM-361/263
BEEDLE-RIES, GEORGE E.	HMH-463
BEELER, WILLIAM R.	HMM-265
BEERY, JACK	HMM-164/165
BEGLEY, JACK	HMM-265
BELFOURE, ED ("FAST E")	HMM-161
BELITSKUS, PATRICK R.	HMM-165
BELL, HARVEY L.	HMM-163
BELL, JAMES A. ("TRIPLE-A")	HMM-363/HML-367

NAME	IN COUNTRY SQDN
BELL, RICHARD ("DING DONG")	HMM-364
BELL, ROD	
BELLARS, CHARLEY	
BENCKENSTEIN, STEPHEN C. ("BECK")	HMM-265
BENDER, LAWRENCE ("BUD")	
BENDER, RICHARD R. ("RICH")	HMM-365
BENET, PETE	HMM-263
BENNER, STEVE	HMM-161/H&MS-17
BENNETT, DAVID C.	HMM-363/H&MS-36
BENNETT, PAUL E.	HMM-165
BENNINGTON, BILL	HMM-164
BENO, JOE	HMM-163
BENSON, DOUG	
BENSON, GARY W. ("HAMMER")	HMM-364/263
BENTON, ROBERT	HMM-362
BERCIER, CHARLES H.	HMM-262
BERG, BILL	
BERG, JIM	HMM-265
BERG, JOHN	HMM-164
BERGMAN, CARL E.	HMM-163
BERGMAN, JACK	HMM-164
BERIL, PAUL	
BERINGER, BOB	VMO-6
BERMUDEZ, FRANK D.	HMM-365/161/463
BERREY, CHUCK	HMM-264
BERRY, GERRY ("BEAR")	HMM-161/165
BERRY, THOMAS P. ("T.P.")	HMM-165/265/364
BERRY, TIM ("ORANGE BLOSSOM")	HMM-164
BETHEL, CHARLES E.	VMO-2
BEUMER, ELDON ("BO")	HMM-362/HML-167
BEVIL, PAUL	HMM-161
BEVIS, DENNIS ("DENNY")	HML-367
BEYER, JAMES J. ("J.J.")	HMM-361/263
BEYMA, DENNIS ("DOC")	HMM-164
BEZMAREVICH, RUDY ("ENFORCER")	
BIANCHINO, RICHARD L.	HMM-364
BIBLE, JERRY D. ("CASPAR T.F.G.")	HMM-363
BIEBERBACH, LEONARD C. ("LEN")	VMO-2
BIERBACH, C.C.	
BIGELOW, TIM	VMO-3
BILTIMIER, CHARLES H. ("CHUCK")	HMM-165
BIMAT, TOM R.	HMM-165
BIRD, MALCOLM	HML-367
BIRT, WESLEY	HMM-364
BIRZER, ED	HMM-164
BISHOP, GARY	VMO-2/HML-167
BISPLINGHOFF, GARY	HMM-365/HMH-463
BJORK, DAVID	HMM-364
BJORNAAS, FORREST R. ("B.J.")	H&MS-16
BLACK, JOHN M.	HML-167
BLACK, RUSSELL M. ("RUSS")	HMM-263
BLACKBIRD, MIKE	HMM-362
BLACKMON, ROBERT	
BLADES, LTGEN. ART	HMM-262
BLAIN, JERALD R.	
BLAINE, TOM	HMM-361
BLAIR, BRENT	HMM-161
BLAKE, TOM	HML-367
BLAKEMAN, WYMAN ("BLAKE")	HMM-161/163
BLANC, DICK ("EAGLE EYE")	HMM-161/163
BLANCHARD, DONALD A. ("DOC")	HMM-165/164
BLANCHFIELD, RICH	HMM-164
BLAND, RICHARD	HMM-364
BLANICH, JIM	HMH-463/361 6
BLANTON, R.D. ("BOB")	HMM-263/HML-367
BLAYLOCK, JAMES	
BLAYTON, OSCAR	VMO-6
BLEDSOE, CARL	VMO-2
BLIVEN, RICHARD L.	HMM-164
BLOCH, BOB	H&MS-16
BLOCK, CHARLES ("CHARLIE")	HMM-163/361/463
BLOOMER, ALAN	HMM-362/H&MS-16
BLOOMFIELD, W.P. JR. ("BILL")	HMM-365/161/H&MS
BLOSSER, TERRY S.	HMM-165
BLUHM, BOB	HMM-163/262
BOATWRIGHT, DAVE	HMM-265
BODE, RON	HMM-161/163
BODEN, JOHN B.	VMO-6
BODI, N.E.	HMM-265
BOEMERMAN, GEORGE ("BOOMER")	HMM-365
BOGG, CHARLES F. ("CHARLIE")	HMM-161/H&MS-16
BOHRMAN, ROBERT ("BAD BOB")	HMH-463
BOICE, MIKE ("PIGPEN")	HMM-165
BOLAN, HERMAN	HMM-262
BOLAND, JIM	HMM-263/362/364
BOLES, ANDREW	HMM-265
BOLGER, BILL	
BOLLER, GARY W.	HML-367/VMO-2
BOLTON, JAMES L. ("TRAPPER")	HMM-161/HMH-462
BOMKAMP, NORM	HMM-165
BONG, TOM	HMM-362
BONNER, DONALD C.	HMM-164
BOOHER, RONALD L. ("LEE")	H&MS-36/16/HMM-364
BOONE, BILL	HML-167
BOONE, JIM ("DAN")	HMM-365/HMH-463
BOONE, KEN	HMM-361
BORDELON, ANDY	
BORODAY, PETER J.	HMM-362
BORRIOS, AMADOR D.	HMM-163/262/VMO-2
BOSBONIS, STEVE ("BOZ")	HMM-261
BOSLER, BOB	
BOSQUEZ, MIKE	HMM-165
BOSSERDET, ROBERT E.	HMM-164
BOSTON, DICK	VMO-6
BOSTWICK, HENRY M.	HMM-265
BOTCH, MIKE	HMM-164
BOTELER, RICHARD E.	HMM-363
BOTTORF, TOM	
BOULEY, JOHN ("J.T.")	VMO-2/HML-167

NAME	IN COUNTRY SQDN
BOULLE, JON A.	HMM-161
BOULOS, JOSEPH ("JOE FRED")	HMH-463
BOULTON, STEVE	HML-167
BOUSQUET, BRUCE P. ("FLEA")	HMM-265
BOWDITCH, TOM ("SHADOW")	HMM-164
BOWDLE, JACK	
BOWERS, ROBERT L. ("BUTCH")	HMM-165
BOWLING, JOE D. ("BANSHEE")	HMM-261
BOWMAN, JERRY	HMM-263
BOWMAN, JOHN W. ("BOWS")	HMM-362/HMH-462
BOYACK, KENT	
BOYCE, AL	HMM-263
BOYD, HERMAN P. ("HORSE POWER")	HMM-161
BOYER, BOB	HMM-263
BOYKIN, KEN	
BOZEMAN, STEPHEN M.	HMM-361
BRAATAN, MGEN. TOM	HMM-263
BRACY, ROBERT A. ("BOB")	HMM-362/VMO-2
BRADBURY, LAWRENCE W. ("BRAD")	H&MS-36/VMO-6
BRADDON, JOHN R.	HMM-364
BRADFORD, BILL	
BRADLEY, CHRIS ("BREATH")	VMO-2
BRADLEY, DAVID D.	HMM-165
BRADLEY, EDWARD ("TED")	HMM-162/362
BRADLEY, KEN	
BRADLEY, THOMAS A.	HMM-361/164
BRADSHAW, MICHAEL E.	HMM-165
BRADY, JAMES J. ("DIAMOND JIM")	HMM-162
BRAKE, BENNIE M. ("BEAR")	HMM-163
BRAMAN, DOUG	HMM-161
BRAND, DON	HMM-165
BRAND, HARRISON	HMM-161
BRAND, LOREN E. ("BRANDY")	HMM-265
BRANDON, DAVID E.	HMM-164
BRANDON, JACK A.	VMO-6
BRANDT, STEVE ("BIG DOG")	HMM-364/263
BRANK, W. SCOTT ("GRIT")	HMM-165/364
BRANNIGAN, JAY	HMM-263
BRANUM, W.D.	HMM-364
BRATTON, ANDY	HMM-262
BRAVARD, STEVE	HMM-262
BRAY, RICHARD P. ("PAT")	HMM-362
BREAZEALE, H. PAYNE III	HMM-263
BREEDING, RANDY	HMM-364/262
BREESE, CHIP	HMM-363
BREIGHNER, MIKE	
BRENNAN, JOHN	
BRENNEMAN, BEN	VMO-2
BRESNAHAN, ART ("BUDDHA")	HMM-164/262
BRESSETTE, GLEN D. SR. ("CHIEF")	HMM-165
BREUSS, TOM	HMM-161
BREWSTER, JERRY V.	HMM-262/364
BREY, JAMES	HMM-164/264
BREY, JIM	HMM-164
BRICKELL, JAY ("PAPPY")	HMM-163
BRIGHT, GARY R.	VMO-6
BRINGHAM, BOB	HMM-362
BRISCOE, EDWARD D. ("ED")	HMM-163/364
BRITTON, LARRY W. ("SLICK")	HMM-364
BRITTON, MICHAEL R.	HMH-461
BROCK, T.C. JR.	HMM-165
BRODERICK, TOM	HML-167
BRODIE, DONALD C.	HMM-164/362
BROKAW, JIM	HMH-463
BRONSON, JOHN	
BRONSON, PAUL	
BROOKINS, ROBERT D. ("SNAKE")	HMH-463
BROOKS, DAVID D.	HMM-165
BROOKS, KENNETH E. ("KEN")	HMM-265
BROOKS, WILLIAM M.	HMM-162/163/362
BROPHY, LARRY	HMM-161
BROSS, NEIL	HMM-164
BROUGHTON, W.C. JR. ("BILL")	
BROWN, ARTHUR	VMO-6
BROWN, BEN	
BROWN, BOB	VMO-2/H&MS-16
BROWN, COSTA	
BROWN, DAVID C.	HMM-362/HML-367
BROWN, GARY LEE	HMM-165
BROWN, JERRY L.	HMM-163
BROWN, STEVE ("BROWNIE")	HMM-262/164
BROWN, WILLIAM W. III ("W.W.")	HMM-164
BROWNFIELD, TOM ("BROWNIE")	HMM-362
BRUDZINSKI, GEORGE	HML-367
BRUNNER, DENNIS	HMM-165
BRUSH, ROY	HMM-364
BRUST, BOB	HMM-361/HML-167
BRUST, DAVID E. ("DAVE")	HMM-263
BRUYERE, JAMES L. JR. ("FRENCHY")	HMM-163/365
BRYAN, REX	HMM-262
BRYANT, DAN ("ACTUAL")	HML-167
BRYANT, DAN ("SIMULATED")	
BRYCE, TOM	HMM-164
BUCCERI, JACK	
BUCKALEW, DANNY K.	HMH-463/361
BUCKINGHAM, SCOTT	HML-167
BUCKMAN, JOE ("GEEK")	HMM-361
BUCKNER, GORDON H.	HMM-362
BUFFINGTON, EARL	HMM-163
BUFTON, EARL L.	HMM-364/163
BULLARD, IKE	HMM-364
BULLOCK, JIM	HMM-165
BULOW, BOB ("BUFF")	HMM-161/263
BUMM, MARK ("BOOMER")	HMM-364/262
BUNTING, DAVE	HMM-362/363
BURBANK, JACKIE Y.	MAG-16
BURGEMEISTER, AL	
BURGESS, GLENN ("SMOKE")	HMM-364
BURGESS, JOHN ("NEW GUY")	HMM-361
BURGESS, MICHAEL R. ("MIKE")	HMM-165
BURGESS, TIM ("BYRD")	HMM-263
BURKE, DAVID	HMM-164
BURKE, JOHN	HMM-363
BURKE, MARIUS JR. ("MAROOSCH")	AIR AMERICA
BURKE, NICHOLAS ("NICK")	VMO-6/HML-367
BURKMIER, RICK ("TROOP")	HMM-263
BURKS, LAGRAND	VMO-2
BURNETT, BOB	
BURNHAM, AI	HMM-263

NAME	IN COUNTRY SQDN
BURNHAM, BOB	HMM-264/364
BURNS, GEORGE	HMM-161/263
BURNS, JACK	HMH-462
BURNS, JOHN G.	HMM-362
BURNS, TOM	VMO-6
BURNWORTH, J.W. ("PAPPY")	HMM-261/H&MS-16
BURRIS, BOB	HMM-164
BURSON, SAMMY R.	HMM-165
BURSTEIN, JEFF ("MOONEY")	H&MS-16/MAG-16
BURT, JIM ("REDEYE")	HMM-365/164/VMO-2
BUSBY, GLENN	HMM-163
BUSH, DALE	HMM-161
BUSHLOW, DAVID M. ("DAVE")	VMO-6
BUSTAMANTE, RAUL ("DIMMER TACO")	HMH-463
BUTLER, JAMES ("SMEDS")	HMM-164/362
BUTLER, JOHN ("SMEDLEY")	HMM-163/264
BUTLER, MIKE	
BUTTS, JERRY	
BYRD, BOBBY	HMM-165
BYRNE, BILL	
BYRNE, G.A.	
BYRNE, REG ("SMOKEY")	HMM-265
BYRNES, JIM ("DIGGER")	HMM-364
— C —	
CADY, ARTHUR R.	HMM-362
CAGLE, BILL	HMM-165
CAHIL, JIM	
CAHRAN, BUDDY	1st MAW
CAIN, ALLAN D.	HMM-362
CAIN, JOHN ("KILLER")	HMM-262
CALDER, ADRIAN R.	MAG-16
CALDON, DAVID ("MARBLE")	
CALDWELL, BILL	HMM-361
CALDWELL, TOM	
CALEF, MARSHALL M. II ("SWAMP")	HMM-361
CALKINS, CHET	HMM-161
CALKINS, R. BARKLEY ("BARK")	HMM-162
CALLISON, GEORGE	HMM-263
CALVERT, JOHN	
CALVERT, MICHAEL	HMM-262
CAMERON, DOUGLAS L.	HMM-263/265
CAMPBELL, DONALD F.	MAG-36
CAMPBELL, GENE A.	HMM-361
CAMPBELL, HARRY J. ("SOUPY")	HMM-265
CAMPBELL, JERRY SANTI	HMM-165
CAMPBELL, JIM	HML-167
CAMPBELL, JIM ("RED BONES")	HMH-462
CAMPOS, RAY	HMH-361/463
CAMPOSE, JOSE	HMM-264
CANADA, RAY	HMM-262
CANFIELD, CLAY	HMM-362
CANINGTON, SHAD	VMO-3
CANNIS, ROBERT M.	
CANNON, C.E. ("CHARLIE")	HMM-364/VMO-6
CANNON, TOM	HMH-462/H&MS-36
CANZONERI, TONY	HML-167
CAPDEPON, HENRY ("CAP")	
CAPEHART, ED	
CARBARY, R.R. ("TIM")	HMM-363
CARBONE, JOHN	HMM-161/262
CARCASIO, JOE	HMM-161
CARD, KEITH M.	HMM-263
CARDINAL, DAVE	
CARE, JIM	HMM-163
CAREY, RICHARD L. ("BENO")	HMM-262/265/461
CARL, MGEN. MARION E.	1st MAW
CARL, PAUL ("P.J.")	HMM-364
CARLEY, GARY G.	HMM-163/164
CARLEY, MICHAEL	HMM-362
CARLIN, RICHARD	
CARLISLE, JOE	HMM-361
CARLON, KEN ("K.C.")	VMO-6
CARLON, TOM	
CARLSON, BILL	
CARLSON, BOB	H&MS-16
CARLSON, ED	HMM-162
CARLSON, ROD	HMM-265/361
CARLSON, WILLIAM C. ("WILL")	
CARLSTRAND, BOB ("BOBO")	HMH-463
CARNAGEY, LARRY	HMM-165
CARNEY, DON	HMM-262
CAROL, GEORGE	
CARON, RON	HMM-161
CAROPRESO, LOUIS	HMM-163
CARPENTER, MICHAEL	HMM-263
CARR, RICHARD F. JR.	HMM-164
CARROLL, GEORGE W.	
CARROLL, JOHN E.	HMM-362
CARROLL, MICHAEL ("CHICKENMAN")	HMM-262
CARROLL, TOM ("T.C.")	HMM-161
CARSON, EMMETT	HMM-364
CARSON, KELLY	HMM-164
CARTER, AARON L. ("SHAKEY JAKE")	HMM-163
CARTER, DAVE ("DELTA TEE")	HMH-463
CARTER, RAY	
CARTWRIGHT, CARL R.	
CARUSELLE, MICHAEL	HMM-365
CASCIANO, CARMINE ("CASH")	HMM-364
CASCIO, BEN R.	HMM-362
CASEY, GEORGE G.	HMH-462
CASHMAN, MIKE	HMM-164
CASTERLINE, DICK	
CASTLE, RONALD	HMM-265
CATALANO, PETE	HMM-263
CATES, ALLEN	HMM-365
CATES, CHARLES	HMM-362
CAUDLE, JERRY W. ("T.C.")	HMM-361
CAUGHEY, FRANK	
CAVAN, JOHN	
CAVEY, DAVID	HMM-161
CEDERHOLM, ROGER ("ZUF")	
CEDERLIND, LEON E. ("LEE")	HMH-463/MAG-16
CENTORE, RICHARD ("RICK")	VMO-2
CERESKO, DICK	VMO-3
CERRA, BERNIE	HMH-463
CHAFFIN, HARRY	HMM-265
CHAMBERS, DWIGHT	HMM-362
CHAMPION, CAL	HMM-362
CHANCEY, JOHN	HMM-364
CHANDLER, RICHARD III ("RICK")	HMH-463
CHANDLER, STEPHEN D.	HMH-463
CHAPMAN, ART	
CHAPMAN, J.R.	
CHAPMAN, JOE	
CHAPMAN, RICHARD ("DICK")	HMM-163/161
CHAPPLE, BENNETT	HMM-165
CHARLTON, ALBERT	VMO-6
CHASE, CHARLIE JR.	
CHENKUS, EDWARD J.	HMM-362
CHERBONNIER, L. MICHAEL	HMM-263
CHESTER, JOHN	VMO-2/6
CHILDERS, LLOYD	
CHILDRESS, DAVID	HMM-161
CHIUMENTO, JOHN	HMM-361/363/165
CHOATE, T.W. ("TOM")	HMM-263
CHOATE, WILLIAM R. ("BILL")	VMO-6/HMM-164/364
CHRISTENSEN, CALVIN T. ("CHRIS")	HMM-362/164
CHRISTIANSEN, JAC	
CHRISTOPHERSON, NORMAN F.	HMM-164
CHURCH, ROY T.	
CHURCHIN, DENNIS ("S. WHIPLASH")	HMM-163/HMH-462
CIEPLIK, TED	VMO-2
CIHAK, BILL ("LT. FUZZ")	HMM-265/MAG-16
CIPOLLA, HENRY J. ("CHIP")	HMM-363/165/463
CISMESIA, AL	HMM-263
CITRANO, JIM	HMM-363
CIVELLI, JOSEPH R.	VMO-2/HMM-263
CLAPP, ARCHIE J.	HMM-362
CLAPPER, JERRY D.	HMM-161/364
CLARK, CHUCK ("PIGGY")	VMO-3/6/HML-367
CLARK, HAL	HMM-163/265/167
CLARK, J.P.	
CLARK, JERRY N.	HMM-365/161
CLARK, JOE	HMM-363
CLARK, JOHN LEWIS	H&MS-16
CLARK, NORM ("SUPERTWIGET")	HMM-164/263/265/364
CLARK, R.L.	HMM-265
CLARKE, MIKE D.	HMM-265
CLASBY, JOHN ("CLAS")	HMM-361
CLAUSEN, RAYMOND MIKE JR.	HMM-364/263
CLAYTON, WILLIAM C.	HML-367
CLEARY, BILL ("R.F.B.")	HMM-263
CLELAND, CARL	
CLEMENENTS, HARRY	
CLEMENS, MICHAEL H. ("MIKE")	VMO-2
CLIFTON, DICK	HMM-363/364/VMO-2
CLINE, DICK	HMM-362
CLINTON, RICHARD B. ("R.C.")	HMM-161
CLOCK, DONALD G.	HMM-164
COADY, ERIC ("J. GREEN GIANT")	HMM-163/361/363
COBB, DANIEL B.	HMM-165/263
COCHRAN, RALPH E.	
COCKMAN, ED	HMM-163
CODAY, BOB	HMH-361/463
CODY, WILLIAM	HMM-161
COFFIN, JIM	
COFFMAN, HAL	HMM-362
COFTY, CHUCK	1st MAW
COKE, JOHN W.	HMM-263
COLBERT, A. BRUCE	HMM-164/165
COLBORN, WAYNE T.	HMH-361
COLE, AL	
COLEY, AL ("DIRTY AL")	HMM-263
COLLIER, WILLIAM F. ("WILCO")	HMM-161/363
COLLINS, B.D.	HMM-265
COLLINS, BEN	HMH-463/H&MS-16
COLLINS, JAMES D. III ("J.D.")	HMM-262
COLLINS, JESSE	MAG-36
COLLINS, RICH	
COLLINS, STAN	HMR-362/VMO-2
COLLUM, JAMES	VMO-6
COMBS, ROGER E. ("ROSCOE")	HMM-364
COMER, D.K.	H&MS-16
COMER, GEORGE M.	HMR-163
COMMON, TIM	HMM-364
CONARD, JACK W.	HML-167
CONLEY, JEROME E.	HMM-463
CONLEY, WILLIAM J. ("WILD BILL")	HMM-163/HMH-463
CONNELL, J.L.	
CONNELLY, JOHN	HMM-164
CONNELLY, PAT	HMM-164
CONNIFF, J.A. III ("JACK")	HMM-162
CONNOLLY, JIM	
CONNORS, WALT	
CONROY, MICHAEL J.	HMM-164
CONROY, THOMAS R. ("TOM")	HMM-265
CONSTANTINE, TOM T.	VMO-6
CONWAY, DARRELL	HMM-363
COOK, DAVID J. ("COOKIE")	VMO-6/HML-167
COOK, GARY J.	HMM-363
COOK, JIM E. ("NUB")	HMM-164
COOK, ROGER ("ROG")	HMM-362/AIR AMERICA
COOKE, MGEN. RICH ("DARK EYES")	VMO-2
COOLMAN, MICHAEL	
COOPER, DENIS V. ("COOP")	HMM-365/264
COOPER, JACK R.	HMM-261/262/265
COOPER, JOHN	HMM-262
COPE, RUSSELL J.	HMM-361
CORATHERS, JOHN ("CAPPY")	
CORDELL, JOHNNY	HMM-161
CORDIN, GENE	
CORLEY, BILL	
CORLEY, RON ("HOSTILE MAN")	VMO-1
CORLISS, MGEN. GREG	HMM-262
CORMIER, C.J.	HMM-362
CORN, CLIFFORD D. ("CLIFF")	HMM-265
CORNETT, C.B.	VMO-2
CORNETTA, RON ("THE HAWK")	HMM-263
CORNFIELD, CHUCK ("FLAKES")	HMH-462/361/161/367
CORRIE, STEVE	HMM-263
CORS, DALE	HMM-262/164
CORYN, ROBERT	
COSIMANO, NINO	
COSTANZA, FRANK	HML-367
COSTELLO, KEITH W. ("MIKE")	HMM-261
COSTLOW, WALTER N.	HMM-362
COTTAM, JOHN ("NO SLACK")	HMH-463
COTTINGHAM, BOB	HMM-264
COUGHLIN, TOM	HMM-361
COUITT, CHARLES	HMM-164
COUNCIL, RON	HMM-264
COUNCILMAN, JOHN D. ("J.D.")	HMM-162
COWLING, JOHN	HMM-364
COWPER, THOMAS J. ("TOM")	HMM-163
COWPERTHWAIT, WILLIAM C.	
COX, GEORGE W.	HMM-364
COX, JOHN ("THE GUNNER")	HMM-165
COX, TONY	
CRAIG, CARL J.	HMM-164
CRAIG, HERB	
CRAIG, LARRY	
CRAIG, O.J.	HMM-165
CRAIG, PHIL	HMM-361
CRAWFORD, DELMAR DAN	MABS-16/HMM-263
CRAWFORD, DICK	HML-367
CRAWFORD, RICK	
CREAL, TOM	HMM-263
CREAMER, C.E. ("FAST EDDY")	HMM-362/HMH-463
CREECH, JIMMIE A. ("ABLE")	HMM-161/263
CRETNEY, WARREN G.	HMM-364/367/VMO-6
CREW, RANDOLPH E. ("RANDY")	HML-367/167
CREWS, BILL	HMM-164
CREWS, MOSS ("CLIFF")	HML-167
CREWS, TERRY	HMM-263
CRITSER, RON	
CRONE, FORREST W. ("SQUIRT")	HMM-361/HML-367
CRONIN, BGEN. JOHN	
CROOKALL, C. ("FAT CHARLIE")	HMM-262
CROSS, DAVID L.	HMM-364
CROSS, J.D.	HMM-161
CROW, BRUCE ("BUZZARD")	HMM-361
CROW, THOMAS V.	HMM-364
CROWE, EDWARD H. ("DINK")	HMH-361/HML-367
CROWE, WILLIAM R. ("BILL")	HMM-364
CRUMPTON, THOMAS D.	HMM-165
CRUSING, JOHN R. ("JACK")	HMM-261
CRUTCHER, BRYAN P.	HMM-263
CRUTCHER, JERRY W. ("TEX")	HMM-363/361
CUCKA, JOHN M.	HMH-461
CUDDY, FRANCIS J. JR. ("FRANK")	HMM-265/367/VMO-6
CULLEN, LAWRENCE V. ("LARRY")	HMM-363/MWSG-17
CULVAHOUSE, PAUL	
CUMPSTON, GEORGE	HMM-265
CUNNINGHAM, GARY ("RAT EYES")	HMM-361
CUNNINGHAM, ROBERT	HMM-262
CUNNINGHAM, ROBERT T. JR ("B.C.")	HMM-163/363
CUNNINGHAM, TOM	HML-367
CURL, NORMAN D.	HMM-161
CURLEY, BILL	HMH-463
CURNOW, FRAN	HMH-463/361
CURTIS, O.W.	HMM-162
CURTIS, TERRY ("HORSENIPPLE")	HMM-165
CYR, JACK S. JR. ("CAPT JACK")	1st MAW
— D —	
DABNEY, TOM	
DACY, VINCE	
DAHLSTROM, STEPHEN E.	HMM-263
DAILEDA, BILL	MAG-36
DAKE, MGEN. TERRANCE R.	HMH-462
DALRYMPLE, DONALD	HMH-463
DALTON, ED	HMM-161
DALTON, FRANK	VMO-6
DALTON, JAMES J. II ("MAD DOG")	HM-265
DALTON, JOHN ("J.J.")	HMM-163/164/165
DALY, MICHAEL	MAG-16
DANDREA, JIM	HMM-164/361
DANIELS, BOBBY	HMH-463
DAOUST, DAVID	HMM-164
DARGER, ROCKY E.	HMM-364/262
DARNELL, DENNIS	VMO-2
DARRAN, DOUGLAS C. ("DOUG")	HMM-363
DAU, JIM	HMM-262
DAUTRIEL, NED	HMM-364
DAVID, PAUL	
DAVIDGE, DEAN G.	VMO-6/HMM-165/364
DAVIDSON, JERRY	HMM-263
DAVIDSON, RICHARD M. ("DAVE")	H&MS-16/HMM-362
DAVIES, SPENCE	HMH-463
DAVIS, ALAN F.	HMM-364/262
DAVIS, ALLAN D. ("A.D.")	HMM-165
DAVIS, BILL ("W.J.")	HMM-164
DAVIS, DEAN	
DAVIS, DELL ("DUCK")	HML-367
DAVIS, EDDIE R.	HMM-361
DAVIS, ELMER H. ("FUDD")	HML-167
DAVIS, GEN. JOHN K.	
DAVIS, GEORGE A.	HMM-165
DAVIS, GLENN C.	VMO-2/HML-167
DAVIS, JAY M. JR.	HMM-163/364/167
DAVIS, JIM ("HOG")	HMM-164
DAVIS, JON	HMM-265
DAVIS, MICHAEL	HMM-165/364
DAVIS, RON D.	VMO-2/HML-167
DAVIS, THOMAS S. ("TOM")	HMM-263
DAVIS, WILLIAM	HMM-262
DAVISON, PATRICK E.	VMO-6
DAW, CHARLES J. ("CHARLIE")	HMM-162/165/263
DAY, EDWARD A.	HMM-163/164/361
DEAL, JOSEPH G. ("PAPA")	HMM-163
DEAN, ALMUS L.	HMH-462/463
DEAN, BRUCE B. ("BUD")	HMM-361
DEAR, EDWIN A.	HML-167
DEATON, JESSE W.	HMM-165/463/H&MS
DeBARGE, GEORGE L.	HMH-462/HMM-361
DeBLANC, DANIEL J. ("DAN")	VMO-2
DeBRINCAT, RON	HMM-364
DeHAAN, ANDRE	HMM-265
DeHAAN, JIM	HMM-165/364
DEIBERT, JOHN	
DeJOURNETTE, DAN	HMM-364
DEKKER, JAMES ("JIM")	HMM-162
DEL GROSSO, SEAN	
DELACQUA, RAY	HMM-363/263/364
DeLAIR, DOUGLAS D. ("DOUG")	HMM-163/362
DELGADO, RICHARD B.	
DELMORE, LARRY	
DeLONG, MGEN. MIKE	HMM-262
DELOTELLE, ROY	HMM-263
DELUTE, DAN	HMM-265
DEMPSEY, ROY	HMM-161
DENHOFF, ROBERT W.	HMM-165
DENISTON, JOHN	HMM-161
DENNIS, JOHN	HMM-265
DENNIS, ROBERT	
DENNIS, RON	HMM-164
DENNISON, R.E.	
DENNY, REX	HMM-161/VMO-2
DENTON, THOMAS	HMA-169/369
DEPUY, JAMES	HMM-161/363
DERBYSHIRE, LESLIE L.	HMM-263
DERRICK, ROBERT A.	HMM-263/163/262
DERYLAK, NORBERT L. ("DIRTY D")	HMM-362
DeSHANE, HARRY G.	HMM-365
DeSIMONE, DAVE ("DIZZY")	HMH-463
DeSNOO, PETER G.	HMH-463/HMM-164
DESPARD, PETE	HMM-161
DeVASIER, MARTIN	HMM-263
DEVINE, DOUG ("PROFILE")	VMO-2
DEVITT, E.F.	
DEVORE, GARY	HMM-164
DeWITT, DONALD L. ("DA FROG")	HMH-463/H&MS-16
DEWSRUP, PAUL	
DEXTER, DOUG ("PIPES")	HMM-361
DEYO, DAVID	HMM-361
DIACONT, GEORGE H.	HMM-265
DIAL, BILL	HMM-164
DICK, RICHARD A.	HMM-165
DICKERSON, A.C. ("AL")	HMM-165
DICKEY, CECIL ("BIG 'D'")	HMM-265/165
DICKEY, LARRY E.	HMH-463
DICKEY, STEVE	HMM-262/164
DICKSEN, AL	HMM-261
DICKSON, TOM	HMM-363
DIEBERT, JOHN	VMO-3
DIEHL, J. RAYMOND II ("RAY")	HMM-163/161
DIETRICH, PETER	HMM-161
DILLON, TAD JR.	
DILULLO, RALPH	
DINNDORF, JERRY	
DIRCK, RON	HMM-162
DISHAW, MIKE	MAG-16/36
DiSTAULO, ANTHONY C. ("CHUCK")	HMM-364
DITMORE, JOHN H.	HMM-161/363
DIX, ELLIOT	HMM-265
DIXON, SAM	HMM-362
DOANE, ED ("FIRE PLUG")	HMM-364/H&MS-36
DOBBRATZ, J.R.	
DODSON, GORDON	
DOI, CLYDE	
DOKTOR, A.M.	HMM-161
DOLATA, THOMAS M. ("DOC")	VMO-2/HMM-161/164
DOLNEY, ED	HMM-265/262
DOLSTEIN, DAVID	
DOMER, O.M.	HMM-164
DOMIN, MIKE	HMH-463
DONAGHY, RICH ("BROOKLYN")	
DONAHUE, AL ("A.J.")	HMM-263
DONALDSON, ZANE H.	HMM-163
DONATO, PETER	
DONNELLY, JOHN J.	HMM-362
DONNELLY, THOMAS	HMM-165
DONOFRIO, ERNEST	HMM-165
DONOGHUE, JOHN P.	H&MS-16/HMM-161
DONOVAN, JOSEPH	HMM-364
DONOVAN, MICHAEL J.	HMM-363
DONOVAN, PAT	HMM-364
DOOLEY, GERALD	HMM-265/364
DORFIELD, C.E.	HMM-263
DORISKI, THEODORE ("SKI")	HMM-162/261/361/165
DORR, DON	HMM-364/HMH-463
DORROH, JEFF	
DOSS, PAUL	
DOUGLAS, DWAYNE	HMM-164
DOUGLAS, JAMES R. ("JIM")	HMH-463
DOULL, WILLIAM G. ("BILL")	HMM-263
DOUST, DICK ("MAD DOG")	HMM-263
DOWNING, JOHN ("J. ANIMAL ZERO")	HMM-361
DOWNUM, CHUCK	HMM-164
DRAKE, DONALD L. ("DON")	HMR-163/VMO [Korea]
DRAPER, BILL	
DREMANN, TIM	HMM-265/262
DREWICZ, DUANE R. ("DREW")	HMM-164
DRIEFER, JOHN	HMM-265
DRISCOLL, JOE	HML-367
DRISCOLL, RICHARD W. ("BARNEY")	HMM-161
DRISGULA, A. RANDALL ("RANDY")	HMM-263
DRISKILL, TERRY	
DRUMM, DAN ("TOO DUMB TO DIE")	HML-167
DRURY, RICHARD L.	VMO-2
DUBINSKI, GERRY	
DUBINSKY, RON	
DuBOSE, FRANK JR.	H&MS-16
DUCKETT, JOE B. ("HONEY BUCKET")	HMM-364/262
DUERR, RONALD E.	VMO-6
DUERRE, GEORGE	VMO-2
DUESING, GREG	HML-167
DUFFY, DENNIS ("DUFF")	HMM-363
DUFRESNE, LEON	HMM-262
DUGAS, LOUIS ("ART")	H&MS-16
DUGGER, ROBERT F. JR.	HMM-161/H&MS-16
DUKE, BILLY E. ("OLD MAN")	HMM-165
DUKE, LEILAND M. JR.	VMO-6
DUKELOW, GORDON	
DULICK, JOHN A. JR.	HMM-362/HMH-463
DULUDE, DANIEL D.	HMM-265/165
DUMAS, PATRICK ("PAT")	VMO-2/HML-367
DUMPHY, W.W. ("WHITEY")	HMM-263
DUNCAN, DAVE	HML-367
DUNCAN, JOHN	HMM-161
DUNEV, PETER	HMM-364
DUNFEE, RALPH E.	
DUNFORD, JOHN	HMM-265

NAME	IN COUNTRY SQDN
DUNIEC, BRIAN J.	HMM-161
DUNLAP, ROBERT W. ("BOB")	HML-367
DUNN, GEORGE	HMM-161
DUNN, LOWELL E. JR.	HMM-165
DUNN, RALPH ("RALPHIE")	HMM-162
DUNN, TOM	VMO-2
DUNNE, WILLIAM ("WILD BILL")	AIR AMERICA
DUNSTAN, CHAUNCEY	HML-367
DuPONT, DELBERT	VMO-6
DURAL, JOHN	HMM-265
DURAND, E. RONALD ("RON")	HMM-361
DURANT, DAVE	
DURBIN, FRANK	HMM-265
DURHAM, RALPH ("PATRICK")	HMM-163
DURNEY, MATT ("DUKE")	HMM-164/364
DURR, JIMMY	HMM-164/364
DUSEBERG, ARNIE	HMM-262
DUTCHER, STUART	
DUTTON, VERNON	HMM-364
DYE, ROBERT	HMM-161
DYER, GLENN	
DYKES, CARSON	HMM-161
— E —	
EAGLES, TOM ("DOC")	HMM-361
EAKIN, RANDALL W. ("RANDY")	HMM-262
EAKLE, JAMES R. ("TUBA MAN")	
EARNEST, DEWEY	HMM-364
EARP, WILLIAM	
EASTER, JOHN J. ("J.J.")	HMM-365
EATON, RICHARD	HMM-161
EBERT, W.R. ("RED")	VMO
ECHEVARRIA, DAVID P. ("ETCH")	HMM-163
ECK, J. STEPHEN	HMM-362
EDDINGS, ROBERT L. ("BUZZ")	HMH-463
EDDY, DALE D. ("WHALE")	HMM-163
EDGAR, ANDREW ("ANDY")	HMM-165
EDMUNDS, TOM	HMM-264
EDWARDS, JERRY K.	
EGAN, EDWARD ("CHICKEN NECK")	HMM-263/163
EGGER, CHARLES ("DOC")	HMM-361
EGLIT, VERN	HMM-261
EHLERT, LTGEN. NORM	VMO-2
EHMER, JIM	HMH-463/ HML-367
EHRHARDT, JACK ("SPARKY")	HMM-364
EICHELBERGER, JOHN M.	HMM-164
EIKENBERY, TOD A. ("IKE")	HMM-162/ VMO-6
EISENHAMMER, JOHN	HMM-361
EISENSON, HANK	
EK, DAVE	HMH-463
EKLUND, BARRY A.	HMH-463
ELAM, DAVID L. ("DAVE")	HMM-161
ELGEE, JOHN	HML-367
ELKINGTON, DON	
ELLIOTT, GENE T.	HMM-364
ELLIS, DOUG	HML-267
ELLIS, FRANK B.	HMM-265
ELLIS, PAUL JR.	
ELLIS, RICHMOND K. JR.	HMH-463
ELMONE, RON	HMM-364
ELOE, ED	
ENEBOE, ED	
ENGEL, DON ("MISSING LINK")	HMM-363
ENGERT, E.	HMM-265
ENGESSER, BOB	
ENGLAND, PHIL	HMM-264
ENGLEHART, BILL	HMM-262
ENGLISH, RICH	HMM-363
ENGLISH, TERRY ("THE GHOST")	HMM-361
ENLOE, DON	
ENOCKSON, JOHN ("JACK")	HMM-263/ 161/VMO-2
ENTWISLE, JIM ("TWEET")	HMM-261/ H&MS-16
EPPERLY, I.L. ("SKIP")	HMM-263
EPSTEIN, GARY	
ERB, STEVE	
ERDMAN, ALAN WAYNE ("BIG E")	HMM-262/263
ERICKSON, DICK	HMM-363
ERICKSON, ERIC	MAG-16
ERLANDSON, RICK	HMM-164/ HMH-462
ERLICH, LAWRENCE B.	
ERRANTE, MARIO	HMH-462
ESCOTTO, RUDOLFO ("SCOTTY")	HMM-365/ 362/462
ESMOND, DON	
ESPINOZA, BARNEY	HMM-364/ 263/H&MS
ESTES, WILLIAM T. ("WILD BILL")	HMM-261/ 361/VMO-2
ETCHEVERRY, LOU ("BLK. TORNADO")	VMO-6/ HMM-364
ETHINGTON, DAVID	
ETTER, LEIGHTON S.	HMM-263
EUSANIO, TOM	HMM-161/164
EVANS, GEORGE G. ("TIP")	VMO-2/ HMM-265
EVANS, KENNETH B. ("TEX")	HMM-364
EVANS, RICHARD A.	HMM-262/265
EVERETT, MIKE	HMM-165
EWERS, NORM	HMM-163
EWING, LANA J. [EARL EWING]	HMM-262/364
EWING, TERRY	HMM-262
EXUM, JACK ("FAT JACK")	HMM-265/262
EYRE, JIM	HMM-161/ 167/VMO-2
— F —	
FAHRNER, R.H. ("RUDY")	HMM-365/161
FALESKIE, JOE	HMM-363
FALKE, ROBERT E. ("BOB")	HMM-163
FALZARANO, V.L. ("VINCE")	HMM-163
FAMIANO, LOUIS	MAG-16
FARIES, TOM ("GOLDS")	HMH-463/ HMM-165
FARKAS, RICHARD W.	HMM-263
FARLEY, JIM	HMM-162/363
FARNSWORTH, FRANCIS ("DICK")	H&MS-16
FARRELL, LEO J.	HMM-265/164
FARRELL, TOM ("RUBE")	HMM-163
FARRELL, TOM ("T.J.")	HMM-263
FARRIS, AUBREY	HMM-262
FARROW, JERRY ("CLUTCH")	HMM-361
FASE, MILES	HMM-363
FATO, FRANK R. ("FOX")	HMM-163/362
FAVOR, BILL	HML-167
FEARS, OSCAR B. JR. ("BO")	HMM-262
FEDORICK, WILLIAM M.	HMM-165/263
FEDOROW, GEORGE ("FEDS")	HMM-362
FEENEY, ROBERT J.	HMM-364/362
FELDER, OTIS A. ("BIG O")	HMM-262
FELDT, JACK ("THE GHOST")	HMM-364
FELL, W.C. ("BILL")	HMM-161
FELTER, JOSEPH L.	HMM-161
FENTON, LARRY ("PHANTOM")	HMM-262
FENWICK, BILL	HML-167
FERG, CHARLES J. ("CHARLIE")	H&MS-16/ HML-367
FERGUS, TOM	HMM-263/161
FERGUSON, ROBERT	HMM-164
FERGUSON, ROBERT L. ("FLASH")	HMM-165/265
FERGUSON, ROLAND A.	VMO-6
FERGUSON, TODD	HMM-364/164
FERNEAU, MIKE	MABS-16/ HMM-164
FERRACANE, LOU	HMM-164
FERRIER, JAMES S.	HMM-362
FERRIN, MICHAEL	HMM-164
FESPERMAN, RAY	
FETTER, CHARLES	
FETZER, STEVE ("FETUS")	HMM-164
FICK, KEITH C.	HMM-365
FIELD, H. BRYSON ("BRY")	HMM-361/ 363/161
FIELD, STEVEN E.	HMM-165
FILINA, JON A.	
FILLEY, DWIGHT	
FIMIANI, CARMEN	HMH-462
FINCHAM, KENT	HMM-262
FINEGAN, BRIAN	HMM-161
FINLEY, ERIC	
FINN, R.	
FINNERTY, MIKE	HMM-161
FISH, KEN ("TUNA")	HMH-463
FISHER, CHUCK	HMM-261
FISHER, ZANE ("Z.B.")	HMM-261
FITZGERALD, TERRY	HMM-163/361
FITZPATRICK, RAYMOND P.	HMR-362/263
FIX, HERB	
FIX, RON ("SAND DOLLAR")	HMM-362/ 363/VMO-2
FLAGG, FRANK	HMM-163/ 261/361/362
FLAHERTY, JOE	HMM-162
FLANAGAN, JOHN	
FLATER, M.E. ("RHETT")	HMM-265
FLEMING, ARCHIE H.	HMM-362
FLEMING, FRANCIS G. ("FRANK")	HMH-361/462
FLETCHER, AL	HMM-164
FLORES, SAMUEL JR. ("SAM")	HMM-262
FLOYD, MAX R. ("TOP")	H&MS-16/ HMM-163
FLYNN, GARY	HMM-263
FOCHT, ROBERT	HMM-164
FOLEY, COLEMAN	HMM-161
FOLLIS, JOE E.	VMO-2/ HMM-263/367
FORBES, R.W. ("BOB")	HMM-262
FORD, JOHN A. ("JACK")	HMM-161
FORD, WAYNE	HMM-161/ 263/462
FORNEY, MIKE	
FORSYTH, ROBERT W. ("BOB")	HMM-263
FORTE, MIKE ("THE BUSH")	HMM-263
FORTINBERRY, HARRY C.	HML-367
FOSTER, BILL	HMM-163
FOSTER, ED	
FOSTER, SAM	
FOWLER, FRED T. ("THUMPER")	HMM-164/165
FOWLER, JERRY	HML-167
FOWLER, MICHAEL D. ("DOC")	HMM-362/364
FOWLER, RADM. JAMES R. ("DOC")	HMM-163
FOWLER, RICHARD M. ("DICK")	HMM-163/161
FOX, ARLAND	HMM-262
FOX, DAVID T. ("FOXIE")	HMM-164/161
FOX, PATRICK	HMM-265
FOX, RAYMOND	VMO-2
FOX, RICHARD ("ROCK")	HMM-165
FOX, WILLIAM T. ("FOXEY")	HMM-364
FRAIN, ADANA ("HONDO")	HMM-365
FRAIOLI, GERRY	VMO-2/6
FRANKLE, JOHN	HMM-261
FRANKLIN, RAY ("WRONGFINGER")	HMH-463
FRANZ, RON ("DOC")	HMM-163/ 362/165/263
FRASER, LARRY	HMM-362
FRASIER, WILLIAM C.	HMM-265
FREED, PHIL	
FREEMAN, ROBERT B.	HML-167
FREESE, GARY	
FREY, GARY	HMM-262
FREY, PHIL	HMM-164/165
FRIDAY, GEORGE A. ("JOE")	HMM-363/361
FRIER, JIM	
FRISENDA, AL	HMM-162/ 364/164
FRISKE, DAVE	HMM-264/364
FRITZ, OTTO	HMM-265
FRITZLER, BOB	HMM-162/361
FROLICK, FRANK	HMM-262
FRY, ROBERT D. ("PIG")	HMM-165
FUHRMANN, BUFF ("BULL")	HMH-462/ HMM-165
FUHRMANN, MIKE	
FUHRMANN, ROBERT O. ("BOB")	VMO-2/ HMM-365/163
FULMER, MARK T.	HMM-361
FULTON, JOHN	HMM-165
FULTON, S.J. ("SAM")	VMO-2
FUNK, DONALD	HMM-362
FUQUA, FRANK	HMM-265
FURLOW, JIM	HMM-263
— G —	
GAGE, BILL	
GAGGERO, LARRY ("GAGS")	HMM-261
GAGNON, RENE	VMO-6
GAGNON, RON ("ROTATING RON")	HMM-161
GAJEWSKI, FRANK	HMM-163
GALBRAITH, JIM	HMM-364
GALE, JOHN F.	
GALE, SIDNEY R. ("GUMBO")	HMM-364/ H&MS-36
GALLAGHER, WALTER J. ("WHEELS")	HMH-462
GALLAWAY, PATRICK	HMM-165
GALLUP, BYRON ("BY")	HMM-361
GALVIN, MIKE	HMM-161
GAMMACK, GREGG	HMM-361
GANEY, JERRY	HMM-161
GANGI, ROGER	HMM-161
GANTZ, BILL	VMO-2
GARCIA, DANIEL	HMM-265
GARCIA, HOWARD	
GARCIA, ISMAEL ("MIKE")	HMM-164
GARCIA, JOHN N.	HMM-161
GARCIA, MANUEL A. ("SKEETER")	HMM-165
GARCIA, RAY	HMM-261
GARCIA, TONY	HMM-364
GARD, GARY	HMM-364
GARDELLA, ROBERT L.	HMM-164/265
GARDNER, DAVID	H&MS-16
GARLETTS, STEVEN J. ("ROCKY")	HMH-463
GARMUS, LARRY	HMM-161
GARNER, GARY ("CB")	VMO-2
GARNER, JERRY	VMO-6
GAROUTTE, JIM	
GAROUTTE, LEE	
GARRETT, LARRY	
GARRISON, ROYCE	
GARTMAN, JERALD B. ("JERRY")	HMM-165
GARVIN, ANCIL B. ("OKIE 1")	HMM-262
GASPARD, CARROLL J. ("FRENCHIE")	HMM-361
GATES, HARDY	HMH-462
GATES, PHIL ("PHILTHY")	HMM-361/262
GATEWOOD, CHARLES L.	HMM-161
GATEWOOD, RON	HMM-262
GATLING, LAWLER	
GATZ, FRED	HMM-161/ HML-367
GAUTHIER, JIM	HMM-163/ 462/463
GAY, JOSEPH	VMO-6
GAYER, DICK ("TACO")	VMO-2
GAYFORD, PETER J.	HMM-161
GEASLIN,	
GEBEL, JOHN R. ("BABY-SON")	HMM-363
GEE, DAVID N.	HML-367
GEER, RUDY	HMM-161/364
GEIS, ROYALL W. ("GOOSE")	HMM-162/263
GEISSER, JOHN M. ("TAC JACK")	HMM-261/ 263/161/163
GEMIGNANI, RALPH D.	HMM-263
GENNARO, VINCENT ("VINNIE")	HMM-165
GENOVA, JOE JR. ("GEKE")	HMM-265
GENTRY, JAMES E. ("JIM")	HMM-364
GENTRY, JAMES R. ("WEASEL")	HMM-162/ 167/367
GENTRY, RAY JR.	VMO/HML
GENTZLER, GREGG ("SATCH")	HMM-164
GERARD, C.G. ("JUG")	VMO-6/ H&MS-39
GERMAN, KENNETH	HMM-363
GETMAN, HOWARD E. ("PREACHER")	HMM-364
GETTES, GERALD L.	HMM-163/ 164/361
GHASTIN, JAMES R.	VMO-6
GIBBONEY, JAMES W.	VMO-2/ HML-167
GIBBONS, JERRY P.	HMM-263
GIBSON, DALE	HMH-463
GIBSON, E.L. ("GIB")	
GIDDINGS, JOHN A. ("DOC")	VMO-2/ HMM-163
GIDEONSE, H.A. ("SANDY")	HMM-261/ 362/463
GIESE, JOHN T.	HML-167
GILBERT, ALLEN D.	HMM-362
GILBERT, E.W.	HMM-265
GILBERT, JOHN G.	VMO-6/ HML-367
GILBERT, RICK	HMM-263
GILLASPIE, ROBERT A. JR.	HMM-262
GILLESPIE, BRUCE	HMM-361
GILLESPIE, GARY L.	HMM-361
GILLESPIE, ROBERT	HMM-163/363
GILLIDETTE, ART	
GILLIGAN, MICHAEL	HMM-263
GILLIGAND, WALT	HMM-164
GILLILAND, WOODY ("COACH")	HML-367
GILLIS, DAVE	HMM-164
GILMAN, DAN	H&MS-17
GILMARTIN, R.T.	
GILMORE, JOSEPH	
GILMORE, MIKE	
GILREATH, JERRY	HMM-164
GILSON, RICHARD	HMM-164
GILSRUD, GARY C.	FAC
GILTON, LELAND R.	HMM-163/ 361/263/161
GINGRICH, ROGER	HMM-262
GITCHO, G. ("SHAKEY GEORGE")	HMM-364
GIULIO, FORONATO R.	H&MS-36
GIVAN, JAMES E. ("JIM")	HMM-364/ HMH-462
GLASER, C. ("CHUNKY MONKEY")	HMM-361
GLEASON, DICK	HMM-263
GLEN, MIKE	
GLIGOR, PETE	HMM-261/362
GLOWCZWSKI, THOMAS ("ALPHABET")	HMH-463/ H&MS/HML
GLYNN, D. MIKE	
GOALEN, FRANCIS ("FRANK")	HMM-163
GOBLE, WOODY	HMM-262
GODDARD, JOEL D.	HMM-361
GODWIN, BOB	HMM-264
GODWIN, ROBERT E.	HMM-164
GOEBEL, JERRY L.	HMM-263
GOETZ, PETE	HMM-263
GOLDBERG, IRV ("GOLDIE")	HMM-165
GOLDEN, JOHN P.	HMM-161
GOLDING, DAVID LEE ("DAVE")	HMM-364/362
GOLDSTON, WILLIAM P. ("GOLDIE")	HMM-265
GOLICH, JOHN X.	
GONNEVILLE, JEAN G.	
GONZALES, JOHN	HMM-164
GONZALES, STEPHEN ("SLOW MO")	HML-167
GOOD, DAVID P. ("D.P.")	HMM-261/362
GOODFALLOW, BERT	HMM-365/ VMO-6
GOODFELLOW, J.H. JR.	HMM-161
GOODSON, BILL	HMM-163/265
GOODWIN, RON ("GOODY T LOUSE")	HMM-361/ VMO-6
GOODWIN, TOM	
GOODWIN, WILLIAM	HMM-164
GORDON, B.F. ("BARF")	
GORDON, BERNIE	HMM-163
GORDON, JAMES T. ("COYOTE")	HMM-362
GORDON, JOHN	
GORMAN, BUTCH	HMM-362/363
GORMAN, GEORGE	HMM-164
GORNEK, JAMES J. ("JIM")	HMM-263
GOTTESMAN, DAN ("HOT BRAKES")	VMO-2
GRABE. GEORGE ("GRABBY")	HMM-361
GRAFF, ART	VMO-3
GRAFMAN, PRESTON	HMM-165
GRAHAM, CURT	HMM-365
GRAHAM, GORDON	HMM-263
GRAHAM, RICK	
GRANATA, PETE	HMM-163
GRASSI, FRANK T.	HMM-363
GRAUL, ANTHONY P.	HMM-165
GRAVELLE, WES	HMM-163
GRAVES, TED	HMM-364
GRAY, COURTLAND P. III ("COURT")	VMO-2
GRAY, EDGAR M.	HMM-164
GRAY, JOE	HMM-264
GRAY, MICHAEL	
GRAY, O.J. ("JUICE")	HMH-463
GRAZIANO, NEAL ("ROCKY")	HMM-361/261
GREEN, BILL ("SPINNER")	HMM-362/ HMH-463
GREEN, DICK	VMO-3
GREEN, MIKE	
GREEN, PAUL	
GREENE, GENE	
GREER, JESSE	HMM-361
GREGOIRE, PAUL	HMM-163/263
GREGORY, BILL ("LT. RATRY")	HMM-163
GREGSON, RICHARD ("RICH")	VMO-3/FAC
GRELL, TIM ("DOC")	HMM-262/ MAG-16/36
GRELSON, DARRELL	HMM-363
GRESHLE, JOHN F.	HMM-364/ 163/H&MS
GRESSLIN, BILL	HMM-263
GRIFFIN, WILLMONT ("MURDOC")	
GRIMMER, JAMES	
GRIMSTEAD, RICK ("BUCK")	HMM-364
GROAH, LARRY	
GRONAU, RONALD ("PAPA SAN")	HMM-161
GRONIGER, BILL	
GROSS, KENNETH L. ("HARRY")	HMM-364/ HMH-462
GROSSFUSS, GEORGE	HMM-161/ 364/H&MS
GROVES, RON	HMM-164/262
GRUBBS, CHARLES E. ("BUD")	HMM-164
GRUNDY, RAY	
GRUSZKA, DIMITRI	HMM-263
GUAY, ROBERT ("BOB")	HMM-261
GUERRERO, GREGORIO S. ("GEORGE")	HMM-262/ 164
GUERTIN, FRED	HMM-265
GUILFORD, ROBIN ("SPARROW")	HMM-163/363
GUINEE, VINCENT J. JR. ("VINCE")	HMM-261/361
GULBRANDSEN, GARY	HMH-363
GULCZYNSKI, KENNETH R.	HMM-263
GULLEDGE, FRANK A. ("UNCLE FRANK")	HMM-364
GULLICKSON, GARY D.	HMM-364
GULLING, LOU	HMM-364
GUNNELS, JOE	
GUNNISS, GORDON ("SCOOP")	HMM-163/262
GUSTAFSON, W.H. ("GUS")	HMM-365/ AIR AMERICA
GUTOWSKI, MARTIN	HMM-363/265
GUY, ROSS	HMM-364/262
GUYER, MARTIN D. ("MARTY")	HMM-361/362
— H —	
HAAG, PATRICIA [MIKE HAAG]	HMM-163/364
HABECK, RICHARD H.	HMM-164
HACHTEL, BOB	HMM-263/362
HACKER, LARRY	HMM-164/165
HACKLEY, BOB	
HACKMAN, BILL ("HACK")	HMM-364/165
HADAR, STEVE	HMM-263
HADDOCK, TOM	
HADZEWYCZ, GEORGE	
HAGERMAN, JESS C.	HMM-363
HAGGERTY, RICHARD K.	
HAGLE, GRANT	
HAINES, RICHARD	VMO-6
HAIRE, DAN	HMM-161
HALDERMAN, PAUL E.	HMM-161/362
HALL, ED	
HALL, FRED	
HALL, JAMES G. ("J.G.")	HMM-364
HALL, JON ("J.R.")	HMM-165
HALL, LARRY ("LURCH")	HMM-263
HALL, ORAMEL E.	HMM-362
HALL, RICHARD H.	HMM-361
HALL, RICK	VMO-2
HALL, TOM M.	HML-367/ HMH-461
HALLAM, MICHAEL R.	HMM-363
HALLEY, MIKE	HMM-261
HALLIDAY, G. FRED ("TAPE")	HMM-163/ 363/H&MS
HALTER, BEN W.	HMR-461
HALVORSEN, WILLIAM ("WILD BILL")	HMM-163
HAMBY, DAVID H.	VMO-2
HAMERA, WALT	
HAMILTON, BOB ("HAM")	HML-367
HAMILTON, CARL	HMM-162/164
HAMILTON, DON E. ("HOLLYWOOD")	HMM-163
HAMILTON, LARRY W. ("DRAGONFLY")	H&MS-16
HAMILTON, TOM	HML-367/167
HAMMER, J.M.	HMM-263
HAMMITT, JIM	
HAMMOND, DENNIS E. ("DUKE")	HMM-363/ 265/163
HANAVAN, MIKE	
HANDLEY, BOB	HMM-165
HANLEY, DALE A.	MAG-16
HANNER, N.F. ("PETE")	
HANNON,	
HANRAHAN, JOHN F.	HMM-364
HANSEN, BRUCE L.	HMM-362/ HMH-463
HANSEN, JACOB A. ("DOC")	HMM-365
HANSON, MAURICE ("MO")	MAG-16
HARDESTY, MICHAEL G.	HMM-263
HARDIN, BOBBY L.	HMM-165
HARDIN, RON	
HARDING, BOB	
HARGRAVE, DONNA [R. HARGRAVE]	HMM-363/ 361/367
HARGROVE, PARKER	HMM-161
HARJUNG, BRUCE	
HARKINS, PATRICK J. JR. ("PORKY")	HMM-161
HARKLESS, RON	HMM-362
HARLAN, ROBERT ("ROB")	HMM-365
HARNLEY, JEFF	HMM-161
HARPER, CHARLES	HMM-161
HARPER, RICHARD ("HARPO")	HMM-265

NAME	IN COUNTRY SQDN
HARR, MICHAEL G. ("MIKE")	HMM-262
HARRIGAN, DAN	HMM-265
HARRINGTON, PHIL	HMM-165
HARRINGTON, R. ("SKINNY")	
HARRIS, CHRIS	MAG-36/39
HARRIS, ED	HMM-165
HARRIS, JIMMIE L.	HMM-364
HARRIS, LARRY	
HARRISON, ROBERT L. ("HARRY")	HMM-262
HARRY, R.T.	HMM-165/265
HARRY, SEXTON	HML-367
HART, BUD	VMO-2/6
HARTER, GARY	HMM-165
HARTFORD, SKIP	
HARTIGAN, BILL	VMO-2
HARTLE, WES	
HARTMAN, DOUG	HMM-165
HARTUNG, DAVID	HMM-164
HARVEY, J.B.D.	VMO-2/ HMM-165
HARVEY, PAT	HMM-161
HARVEY, ROBERT L.	HMM-265
HASSON, WAYNE	HMM-361
HATCH, BILL ("BOOBY")	HMM-262
HATCH, JIM	HMM-161
HATCHER, GARRET ("FAT HATCH")	HMM-363/ 362/164
HATELY, JIM ("BOY")	HMM-361
HATRICK, DAVID	HMM-164
HATTON, RONALD ("MAD HATTER")	HMM-361/ 362/MWSG
HAUF, JIM	HMM-361
HAVARD, JERRY	HMM-164
HAWES, GEORGE ("SCARFACE 55")	HML-367
HAWKINS, HERB ("HAWKEYE")	HMH-463
HAWKINS, ROBERT W. ("HAWK")	HML-269
HAWTHORNE, DAN	
HAX, JOHN	HMM-163
HAYDEN, BOYD	HMM-261
HAYES, CHARLES ("CHUCK")	HMM-265/ VMO-6
HAYES, CLIFF ("GABBY")	H&MS-36/ HMM-364
HAYES, KYLE ("GABBY")	HMM-265
HAZELBAKER, V. ("WAYNE")	VMO-2
HEACOCK, F. JOEL ("JOSE")	HMM-263
HEALD, RON	HMM-362/ VMO-6
HEALING, NICK J. ("LURCH")	HMM-262
HEALY, JOE	VMO-2
HEALY, JOHN	
HEARON, LARRY ("ROOKIE II")	HMM-265
HEARTLAND, WILLIAM	HMM-262
HEATLEY, ROBERT L.	
HEBERT, GARY T.	VMO-6
HEDDERMAN, DAVE	
HEER, DAN ("HAIR")	HML-167
HEFFERNAN, J. ("HUGE")	HMH-462
HEIBERG,	
HEIFNER, TOM ("CAPTAIN PERFECT")	HMM-265/263
HEIM, DONALD C.	HMM-165
HEIMAN, PETE	
HEIN, ROBERT	HMM-163
HEINLE, EDMUND E. ("ED")	HMM-163/361
HEINS, F. ("WHISPERING FRANK")	HMM-162/ HMH-463
HEINZEL, JOHN R.	HML-167/367
HELLER, RON	HMM-362/ HMH-462
HELLRIEGEL, JOHN	HMM-261
HELM, KURT G.	HMM-265
HEMMING, PHIL ("FILTHY PHIL")	HMM-364/ 263/165
HENDERSON, JACK L.	HMM-163
HENDERSON, JIM	HMM-261
HENDRICKSON, LARRY E.	HMM-164/ 165/265
HENDRIE, RICHARD E. ("CLUTCH")	VMO-2
HENNESSEY, BILL	HMM-161
HENRY, JIM	
HENRY, ROBERT	VMO-6
HENRY, ROBERT L. ("HANK")	VMO-6
HENSEN, HALE	HMM-263
HENSHALL, ERIC B. ("HURRICANE")	1st MARDIV
HERBERG, RICHARD JR. ("RICH")	HMM-262
HERLOCKER, ALAN	HMM-161
HERMAN, DONALD L.	HMM-164
HERMAN, ROGER A. ("DUKE")	HMM-361
HERN, CHARLES R.	VMO-6
HERNANDEZ, DAVID	HMM-165
HERNANDEZ, ENRIQUE	HMM-361
HERNANDEZ, H. ("CRAZY CUBAN")	VMO-1
HERNANDEZ, L.E.	HMH-463
HERRMAN, RALPH V.	HMM-163
HERRON, ("BUCK")	
HESTER, WILLIAM S.	HMM-265
HETH, BASIL	HMM-265
HETRICK, DAVID	HMM-164
HEWES, WILLIAM T. ("TOM")	HMM-362/ H&MS-36
HEYER, HENRY T. ("TOM")	HMM-362
HEYSER, DONALD	HMM-265
HIAN, ROBERT A.	HMM-164
HICKS, PRESTON E. ("PETE")	HMM-265
HIGGINS, MIKE ("BEAK")	HMM-161
HIGGS, W.W. ("BILL")	
HIGH, BILL	HMM-263/ H&MS-16
HILL, EDWARD S. ("ED")	HMM-265
HILL, FRANK G.	HMM-163
HILL, GARY	
HILL, GEORGE	HMM-164
HILL, TOM	HMM-363
HILLE, KLAUS P.	VMO-6
HILLER, RUSS	
HILMANDOLAR, BUSTER	VMO-6
HILTIBRAN, JACK J.	HMM-164
HILTON, BILLIE	VMO-6
HILTON, J.D. ("J. DOG")	HMH-463
HINES, KEN	
HINGA, GERALD R.	HMM-165
HINKLE, ED	HMH-462
HINTON, ALLYN J. JR. ("BIG AL")	HML-167
HINTZ, DICK	HMM-163
HINTZ, JIM ("GUNNY")	HMM-364
HIPPERT, JIM	HMM-263
HIPPNER, RICH ("HIP")	HMM-261
HIRE, ED	HMM-363
HIRSCH, DON	HMM-362
HITCHCOCK, BILL	HMM-264
HOAG, JACK	HMM-161
HOBBS, ROBERT W. ("BOB")	HMM-261
HOBBS, VIC	HMM-264
HOCEVAR, FRITZ	HMM-263
HOCKADAY, JACK	
HODGE, MAC	HMH-463
HODGEN, DON	HMM-264/364
HODGINS, JERRY ("GUNNY")	HMM-164
HODGSON, JIM	HMM-265
HOEKSTRA, JIM	MACV SOG
HOEN, G.G.	
HOERTZ, JIM	MAG-16
HOESCH, MARTY	HMM-164
HOFER, LESLIE	HMM-365
HOFF, WILLIAM	
HOFFMAN, BRUCE	VMO-2
HOGAN, JACK ("GOAT")	HMM-167
HOGG, HIRAM A.	HMM-363/ 265/462
HOGUE, JIM	HML-167
HOGUE, JOHN	HMM-262
HOHMAN, RUSSELL	HMM-362
HOLLAND, DAVID K.	HMM-163
HOLLAND, JAMES ("DUTCH")	VMO-3
HOLLAND, MICHAEL ("RAGMAN")	HMM-263
HOLLIFIELD, JACK E. ("HOLLY")	H&MS-16/ HMM-163
HOLLIS, JAMES A. ("SHAKEY")	HMM-165
HOLLISTER, NEAL F.	HMM-164
HOLM, JOHN	
HOLM, KENNETH L. ("GOMER")	HMM-362
HOLMAN, RUSS	
HOLMES, ALLAN ("BIG AL")	HMM-164/165
HOLMES, DON	HMM-161/363
HOLMES, JAMES C.	H&MS/HMM- 265/463
HOLMES, WILLIAM T. ("T")	HMM-263
HOLSTEAD, G.H.	HMM-265
HOLT, JIM	
HOME, RALPH	HMM-262
HONEYCUTT, F.H. ("BO")	
HOOTEN, LEROY J.	HMM-165
HOOTON, DICK	HMM-361/ 161/363/362
HOPKINS, ROY	HML-367
HOPKINS, THOMAS ("HOPPY")	VMO-6
HORNE, EDWARD A. ("ED")	HML-367
HORNYAK, CHARLES R.	HMM-265
HORSTER, ALEX ("ACK ACK")	HMM-263
HORTON, ANSLEY S. ("ANS")	HMM-162
HORTON, GARY	HMM-164
HORTON, SAMUEL M. ("SAM")	HMR(L)-163
HORVE, ODDMUND L. ("HORVIS")	HMM-163/ H&MS-16
HOSHINO, HENRY ("HANK")	HMM-364
HOUGHTON, RICHARD L. JR.	HMM-362
HOULE, GEORGE A.	HMM-361
HOUSE, ED	HMM-164
HOUSTON, ROBERT C. ("BOB")	VMO-3/ HML-367
HOWARD, DONALD L.	
HOWARD, JAMES E.	HML-167
HOWARD, JERRY	HMM-164
HOWARD, MORGAN ("MUGS")	
HOWARD, PHIL ("THE HAWK")	VMO-2/6
HOWARD, ROBERT L.	HMM-165
HOWARD, TOM	HMH-461
HOWELL, JACK	
HOWELL, P.E.	HMM-265
HOWERTON, ROBERT E. ("BOB")	HMM-163
HOWLAND, HARRY	HMM-161
HOXTON, A.R. III ("ARCH")	HMM-164/ FAC
HOYLE, ROBERT A. ("BOB")	VMO-6
HRUSCH, DAVID	HMM-263
HRUSKA, JAN ("ROOSTER")	
HUBBARD, DONALD	HML-167
HUBBARD, G. RICHARD ("DICK")	HMM-263/365
HUBBELL, JERRY	VMO-3
HUBER, CHARLES M.	1st MAW
HUBER, ROBERT A.	HMM-164/ 165/463
HUCKEMEYER, M.R.	HMM-263
HUCKLE, RICHARD	HML-367
HUDSON, WINSTON ("HUD")	HMM-262
HUEBNER, A.C. ("TONY")	HMM-363
HUFFCUT, BILL ("THE ICEMAN")	HMM-263/ 363/VMO-6
HUFFORD, JOHN ("HUFF")	HMM-164
HUGHES, GARY	HML-167
HUGHES, GREGORY	HMM-361/263
HUGHES, JAMES L. ("HUGGS")	HMM-162/ 165/H&MS
HUGHES, LOWRY ("HIPPO")	HMM-161
HUGHES, ROBERT C. ("HUGGLES")	HMM-163
HUMMEL, TOM	HMM-165
HUMMER, JOHN R.	HMM-361/ 263/VMO-2
HUMPHREY, WILLIAM A.	HMA-369
HUNDLEY, THOMAS	HMM-265
HUNNEYMAN, ED	HMM-362/ 363/364
HUNT, HARRY	HMR-263/362
HUNT, JOEL	HMM-361
HUNT, JOHN E. ("ACE")	HMM-265
HUNT, PAUL O.	H&MS-16
HUNT, ROGER ("YORK")	HMM-165
HUNTSBERGER, ROBERT B.	
HURLBUT, F. WAYNE ("HURLY")	HMM-163
HURLEY, RAY D.	HMM-161
HURME, SEPPO I. ("SEP")	HMM-361/ 167/VMO-2
HUSS, TED	HMH-463
HUST, LARRY R.	HMM-365
HUTCHINS, ROBERT A. ("HUTCH")	HMM-261
HUTCHINSON, G. CALVIN ("CAL")	VMO-2/6
HUTCHISON, JOHN	HMM-163/ H&MS-16
HUTH, LAWRENCE L.	HMM-361
HUTTON,	
HYDE, STEVE	HMM-164
HYMAN, PETE	VMO-3

— I —

NAME	IN COUNTRY SQDN
IAROSIS, THOMAS	HMM-164
IHLI, LEO J.	
ILZHOEFER, BOB ("ILZ")	HMM-261/ 163/164
INGRAM, DOUG	
INGVOLDSTAD, ORLANDO ("LANNY")	VMO-6/2
INMAN, B.T. ("DUMP TRUCK")	HMM-361/363
IRWIN, ED	HMM-161

— J —

NAME	IN COUNTRY SQDN
JACKS, SAM ("THE KID")	HMM-165
JACKSON, GEORGE	
JACKSON, LARRY ("BUDDHA BODY")	HMM-265/164
JACKSON, MARK	HMM-361/163
JACKSON, REED	HMM-265
JACKSON, ROBERT	HMM-164
JACKSON, WILLIAM J.	
JACOBS, JOHN R. ("JAKE")	HMM-365
JACOBS, ROBERT E. ("JAKE")	
JACOBSON, ALLEN ("JAKE")	HMM-263
JACOBSON, LES	
JACOBSON, NEAL E.	H&MS-16
JACQUES, RONALD E.	HMM-361
JAEGER, JIM ("JIMMIE")	VMO-6
JAMES, TOMMY L.	VMO-2
JANNING, WILLIAM	HMM-262
JANOVICZ, MIKE	
JANSEN, RANDALL J.	HMM-363
JAQUISH, MIKE	HMM-262
JARRETT, PAUL ("KIP")	HMM-163
JARUSINSKY, JIM	MABS-16
JARVIS, HAROLD ("JARVY JARV")	HMM-263
JARVIS, JAMES W.	HMM-165
JASMINE, RICHARD	
JASSICA, R.A.	HMM-263
JEALOUS, BART	HMM-362
JEFFERSON, GORDON ("GORDO")	HMM-164
JEFFERSON, JAMES	MAG-36
JEFFRYES, JIM	
JELLISON, JEROME C. ("JELLY")	HMM-262/164
JENISON, P.B.	HMM-164
JENKINS, STUART D.	HMH-361/463
JENNINGS, PHIL	
JENNINGS, RONALD	VMO-2
JENSEN, DUANE S.	HMM-265/364
JERNIGAN, JOHN C.	HMM-361
JEWELL, JACK	
JIVIDEN, LARRY	HMM-364/165
JOBE, TONY	HMM-161
JOHANNESEN, BOB ("PACK RAT")	HML-167
JOHNS, DAVE	
JOHNSON, BRUCE W. ("B.W.")	HMM-265
JOHNSON, C.E.	HMM-261
JOHNSON, DAVE	HML-167
JOHNSON, G.L.	
JOHNSON, GARY	
JOHNSON, JAMES C.	HMH-463
JOHNSON, JERRY ("OINK CHASE")	HMM-262
JOHNSON, JOHN ("JOHNNY")	HMM-163/VMO-2
JOHNSON, JOHN E. JR. ("J.J.")	HMM-164
JOHNSON, KEN	
JOHNSON, KENNETH H. ("CHUCK")	HMM-265/262
JOHNSON, LARRY N.	HMA-369
JOHNSON, MICHAEL L.	HMM-161
JOHNSON, RAY I.	HMM-165
JOHNSON, ROBERT H. ("HEAD")	HMM-165
JOHNSON, SCOTT ("SCOTTY")	HMM-165
JOHNSON, WARD	HMM-164
JOHNSTON, CHUCK ("DOC")	HMM-261
JOHNSTON, PAUL	HMM-262/164
JONES, BOB ("DOC")	HMM-164/362
JONES, BUD ("HONEZ")	HMM-162
JONES, CARL ("FIRE TEAM")	1st MAW
JONES, DALE	HMM-365
JONES, DAVID A.	HMM-362/ HMH-463
JONES, GARY	
JONES, GRANT	HML-367
JONES, HAROLD W. ("JONESY")	HMM-262
JONES, JOHN C.	HMM-165
JONES, JOHN J. ("J.J.")	HMM-164
JONES, P.J. ("HOSTAGE PJ")	H&MS-16
JONES, RALPH E. II	HMM-361
JONES, RICHARD	HMM-362/163
JONES, ROBERT A.	HMH-463
JONES, WILLIAM	
JONES, WILLIAM D. ("W.D.")	HMM-163
JORDAN, DALE	
JORGENSEN, STAN	HMM-165
JOYCE, DICK	
JUBA, CRAIG	HMH-463
JULIAN, WAYNE A.	HMM-262

— K —

NAME	IN COUNTRY SQDN
KAAPU, KEKOA	
KABA, RON	HML-367
KAHLER, BILLY J. ("BILL")	HMM-161/ HMH-462
KAIER, DAVID J.	
KAISER, GREG	
KAISER, LOUIS E. ("GUNNER")	HMM-263
KAISER, RUDY ("NAILS")	HMM-161/364
KALATA, RICHARD	HMM-262
KALMECK, MARTY	HMM-164
KAMMAN, WAYNE	
KANALEY, TOM	HMM-361
KANLOCK, BRUCE	HMM-161
KAPETAN, NICK J.	HMM-362
KARAMARKOVICH, MGEN. GEORGE	
KARLIN, WAYNE S.	HMM-164/ 263/161
KAROW, BOB	HMH-361
KARR, JAMES E. ("JIM")	
KARR, LLOYD N. ("LITTLE TOP")	HMR-362/ 262/164
KATZMARK, MICHAEL	
KAWALEK, DENNIS ("D.K.")	HMM-163/361
KECK, KEN	
KECKLER, RICH ("KECK")	HMM-364
KEEFE, BOB	
KEEGAN, MICHAEL G. ("MIKE")	HMM-265/262
KEELE, M. DUANE	HMM-161/163
KEELER, BOB ("JUNKMAN")	H&MS-16
KEEN, JIM	HML-367
KEIDEL, DALE E.	HMM-361
KEITH, WES ("FLUFF")	VMO-2
KELLEHER, BOB	
KELLENBERGER, BILL	HMM-363/161
KELLER, BILL	HMM-362/263
KELLER, BOB	HMM-361/ VMO-2
KELLER, FLOYD	HMM-262
KELLER, LTGEN. ROBERT P.	A.W.C. 1st MAW
KELLER, NOEL	HMM-263
KELLEY, RAY JR. ("MACHINE GUN")	
KELLEY, WALT	HMM-262
KELLEY, WILLIAM J. ("DOC")	HMM-262/ MABS
KELLOGG, GORDON	HMM-262
KELLY, BILL ("KEG")	HMH-462/362/ H&MS
KELLY, GERRY	HMM-261
KELLY, KEN	
KELLY, MICHAEL	HMM-164
KELLY, PAUL L.	HMH-463
KELLY, ROGER	
KELLY, SAM C. III	HMM-265
KELLY, SCOTT ("SCOTTIE")	HMM-362/ 163/364
KELLY, TOM ("T.J.")	HMM-361
KEMNA, DAVE	HMM-362
KEMPFFER, BILL	HMM-263
KENDALL, STU ("OSCAR")	HMM-361
KENDALL, W.H. ("KEN")	HMM-361/ HML-367
KENNEDY, DAVID	VMO-2
KENNEDY, FRANK	HMH-463
KENNEDY, HERBERT	HMM-164
KENNEDY, J.J.	
KENNEDY, JIM	HMM-262
KENNEDY, THOMAS H.	HMM-164
KENNETT, W.M. ("MIKE")	HMM-362
KENNEY, BOB ("TOO TALL")	HMH-462
KENNY, JIM ("HOOK SLIDE")	HMM-261
KENNY, PAT ("SWIFT CHUCK")	HMM-364
KENT, ARTHUR L.	H&MS-16
KENYON, TOM	
KEOWN, DAVE	HMM-362/ VMO-2
KERINS, AL	HMM-261
KERR, GARY	HMM-265
KERR, THOMAS J.	HMM-164
KERWIN, FRANCIS J. JR. ("FRANK")	HMM-364/363
KESSLER, JAMES R. ("JIM")	MAG-16/ MAG-36
KESTLER, ROBERT ("BOB")	HMM-163
KETCHUM, JOHN S.	VMO-3
KEY, JOHN HENRY	VMO-2/ HML-167
KEYES(Frederiksen),TED ("D. CHZ.")	HMH-463
KEYFEL, AL	HMM-264/364
KIDDER, ART ("HOSTAGE KIDD")	VMO-2
KIERSH, ERIC	HMM-165
KILBANE, BOB ("KILLER")	HMM-261
KILGORE, BOB	
KILMER, ERICH	
KIMAK, CHARLES	HMM-161
KINDEN, JIM	
KINDRED, GARRY	MABS-16
KING, CLAUDE ("SCARFACE-33")	HML-367
KING, EDWARD E.	HMM-362
KING, ETHMER W.	HMM-164
KING, JOHN C. ("ACE LONG")	HMM-261
KING, PATRICK ("PEJAY")	HML-167
KING, PETE	HMM-362
KING, WILLIAM C.	HMM-262
KINGSTON, BILL ("WILLY GROSS")	HMM-263
KINLOCH, BRUCE G.	HMM-161/163
KINSMAN, STAN	HMM-161
KIRBY, E.K.	HMM-263
KIRBY, TOM	
KIRK, ALEXANDER ("AL")	HMM-262/164
KIRK, NORMAN T.	HMM-161
KIRKMAN, TIM ("CHIEF")	HMM-361
KIRKPATRICK, ARCHIE L.	MABS-36
KIRKPATRICK, ROBERT	HMM-163
KIRSCHNIK, JAMES L.	HMM-262
KISELEWSKY, RICHARD ("RICH")	HMM-364
KISER, JOHN W.	VMO-2
KISSLING, BOB ("HUNTER")	HMM-163/362
KLAAS, LOUIS ("LOUIE")	
KLAHN, FRANK	
KLEPPSATTEL, FRED	
KLING, DAVID A.	HMM-161
KLINGLER, DON	HMM-165
KMIEC, WALTER J.	VMO-6
KNAPP, RANDY	HMM-161
KNECHTEL, STEVE	HMM-163/362
KNESE, PETER G. ("PETE")	HMM-262
KNOWLES, TOM ("MAGNET ASS")	HMM-161/ VMO-2
KNOX, JAMES H.	VMO-3
KNOX, JIM	
KNUCHEL, ROGER L.	HMM-165
KOCH, DAN W.	
KOEPPE, RICHARD H. (RADAR)	HMM-362
KOHANOWICH, ALBERT J.	VMO-6
KOHLER, EDWARD B. JR.	HMM-264/161
KOLBINSKI, JAMES	HMM-361
KOLER, JOE	HMM-365
KOLLER, JOE	HMM-162
KONTRABECKI, GEORGE A.	HMM-161
KONVINSKI, GEORGE	HMM-164
KOONS, JERRY L.	HMM-164
KORAN, STEVE	HMM-263
KOROSEC, ROBERT G. ("BOOMER")	HMM-265/364
KORTE, JIM	VMO-3
KOSINSKI, JOHN L. ("CRASH")	HMM-364
KOSS, JAMES A.	HMH-462
KOWALK, ROBERT J. ("B.J.")	VMO-6/ HML-367
KOZAIN, LARRY	HMR/VMO/ HMM
KOZAR, JOSEPH P. JR. ("KOZY")	MAG-36
KOZLESKY, JOE	HMM-162/ 363/163
KOZOBARICH, LARRY	HMM-263
KRAGES, BERT	VMO-3/HML- 367/H&MS
KRAHE, STEPHEN	HMM-263
KRANE, KJELL	
KRANKUS, JOHN	MACG-14
KREGER, LEN ("DEVIL DOG #1")	VMO-2/ HMM-265
KREKELER, RANDALL D. ("RANDY")	HMM-161/164
KRENK, EDDY	HMM-263
KRILLA, LARRY M.	HMM-163
KRISTAPOVICH, JAMES P. ("KRIS")	HMM-164
KROH, NORM	
KRUGER, GENE	HMM-262
KRUM, CHARLES C.	HMM-265
KRUSE, MICHAEL A.	HMM-262
KRYWKO, WALLY	HMM-364/ HML-167
KUBECK, STANLEY	HMM-263
KUCI, MGEN. DICK	HMM-361
KUCZERO, TED	HMM-365
KUEHN, DONALD ("DON")	HMM-262
KUETON, MIKE	HMH-463
KUFELDT, ED	VMO-6
KUHEIM, JOHN E.	HMM-165
KUHLMEYER, BILL	HMH-463
KUHN, DAN	

NAME	IN COUNTRY SQDN
KUHN, KEN	HMM-364
KUHNS, JOHN ("KUHNSY")	HMM-263
KUKLOK, KEVIN ("KUK")	HML-367
KULICK, EDWARD J. ("ED")	HMM-165
KUN, ERNEST ("ERNIE")	HMM-364
KUNKEL, HENRY	HMM-263
KUSKE, GREGORY W. ("CUSS")	HMM-165
KUX, STEVE ("INJUN")	HMM-262/364
KUYKENDALL, ED ("DUTCHMAN")	VMO-2
KYLLO, KELLAN ("K-BAR")	HMM-262
— L —	
LABRIE, LARRY	HMM-364
LADD, BOBBY T.	HMM-262
LaFOUNTAINE, NORM ("FRENCHY")	HML-167
LAIRD, JIM ("TATOR")	HMM-362
LAKE, BRUCE	HMM-265
LAMASCUS, ZANE ("ZEKE")	HMM-161/362
LAMB, FLOYD	HMM-164
LAMB, H.B. ("HANK")	HMM-364
LAMBERT, DONALD	HMM-161
LaMONTAGNE, ROBERT	
LANCASTER, ROB	HMM-363
LANDAU, O.J. ("JUICE")	
LANDERS, RICK	HMM-161
LANDFORD, WILLIAM	HMM-164
LANDWEHR, DENNIS ("POPEYE")	HMM-263
LANE, M.V. ("SKIP")	
LANG, JAMES	HMM-163
LANGE, ROBERT ("DOC")	HMM-164
LANGENFELD, RICHARD	HMM-265
LANGLEY, ED	
LANGLOIS, DAVE	HMM-164/165/262
LANHAM, CARLYLE ("C.T.")	H&MS-16
LaPLANTE, DAVID W. ("BUSH")	HMM-262/H&MS-16
LARIMER, KENT ("DASH-2")	
LARIVIERE, DONALD G.	HMM-164
LaROCCA, AL ("SILK")	HMM-361
LaROUETTE, A.G. JR. ("FRENCHIE")	HMM-162/264/361
LARSEN, ERIK	VMO-2
LARSEN, LEONARD S.	HMM-161/163/461
LARSEN, RON	
LARSON, DUANE D. ("SWEDE")	HMM-163/H&MS-36
LARSON, GARY M.	HMM-164
LARSON, LARRY ("LARS")	HMH-463
LARSON, LEWIS C. ("LOW PASS LEW")	VMO-2
LaRUE, HALLER G.	HMM-164
LaRUE, JAY	HMM-364
LASETER, JAMES W. ("JIM")	HMM-361/HML-167
LaSHOMB, GARY L. ("LASH")	HMM-362/263
LASSITER, JOHN R. ("BOB")	MAG-16
LATHERY, W.M. ("BILL")	HMM-262
LATON, ROGER	HMM-263
LATTIMER, JIM ("LATTS")	VMO-2
LAUBACH, DAVID W.	HMM-165
LAURITSEN, ROBERT	
LAUX, PETER	HML-367
LaVALLE, WAYNE T.	HMM-265
LaVIANO, LINDEN T.	HMM-362
LaVIOLETTE, DAVID F. ("FALCON")	HMM-363/HML-167
LAWSON, HERBERT F. ("HERBY")	HMM-261/262
LAWSON, JOHN H.	MABS-36
LAWSON, RALPH D.	HMM-364/363/463
LAYMAN, MICHAEL R. ("MIKE")	HMM-365
LAYTON, DAVID G.	HMM-164
LEA, CHARLES DEAN ("CHUCK")	HMM-362/263/VMO-6
LEACH, EDWARD L.	HMM-163
LEAHY, MIKE ("POGUE")	HMM-363/161
LEAL, ABELARDO	HMM-161
LEAMING, GEORGE	HML-167
LEAR, GERALD	HMM-265
LEARNARD, CLIFFORD F.	H&MS-36
LEARNED, DAVE	HMM-363/364
LEARY, ALBERT D. JR. ("AL")	HMR-161/VMO [Korea]
LeBLANC, DONAT J. ("DAN")	HMM-363
LeBLANC, GARY	MAG-36
LECKY, HUGH ("JERRY")	HMM-261
LEDBETTER, WALT ("PINCHPECKER")	HMM-263/365
LEDET, JOHN J. ("FRENCHY")	H&MS-16
LEE, BARRY ("FUBAR")	HMM-161
LEE, EUGENE	HMM-161
LEE, GREG	
LEE, JAMES FRANKLIN JR. ("SKIP")	HMM-265
LEE, PETER B.	
LEEK, FRED	HMM-364
LEGAS, DAVE	HMM-364
LEGER, KEN	HMM-165
LEGG, JOHN	HMM-263
LEHMAN, RONALD G. ("RON")	HMM-165
LEIBY, TOM ("HABU")	HMM-261
LEIGHTON, DAVE ("THE GREY FOX")	HMM-164/362
LEIGHTON, JOHN	
LeJEUNE, JAMES D. ("SWAMPY")	VMO-6
LELASH, BILL	HMM-362/164
LEMASTER, RON	HMM-164
LEMKE, DANIEL	HMM-165
LeMOINE, NED	HMM-262
LENTZ, LARRY	HMM-263
LEONARD, ALBERT E. ("BUD")	HMM-263
LEONARD, JOHN J.	HMM-161
LERAN, BOB	HMM-263
LESSARD, JOHN	HMM-364
LEU, R. ("CHIEF")	
LEVAN, BOB	
LEVARDIE, JAMES A.	VMO-6
LEVIN, AL ("FLIGHT QUACK")	HMM-265/361
LEVY, ROBERT L. ("BABY CAKES")	HMM-164
LEWER, DICK	HMM-361/VMO-2
LEWIS, CHARLES W. ("CHUCK")	HMM-165
LEWIS, FLOYD C. ("LEW")	HMM-161/263
LEWIS, TYRONE	HMM-263
LEWZADER, MICHAEL G.	HML-167
LIBRA, DEAN	HMH-463
LIEBER, JACE	VMO-6/HML-367
LIEM, RON	HMM-163
LILLEY, HOWARD ("DOC")	HMM-261
LIND, A.W. ("DUKE")	VMO-2
LIND, RONALD J. ("R.J.")	HMM-165/265
LINDEMAN, GEORGE	
LINDLEY, MALCOLM G.	

NAME	IN COUNTRY SQDN
LINDQUIST, DAN	HMM-165
LINDSAY, BILL	HMM-164
LINDSEY, TOM	HMM-265
LINEBAUGH, STEVE	HMM-164/262
LINEBERGER, HOWARD ("PAUL")	HML-167
LINKE, BILL	
LINKES, LEONARD ("SNOOPY")	HMM-163/VMO-2
LINKOUS, CLAY E. ("DOC")	HMM-364
LIPKING, MIKE ("RUBE LIPKINSKI")	HMM-165
LIPPUS, FRANK	HMM-265
LIPSCOMBE, JOHN	HMM-163
LITTLEPAGE, BILL	
LIUSKA, DAVID A. ("LUKE")	HMM-264
LIVESEY, W.A.	HMM-361
LLOYD, BOB	HMM-262
LLOYD, CALVIN A. II ("CAL")	HMM-362/263/VMO-2
LOBDELL, GUY	HMM-263
LOCKE, FREDERICK A. ("FRED")	HMM-265/364
LOCKRIDGE, GARY	VMO-2
LOCKWOOD, HAROLD E.	HMH 463
LODGE, JACK	HMM-362
LOFTIN, KREIG	HMM-162
LOGAN, G.L. ("RED")	HMM-262/265
LOGSDON, R.C. ("YOGI")	HMH-462
LOGUE, KEN ("K.D.")	HMM-362
LOGUE, WILLIAM ("BILL")	HMM-161/H&MS
LOHNSEN, DAVID	HMM-362
LONDON, JOHN H.	
LONG, NORMAN A.	HMM-364
LONGDIN, JOHN F. JR. ("FANGDIN")	HMM-362/361
LONGHURST, BILL ("LONG")	HMR(L)-362
LONGSWORTH, TOM	HML-367
LOOMIS, GARY	HMM-364/165/H&MS
LOOP, JAMES	HMM-161
LOPES, GORDON D.	AIR AMERICA
LOPEZ, JUAN R.	HMM-165
LOQUE, WILLIAM F.	HMM-161
LOSEY, JAMES L.	HMM-362
LOTTMAN, W.J.	
LOUD, VERN JR.	HMM-164
LOUDER, RICK	HMM-161
LOVE, CHARLES T. ("CHARLEY")	HML-367
LOVE, DENNIS D.	HMH-462
LOVE, MARCELLUS	HMM-164
LOWRY, DWAYNE	
LOZITO, PAUL	
LOZOYA, ROBERT	VMO-6
LUBY, JACK	HMM-361
LUCAS, GARY A.	HMM-265
LUCAS, WILLIAM	HMM-364
LUCIA, JOSEPH S.	HMM-165
LUCKEY, STEVE	
LUCKIE, MACK ("LUCKY")	HMH-463
LUDWIG, JOHN	HMM-261
LUHRSEN, DAVE ("LURCH")	HMM-362
LUI, CALVIN W. ("CAL")	HMM-161
LUKS, RON ("RANK RON")	HMM-263
LUND, BOB	VMO-6
LUPPUS, FRANK E.	HMM-265
LUTES, MORRIS ("MOOSE")	HMH-463
LUTHER, GENE	HMM-265
LYCAN, GEORGE	HMM-364
LYMAN, MARTY	HMM-262
LYNCH, GARY	HMM-265
LYON, BILL	HMM-161
LYONS, DON	HMM-163
— M —	
MAAS, BERTRAM A. ("BERT")	VMO-6/H&MS-36
MABEY, BOB	HML-367
MACAULAY, AL	HMM-161/165
MacFARLANE, JOHN L.	HMM-265
MacINTYRE, KEN	HMM-165
MacINTYRE, RICHARD	HMM-262
MACK, BOB	
MacKAY, STEPHEN A.	HMM-365
MacKELLER, DAN	HMM-164
MACKIE, KEN	HMM-261/165
MACKLIN, KEVIN T.	HMM-165
MacRAE, DUNCAN III ("PIG")	HMM-263
MADDEN, MIKE ("MAD DOG")	HMM-161/H&MS-16
MADDEN, PATRICK	HMM-262
MADDEN, TOM	HMM-165
MADDOCKS, CHARLES ("MAD DOG")	VMO-2
MADRI, RICH	HMM-361
MAESTRINI, DAVID	VMO-6
MAGALLANES, FRANK A. ("MAGGIE")	
MAHALICH, NORM	VMO-2
MAHERN, R.E.	
MAHONEY, JIM	HMM-164
MAHONEY, P.T.	
MAIZ, JACK J.	MABS-36
MAJOR, BRUCE	HMM-265
MAJORS, DAVID	
MAKI, SAMUEL A. ("SAMMY")	
MALEY, BOB	HMM-264/364
MALIK, EDWARD C. ("CHIP")	HMM-263
MALKERSON, SHERMAN P.	HMM-164
MALONE, GENE A. JR.	HMM-164
MALONE, GLENN R.	HMM-165
MALONEY, CLIFFORD	VMO-6
MALONEY, LTGEN. BILL	VMO-6
MALOY, STANLEY J.	HMM-265
MALTBY, JOHN B.	HMR-361
MANDEL, BRUCE	HMM-361
MANDITCH, DAN	HML-167
MANIFOLD, JAMES W.	
MANISCOLA, TONY ("THE COUNT")	MABS-36
MANN, BENNIE	HMM-163
MANN, JACK ("PEACH")	HMM-163/263
MANN, TERRY W.	HMM-362
MANNING, C.H. ("HUD")	HMM-165/HMM-364
MANNSCHRECK, CRAIG ("CRAIGO")	
MANSHIP, EDWARD S. ("STAN")	HMR-162/362/161
MANTON, WILLIAM A. ("WILL")	HMM-263
MANZ, DOUG	HMM-165
MARCH, BILL	HML-167
MARINUCCI, MIKE ("MAD")	HMH-463
MARQUARD, PAUL W.	VMO-2
MARQUETTE, EUGENE ("GENO")	VMO-6/HMM-263/364
MARSH, LARRY C. ("SWAMP")	HMM-361
MARSHALL, MIKE	AIR AMERICA
MARSHALL, ROBERT ("BOB")	HMM-364
MARTELL, DAVE	HMM-263

NAME	IN COUNTRY SQDN
MARTENS, EDWARD C. JR. ("CHRIS")	HMM-261/164/165/265
MARTIN, BILL C.	HMM-365
MARTIN, DAVID	HMM-161/262/364
MARTIN, DICK ("MOOSE")	HMM-261
MARTIN, DOUG	VMO-2
MARTIN, EARL	
MARTIN, GARY	HMM-265
MARTIN, HENRY	HMM-262
MARTIN, JERRY	HMM-263
MARTIN, JERRY	
MARTIN, JIM	HMM-164
MARTIN, JIM ("JAY EM")	HMM-263/H&MS-16
MARTIN, M. c/o V. Zelent [V2M4N8]	HML-167
MARTIN, RICH	
MARTIN, STEVE	
MARTINDALE, JIM ("BUSHMAN")	HMM-265
MARTINDALE, TOM ("MARTY")	HMM-161/263
MARTINEZ, ANTONIO	
MARTINEZ, DAVID A. ("DOC")	HMM-163/262
MARTINEZ, F. ("MARTY")	HMM-164
MARTINEZ, VIC ("BOOGALOO")	VMO-3
MASON, KURT ("SMOKE")	VMO-2
MASON, ROBERT B. ("BOB")	
MASSARI, KERRY M.	HMM-161/VMO-2
MASSEY, BILLY R.	HMM-165
MASSEY, HAROLD E. ("GENE")	HMM-161
MASSEY, ROBERT ("SKIP")	VMO-2/HMA-367
MASTERS, GEORGE	
MASTERS, JOHN	VMO-6
MASTRINI, DAVID	VMO-6
MATEUS, RAUL E.	HMM-163
MATHESON, BRUCE J.	MAG-36
MATHEWS, ROBERT A. ("BOB")	H&MS-16
MATHIEU, DENNIS M.	H&MS/VMO-6
MATTHEWS, JIM	VMO-6/H&MS
MATTHEWS, PETER F.	
MATTISON, JEFF ("BAMBI")	HMM-364
MAUPIN, WES	HMM-165
MAURER, JOHN G. ("HOOKER")	HMM-364/H&MS
MAVRELIS, PETE	
MAXWELL, JEWELL C. JR. ("CLINT")	H&MS-16
MAXWELL, JOHN T. ("THE RED MAX")	HMM-265
MAXWELL, STEVE	HMH-463
MAY, JAMES E.	HMM-164
MAY, LEE	
MAYES, ROBERT ("BOB")	HMM-363
MAYNARD, JOHN C. ("HAPPY JACK")	VMO-2/HMH-463
MAYNE, GERALD C. ("JERRY")	HMM-365/163/164
MAYS, HERBIE	HMM-164
MAYTON, JAMES A. ("DOC")	VMO-2/MAG-16/36
MAZE, ISAAC D.	HMM-365
MAZON, LEN	HMM-161/263
MAZURAK, PETER A. ("DOC")	HMM-165/263
McADAMS, GREG	VMO-6
McALEAVEY, BRUCE C.	HMM-164/362
McALLISTER, T.C.	HMM-265
McAMIS, RONALD W. ("MAC")	HMM-262
McBEE, BAILEY H.	HMM-263
McBRIDE, CHUCK	
McBRIDE, THOMAS E. ("MAC")	HMM-161/HMH-462
McBROOM, DENNIS	VMO-6
McCABE, MARVIN ("MAC")	HMM-263
McCALL, LYN	VMO-2/HML-367
McCALLIE, PAUL T.	HMM-263
McCALLUM, LEO	
McCARTHY, TOM	HMM-265/364
McCARTY, DAN A.	VMO-2
McCASLAND, LOUIS P. JR. ("LOU")	HMM-162
McCHARGUE, PHIL D.	HMM-265
McCLAIN, MIKE	HML-167
McCLEARY, BRYCE	HMM-164
McCLEES, JERRY	HMH-463
McCLELLAN, ROBERT ("MacATTACK")	HMM-161/VMO-2
McCLUNG, DARRELL	HMM-363
McCLUSKEY, W.C. ("BILL")	HMM-362
McCOMBS, WILLIAM S.	HMH-463
McCORD, RICHARD	HMM-364
McCORKLE, BGEN. FRED ("ASSASSIN")	HMM-262
McCORMICK, DANIEL	HMM-164
McCORMICK, HERB	HMM-264
McCOY, KEN	HMM-161
McCRARY, J.R.	HMM-365
McCREARY, BRICE ("MAC")	HMM-164/165
McCRORY, FREDERICK	HMM-265
McDERMOTT, TOM	HMH-361/463
McDONALD, LARRY	HMM-265/164
McDONALD, MARK ("WORM")	HMM-164
McDONALD, TOM	HML-367
McDONOUGH, LELAND ("BRUNO")	HMM-361
McDYRE, DAN ("SAVAGE")	HMM-165/MWHS-1
McEACHRAN, ROBERT B. ("BOB")	HMM-365
McELWEE, MICHAEL M. ("MIKE")	HMM-161
McFALL, MILTON L.	HMM-164
McGADDEN, JOSEPH R. ("TIO JOSE")	MAG-16
McGAMA, DOYLE	
McGANIGLE, CHUCK	HMM-161/263
McGAW, WILLIAM A. ("BILL")	VMO-6
McGEE, JAMES B. ("3D McGEE")	VMO-2/1
McGEORGE, WILLIAM E. ("BILL")	VMO-2
McGINLEY, BERNIE	HMM-164
McGINN, BOB	
McGINN, GERALD ("MAC")	HMM-165
McGINN, JIM	HML-367
McGLOTHLIN, DONALD H.	HMM-262
McGOVERN, ROB	HMM-262
McGOWAN, TERRY R.	
McGREW, ERIC	HMM-262
McGUFFY, TOM	HMM-262
McILVAIN, BOB	HMM-163
McINTIRE, VIRGIL	
McINTOSH, RANDY	
McJILTON, JACK	HMM-364
McJOYNER, E.M. JR.	MAG-16
McKAY, JAMES E. ("JIM")	VMO-2
McKEAN, L.	HMM-161
McKEE, GEORGE	HMM-365
McKELLER, JOE	HMM-161
McKENNEY, JAMES B. ("BONES")	1st MAW
McKERNAN, JOHNNY D. ("MACK")	VMO-6

NAME	IN COUNTRY SQDN
McKIERNAN, BOB	
McKINNEY, CHARLES	HMM-365
McKITRICK, R.	HMM-164
McKNIGHT, DANIEL P. ("DAN")	HMH-361/463
McKNIGHT, TOM ("MAX GROSS")	HMM-362
McLEAN, FRED	HML-367
McLEARY, DAVID	
McMAHON, DICK ("PIGPEN")	HMM-364
McMANAWAY, JIM	HMM-361
McMENAMIN, ("MAC")	
McMORROW, JOHN	
McNEESE, ROBERT B. JR.	
McPARTLIN, FRED ("McP")	VMO-1
McPHERON, DON	HMM-265
McQUEEN, ROBERT P. JR.	HMM-363
McSHANE, WILLIAM J. ("WILLY MAC")	HMM-361
McSORLEY, DAVE ("SMILEY")	HMM-364
McVAY, DON	HMM-362/263/VMO-2
MEADOR, ROBERT D. ("BOB")	HML-367
MEARS, BRUCE R. ("DOC")	H&MS-16
MEDLEY, VIRGIL ("ZIPPER")	VMO-6
MEDLIN, DUANE ("D.J.")	HMM-263/262
MEDLIN, LARRY	HMM-165
MEHARG, BEN	VMO-2
MEHIRTER, JIM	
MEIDELL, LARRY	HMM-165
MELANCON, LENNY	HMM-361/364
MELAND, QUINT ("GOOSE")	HMM-361
MELENDEZ, JOSE A.	VMO-2
MELIN, MIKE ("MUSH")	HMM-361/362
MELL, FRANK E. ("MEL")	VMO-2/H&MS-16
MENDENHALL, HERB	HMM-265
MENDEZ, RONALD D.	HML-167
MERINO, LLOYD J.	HMM-165
MERRILL,	
MERRITT, BILL	HMM-265
MERRITT, JERRY	
MERRITT, JOHN W.	HMM-161
MESKAN, DONALD W.	HMM-362
MESSICK, JOHN M.	HMM-363
MEYDAD, RICHARD	
MEYER, HAROLD ("HAL")	HMM-365
MEYER, RONALD H.	HMM-262
MEYERS, ("TANK")	
MEYERS, DANNY ("HOLLYWOOD")	HML-367
MEYERS, DAVID	HMM-165
MEYERS, DICK	
MICHAELS, RAY	
MICKLES, RAY C.	HMM-361/161
MILES, WILLIAM H. ("BULLETHEAD")	VMO-6
MILLER, AL	
MILLER, BILL [RICHARD R. MILLER]	HMM-164
MILLER, BOB ("BLUEHAWK")	HMM-263
MILLER, F.W. ("SKIP")	HMM-163
MILLER, HUEY	HML-367
MILLER, JOE E.	HMH-463
MILLER, MGEN. DONALD E.P. ("DEP")	H&MS-16/VMO-2
MILLER, NORM	VMO-2
MILLER, O.J.	
MILLER, PAUL	HMM-361
MILLER, PHIL	HMM-263
MILLER, RAYMOND E. ("GENE")	HMM-161/263
MILLER, RICHARD	VMO-2
MILLER, STEVE	VMO-2
MILLER, T.O.	
MILLER, THOMAS F. ("CHESTY")	HML-167
MILLER, WALLACE B. ("WALLY")	HMM-262
MILLETTE, DENNIS J.	HMM-263/162
MILLHOUSE, TIM	
MILLNER, FRANK	
MILLS, FRANCIS ("HAL")	HMM-262
MILLS, GORDON R. ("BOB")	HMM-364/265/164/165
MILLS, HARRY R. ("BOB")	HMM-163/265/164
MILNER, WILLIAM	HMM-161
MINCEY, JAMES S. ("MAGNET ASS")	HML-167
MINICK, DAVE	HMM-164/HMH-463
MISHKA, BOB	HMM-363
MITCHELL, CHARLES	HMM-164
MITCHELL, EUGENE	
MITCHELL, JOE	
MITCHELL, MACK E.	HMM-162
MITCHELL, NORMAN L. ("NORM")	HMH-463/462
MITCHELL, RICHARD F. ("MITCH")	HMM-262
MITCHELL, ROBERT	HMM-165
MITSIN, LEO	VMO-6
MIX, DAVID	
MODGLIN, CARROL W.	HMM-461/162/163
MODZELEWSKI, ED	HMM-362
MOE, MAURICE ("MAURY")	HMM-364
MOFFETT, DON	
MOFFETT, JOHN	HML-367
MOIST, JOHN	
MOLDOVAN, KENNETH R. ("CHIEF")	H&MS-16
MOLGART, ROY	HMM-365
MONCH, CHARLIE	HMM-365
MONK, BOB	HMM-161
MONK, GARY ("CATFISH")	HMM-364
MONROE, AL	VMO-6
MONROE, JOHN	HMM-164
MONROE, TERRY S.	HMM-365
MONSERRATE, LARRY ("PAPPY")	HMM-265
MONTAGUE, PAUL ("COUNT")	HMM-364/264
MONTE, ANTHONY F.	HMM-364
MONTELEONE, JOE	
MONTGOMERY, J.R. ("MONTY")	HMM-265/MAG-36
MONTOYA, PAUL	HML-167
MOODY, PAUL	HMM-265
MOON, DONALD E.	H&MS-36
MOORE, DAVID J. ("DAVE")	HMM-163/362/164
MOORE, ERNEST O.	HMM-164
MOORE, FRANK ("COLORADO")	HMM-363/167/H&MS
MOORE, HARVEY ("BASS")	
MOORE, JACK	VMO-2
MOORE, JOHN A.	HMM-163/361
MOORE, LEON M. JR.	HMM-162
MOORE, PAUL	H&MS-16/HMH-462
MOORE, R.C. ("DANCING BEAR")	HMM-364
MOREAU, PAUL L. ("EVIL")	HMM-362/165/VMO-6
MOREY, A.W. ("PALADIN")	HMH-463/361
MORGAN, BOB ("STONEY")	HMM-263/163/VMO-6

NAME	IN COUNTRY SQDN
MORGAN, EVAN H. JR.	HMM-263
MORGAN, JAMES ("JIM")	HMM-163
MORGAN, ROBERT Y. ("BOB")	VMO-3
MORGENSTERN, JOHN	HMM-265/161
MORGON, ROBERT Y.	VMO-3
MORHARDT, JEFFREY C.	HMM-364/263
MORIARTY, JAMES R. ("JIM")	VMO-2
MORIN, RON ("MOE")	HMM-363
MORLEY, JAMES E. JR. ("JAY")	HMM-261/362
MOROSKY, RONALD W. ("SKI")	HMM-161/162
MORRIS, DAVE	HMM-263
MORRIS, DREW	HMM-262
MORRIS, JOHN R.	VMO-3
MORRIS, M.L. JR. ("SONNY")	HMM-361
MORRIS, SAM ("SAFETY SAM")	HMM-263/165
MORRISON, MIKE	
MORRISON, MIKE ("M&M")	HMM-161/263
MORRISON, TOM	HMM-161
MORRONGIELLO, C.J. ("SANDY")	HMM-161
MORSCHAUSER, MIKE	
MORSHEAD, JOHN W.	MABS-16/ HMM-263
MOSER, GARY W. ("MOE")	HMM-161/265
MOSER, STEVEN	HMM-265
MOSKOSKY, JOE	HMM-164
MOTT, KEN ("KAHUNA")	HMM-361
MOTTER, WILLIAM	HMM-164/265
MOTZ, DONALD F. JR. ("DON")	HMH-463
MOUNTAIN, PAUL E.	HMM-164
MOWRY, EDWARD F. ("EASY ED")	HMM-262
MULLANE, TOM	HMM-263/ 364/H&MS
MULLEN, GERALD ("MOON")	HMM-161/364
MULLEN, HERBERT L. ("MOON")	HMM-262/165
MULLEN, MIKE ("MOON")	HMM-262/ HMA-369
MULROY, JIM ("J.P. LUMBER CO.")	HMM-361/363
MUMFORD, C. ALLEN ("AL")	VMO-2
MUNOZ, ANSELMO	HMM-161
MUNOZ, JUAN R. JR. ("OOPS")	HMM-164/263
MUNTER, WELDON ("MUNT")	HMM-362/ VMO-3
MURPHY, BILL ("SUITCASE")	VMO-3
MURPHY, DAVE	HMM-265/161
MURPHY, ED ("MURPH")	MAG-16
MURPHY, JERRY	
MURPHY, JOHN T. ("MURPH")	HMM-161/ 363/VMO-2
MURPHY, THOMAS J. ("TOM")	HMM-165
MURPHY, WILLIAM	HMM-262
MURRAY, CHRIS	HMM-263
MURRAY, CURT	VMO-6
MURRAY, JAMES	
MURRAY, NORB ("DANCING BEAR")	HMM-163/161
MURRAY, STEVE ("ONLY STEWS")	HML-367/167/ H&MS
MURRAY, THOMAS C. ("T.C.")	HMM-263
MURRAY, WILLIAM A. ("BILL")	HMH-463
MURREN, LOREN ("LEW")	HMM-165
MUSICH, DICK	
MUSSO, MICHAEL ("MOOSE")	HMM-361
MUTH, CHUCK	HMM-161
MYATT, RAMSEY	HMM-361/362
MYERS, CONNIE D.	HMM-165
MYKING, LARRY R.	HMM-163
— N —	
NACCITELLI, ANDY	
NADEAU, A.J.	MAG-16
NADLER, NEIL	HMM-161/164
NALE, BURT	
NALL, JIM ("MAX GROSS")	HMM-262/ H&MS-36
NANCE, H.T.	HMM-162/ 164/VMO-2
NASH, ARTHUR H. ("ART")	HMM-163/362
NASH, OTIS	HMM-362
NASSER, RAYMOND	HMM-262
NAY, DAVID R. ("PELT")	HMM-361
NEALY, JOHN ("SNAKE")	HMM-361
NEBEL, R.M. ("THE BLUE MAX")	HMM-363/361
NEEDHAM, MIKE	
NEES, WILLIAM E. ("BILL")	HMM-262
NEFF, ROBERT L. ("SILVER FOX")	HMM-162/ 261/VMO-6
NEIGHBOUR, WALTER T. ("WALT")	MAG-36
NEIL, BGEN MIKE	AO/1st MAW
NELBACH, A.A.	
NELSON, A. MONTE	HMM-262/163
NELSON, BILL ("FUELS")	HMM-262
NELSON, BOB	
NELSON, DAVE ("NELLY-BELLS")	HMM-163/ 364/262
NELSON, H.E.	HMM-263
NELSON, J.A.	
NELSON, M.R. ("ROCKY")	HMM-262
NELSON, MGEN. RONALD K. ("RON")	
NELSON, RALPH S.	HMM-265
NELSON, RON	HMH-462
NEMETZ, P.T.	HMM-362
NERBURN, BILL	
NESBIT, CHARLES L. ("CHARLEY")	MAG-36
NeSMITH, JOE	HMM-165
NEWBERRY, JOHN	
NEWCOMB, FRED	HMM-265
NEWLON, DAVE	
NEWTON, H.W.	
NICHOL, B.J. ("BUD")	HMM-361/363
NICHOLAS, RALPH ("GUNNY NICK")	HMM-163
NICHOLS, DONALD	HMM-262/364
NICHOLS, ROBERT	HMM-163
NICHOLSON, R.B. ("NICK")	HMM-361
NICK, PAUL M.	HMM-165/265
NICKELE, J. ("JOE SHIT THE RAGMAN")	HMH-463
NIESEN, JEWELL [TINY NIESEN]	HMM-161
NIKOLAI, LEONARD ("THE WEDGE")	HMM-365
NILSON, GEORGE ("CRISPY CRITTER")	HMM-163
NITCHMAN, ALAN C. ("NITCH")	HMM-362
NOBLE, JOHN L. JR.	HMM-361
NOBLETTE, KENNETH B.	HMM-262
NOLAN, JOHN L.	HMM-362
NOLL, ERNIE ("NGUYEN")	HMH-463
NOON, JAMES	HMM-164
NORCOTT, ROBERT B. ("BOBBY")	HMM-262
NORMAN, CHARLES M. ("MIKE")	HMM-261/165
NORMAN, JOHN F. ("HAWG")	HMM-163/ 362/VMO-2
NORRIS, AVERY	HMM-265
NORRIS, PHIL	
NORTON, GARY	VMO-2
NORTON, JAMES D. ("SMOKY")	HMM-362/ 363/367

NAME	IN COUNTRY SQDN
NORTON, JERRY	HMM-363
NORTON, RAYMOND J. ("RAY")	HMM-161/263
NOTEBOOM, DENNIS	HMM-362
NOTTINGHAM, CHARLES R.	HMM-165
NOVAK, JOE ("MOMS")	HMM-162/ HMH-462
NOVICKIS, ANDY	
NOVOTNAK, JOHN	HML-167
NOWAK, ED	HMM-361
NOWOTNY, CHUCK	HMM-164
NUGENT, JIM	HMM-364
NUTTER, DAVE W.	HMM-263
NYE, BOB	HMM-163
NYLANDER, LELAND ("JIM")	HMM-164
NYQUIST, GLENN	
— O —	
O'BRIAN, JIM	
O'BRIEN, DENNIS M. ("O'BIE")	HMM-265/165
O'BRIEN, RICHARD	MAG-16
O'BRYAN, BOB	VMO-3
O'CONNELL, GEORGE J. JR.	HMM-163
O'CONNOR, ED ("EROC")	HMM-261/ 363/364
O'CONNOR, JAMES	
O'CONNOR, TIM	
O'CONNOR, TOM	
O'DONNELL, BERNIE M.	HMM-265
O'DONNELL, PATRICK L.	HMM-165
O'HARA, NORB	
O'KELLEY, JIM	
O'KENNON, BOB ("COBRA")	HML-267
O'MALLEY, TOM JR.	HMM-262
O'MEARA, JOHN ("JOHN O")	HML-367
O'NEIL, BOB ("IRISH")	HMM-163/362
O'NEIL, JERRY ("DOC")	
O'NEIL, LOUIE ("THE RAT")	HMM-264
O'NEILL, BRIAN	HMM-263
O'NEILL, RAY	
O'TOOLE, TIM	HMM-363
OBERHOLTZER, BENJAMIN	VMO-2
OBERLANDER, DON ("OBIE")	HMM-361
OBERLIES, DARRELL	H&MS-16
OCKULY, EUGENE J. ("GENESAN")	HMM-263/ HMH-462
ODDO, BILL	MAG-16
ODGERS, GERALD ("JERRY")	HMM-263
ODOM, JOHN R. III ("BOB")	HMM-161
OFFILL, THOMAS C.	
OGLE, JERRY	
OLDHAM, WADE ("SUPER CHICKEN")	HMH-463
OLE, DICK	
OLITSKY, HARVEY ("HARV")	HMM-263/ 161/163
OLIVER, JOSEPH	HMM-163
OLIVO, TOM	
OLSEN, LARRY	
OLSHESSKI,	
OLSON, CHARLES M. ("OLE")	HMM-362
OLSON, CURT	
OLSON, DOUG	HMM-164
OLSON, GLENN ("ANDY")	VMO-6/ VMO-2
OLSON, KEN ("OLY")	HMM-161/265
OLSON, MARTIN G. ("MARTY")	HMM-164
OLSON, MYRON ("MOSTLY OLSON")	HMM-263/364
OLSON, REID	HMM-261
OLSON, THOMAS C.	HMM-165
OLY, DICK	
ONDRICK, ROBERT M. ("HONDO")	HML-167
OPEAN, MIKE	HMM-361
OPOCENSKY, JARDO JR. ("SKI")	VMO-2
ORAHOOD, DOUG	HMM-364
ORCUTT, ALLEN G.	HMM-164
OREY, RALPH B.	HMM-262
ORKISH, JOHN M.	HMM-364/ 163/463
ORR, THOMAS	
ORTT, DOUGLAS J.	HMM-364
OSBON, KEN ("OZ")	HML-167
OSBORNE, RON	VMO-2/ HMA-369
OSKADA, DANIEL	HMM-164
OSSERMAN, STAN	HMM-161
OSTER, WAYNE	HMM-164
OSTROWSKI, CONRAD J. ("SKI")	HMM-165
OTT, ROBERT F.	HMM-161/263
OTTO, GEORGE	HMM-165/262
OTTO, JIM ("TOOT")	HMM-261
OUBRE, JOHN A.	HMM-361/ MABS-16
OUELLETTE, GARY L.	HMM-165
OUELLETTE, ROGER E.	HMM-361
OVERSTREET, COY	HMM-161/361
OWEN, PATRICK ("PAT")	HML-367
OWENS, DAVID	HMM-263/161
OWENS, DAVID J. ("DAVE")	HMM-364
OWENS, EARL	HMM-164
OWENS, TED	
OWENS, WILLIAM M.	HMM-264
OWLLETT,	
— P —	
PABST, ROBERT C.	HMM-164
PACK, TOM	HMM-265
PADELETTI, JOE	HMM-163/164
PAEGEL, WALT	HMM-265
PAETZNICK, JAMES H.	HMM-362
PAGE, DOUG ("WINDY")	VMO-2/6
PAGE, JACK	HMM-262
PAGE, JACK E.	HMM-262
PAGE, MICHAEL	HMM-164
PAGE, ROBERT W.	
PAGLIONE, CARMEN	MAG-39
PALERMO, CARLO	HMM-164
PALMASON, STEVE ("BLUNDER")	VMO-6
PALMER, CHUCK	HMM-161
PALMER, ERNEST ("BURT")	HMM-362/ 265/165
PALMER, LEONARD E. JR.	HMH-462
PALMER, ROGER	HMM-261
PANSKA, JIM	HMM-161
PAPINEAU, TOM ("PAPPY")	HMM-164
PARADISE, LOUIS	HMM-365
PARDOE, CRAIG	
PARKER, ANDREW JR.	HMM-265
PARKER, GARY ("RED")	HST - KHE SANH
PARKER, PAUL	HMM-263
PARKS, WALTER ("WALT")	HMM-165
PARNELL, BGEN. EDWARD A.	HMM-161
PARNHAM, JOHN ("STUMPY")	HMM-262

NAME	IN COUNTRY SQDN
PARR, JOE	HML-367
PARR, MICHAEL J. ("JACK")	HMM-164
PARRY, DALE T.	HMM-163/165
PARSONS, JIM ("PUDGE")	HMM-364
PARSONS, LARRY	
PARTEE, RICHARD ("DICK")	
PASKEVICH, TONY	VMO-2
PATE, JERRY ("HOME BREW")	HMM-261/262
PATRAITIS, CHET	H&MS-16/ MABS-16
PATRICK, G. LANE	HMM-165
PATRICK, K.W. ("PAT")	HMM-263
PATTERSON, FRED D. JR. ("PAT")	HMM-361
PATTERSON, KELLY	HMM-261
PATTON, CAREY ("PAT")	HMM-262
PATZKE, WILLIAM R. ("P-SAN")	HMM-363/161
PAULSON, E. W. ("BILL")	HMM-263/ VMO-2
PAVAN, DAN	HMM-364/ 362/462
PAWLING, DOUGLAS H. ("DOUGOUT")	HMM-162
PAYNE, COURTNEY B. ("COURT")	HMM-163/ 364/463
PAYNE, PAUL	HMM-263
PAYTON, LEE ("LIMA LIMA")	HMM-165/263
PEACOCK, CLYDE	HMM-161
PEACOCK, MARV	HMM-264/364
PEARCY, LARRY	HMM-161/ 262/364
PEARD, ROGER W. ("REDEYE")	HMH-463/ H&MS-16
PEARSON, MONTY	HMM-164
PEASLEY, DICK	
PECHALONIS, STANLEY J. ("STAN")	HMM-263/161
PECORARO, TONY ("PECK")	VMO-6
PELHAM, WAYNE ("BEACH")	HMH-463
PELKEY, JIM	
PENCEK, BARRY	HML-367
PENDER, DAN ("SCARFACE 42")	HML-367
PENLAND, RONALD C.	HMM-164
PENN, THOMAS L. ("SLICK")	VMO-2
PENNING, FRED ("FRIAR")	HMM-161
PENNINGER, RON	
PENNINGER, SAM	VMO-6
PENTZ, ED ("SKIP")	HMM-263
PEPPLE, MIKE ("PEP")	HMM-165/265
PERCIVAL, WILLIAM F. ("WILD BILL")	VMO-3/ HML-367
PEREIRA, DAVE ("PORTAGEE")	HMM-364
PEREZ, TONY ("SIERRA SIERRA")	HMM-361
PERKINS, G.T.	
PERKINS, JAMES W. ("JIM")	HMM-364
PERKINS, MARVIN D.	
PERRINE, JOHN	
PERROW, RON	
PERRY, C. ROBERT ("BOB")	HMM-165
PERRY, HANK ("HAWK")	HML-367
PERRY, ROBERT E. ("LURCH")	VMO-3
PERRY, RON	VMO-3
PERRYMAN, JAMES ("JIM")	HMM-362/ 262/VMO-6
PERSINGER, PAUL ("FANG")	HMM-165
PERSKY, DON ("DEUCE")	HMH-463
PETERS, ALEXANDER R. ("BOB")	HMR-361
PETERSEN, FRED ("PETE")	HMM-363
PETERSEN, KENT G. ("PETE")	HMM-362/ HML-367
PETERSON, BILL ("W.P.")	
PETERSON, CARL	
PETERSON, DAN	
PETERSON, KELSEY	HMM-364
PETERSON, ROGER ("PETE")	VMO-6
PETRAS, TED	HMM-164
PETREE, D.E. ("PETE")	HMH-463
PETRO, JOHN J. SR.	HMM-161
PETROUSKY, JAMES A. ("JIM")	VMO-6
PETTEYS, DAVID M. ("DELTA MIKE")	HMM-165/265
PETTIS, RON	HMM-365
PETTIS, WADE A.	HMM-263
PFALZGRAFF, MIKE	HMM-164
PHELPS, JOHN	VMO-6
PHIFFER, SAM	
PHILBRICK, ALFRED G. ("AL")	HMM-161
PHILLIPS, B.G. ("BILL")	HMM-163/ 161/H&MS
PHILLIPS, BILL	HMM-164
PHILLIPS, BOB	HMM-161/363
PHILLIPS, BRUCE M.	HMM-361
PHILLIPS, MGEN. RICK	HMH-463/ MABS-16
PHILLIPS, MICHAEL W.	HMM-364
PIATT, JERRY	HMM-262/ 263/265
PICKETT, JEFF	VMO-6
PICONE, ART ("PINECONE")	HMH-361/463
PIDOCK, GARY	HMM-364
PIERSON, JOHN H. ("GOOFY")	HMM-363/364
PIMENTAL, RICHARD ("BULL")	HMM-362
PINCKARD, W.H. ("BUZZ")	HMM-361/362
PINGLE, HOWARD	
PINKERTON, WALTER E. JR. ("PINKY")	HML-367
PINSCHENAT, BARNEY ("B.J.")	
PINSON, JOE	
PINSON, RAY	HMM-262
PIPA, JACK ("BLADES")	VMO-2/3/ HMM-364
PIPER, ART	VMO-6
PIRNIE, DAVID M.	HMM-362
PIRTLE, GORDON	HMM-164
PITMAN, LTGEN. CHUCK	HMM-265
PITTMAN, TED	HMM-265
PITTS, KEN	HMM-262
PITZL, GERRY	
PIXLEY, EDWIN ("ED")	HML-367/ VMO-3
PIXTON, MARVIN F. III ("ROCKET")	VMO-2/ HML-167
PLASTERER, MGEN. ROSS S.	HMM-162/ 165/167
PLATE, ROGER	HMM-364
PLATTE, CURT	HMM-365
PLETCHER, JERRY	HMM-364
PLEVA, JAMES F.	HMM-165
PLUMMER, J.R.	HMM-164
PLUNKETT, CHARLES L. ("CHARLEY")	VMO-2
POE, WILLIAM JR.	HMM-265
POGANY, C.J. ("POGY")	HMH-463
POHL, PAUL A.	HMM-365
POINDEXTER, NEAL ("SARGE")	HMM-161/ HML-167
POLESKI, MARTIN LEE ("MAD DOG")	H&MS-36/ HMM-361

NAME	IN COUNTRY SQDN
POLHEMUS, DICK	HMM-263
POLHEMUS, KEN	
POLI, JOSEPH ("JOE")	HMM-165
POLLEY, DENNIS	HMM-365
POMEROY, ROBERT W. ("GROOVY")	HMM-261
POMPI, WILLIAM	HMM-263
PONNWITZ, AL	HMM-165
POOLE, SAMUEL T. JR.	HMM-364
POPE, BILL	HMM-263
POPE, FRED	HMM-363/ H&HS
POPE, HAL	HMM-361
POPE, JERRY M.	VMO-6
POPP, JOHN A.	HMM-164
PORISCH, KARL	HMM-263
PORTER, GARY LEROY ("TAILSPIN")	HMM-263/ HML-367
PORTER, MERVIN	HMM-261
PORTER, PETER A.	VMO-6
PORTER, ROBERT D. ("BOB")	HMM-265/164
PORZIO, LEN ("BROOKLYN")	HMM-361/ 263/VMO-6
POWELL, NORRIS	
POWELL, TERRY H.	HMM-161
PRATT, FRED	HMM-265/161
PRATT, GEORGE	
PRATT, ORVID W. ("ORV")	HMM-163
PREDOVIC, DAN ("MATTRESS BACK")	HMH-463
PRESSLY, JAMES P.	HMM-164
PRESSNALL, LARRY	
PRESSON, BOB	
PRESTON, STEVE	HMM-163
PRIBANIC, GEORGE ("SKIP")	
PRICE, DENNY	VMO-6
PRICE, JAMES W. JR. ("JIM")	HMM-263/363
PRICE, WILLIAM G. ("WILLY")	HMM-263
PRIEST, JACK D.	HMM-162/261
PRIETO, JUAN	
PRIGGE, ROY J.	HMM-262
PRIMEAUX, LAMOTTE	HMM-164
PROCTOR, DON	HMM-262
PROUT, BRUCE	
PROUTEY, DON ("DOC")	HMM-163/ 262/MAG-36
PROVEN, DONN R. ("DOC")	HMM-362
PRUDEN, JAMES R. JR.	HMM-164
PUGH, JESS	HMM-362/ HMH-461
PURCELL, ROBERT D. ("BOB")	VMO-6
PURDY, JAMES ("DARYLL")	HMM-262
PYLE, HAROLD F. JR. ("ERNIE")	HMM-162
— Q —	
QUACKENBUSH, JOHN ("DUCKWEED")	HMM-164
QUADRINI, BARBRA [F. QUADRINI]	HMR-362
QUALE, RAYMOND A. JR.	
QUINLAN, MOSS	HMM-364
QUINN, MGEN. W.R. ("BILL")	1st MAW
QUINTERO, PETE	HMM-365
— R —	
RACHAL, JOSEPH	VMO-2/6
RAGIN, MICHAEL	
RAINES, T.E.	VMO-2
RAINEY, GARY D.	HMM-262
RAINVILLE, GENE	HMM-365
RALPH, DICK	
RAMIREZ, BARRY P.	HMM-262
RAMSBURY, H.R. ("RAMJET")	HMM-263
RAMSEY, CDR. IRA E. ("BLONDIE")	HMM-363
RAMSEY, GARY	HMM-261
RANDALL, JOSEPH J. JR. ("J.J.")	HMM-262
RANDALL, RUSS	VMO-2
RANKIN, BILL	VMO-6
RANSOM, AL ("ANIMAL")	1st MAW
RARICK, ROBERT W. ("BUFFALO")	HML-167
RASCO, THOMAS	HMM-161/363
RASK, ART ("MWK")	HMM-163
RASNICK, HARVEY R. ("RAZZ")	HMM-265
RATCLIFFE, BOB ("RAT")	HMM-263
RATTERER, RICHARD A.	HMM-164
RAUPP, DOUG	HMH-463
RAUWALD, THOMAS C.	HMM-362/363
RAWLINGS, ART	HMR-161
RAY, DAVE	VMO-6/ HML-367
RAY, JOHN PARKS ("KYTU")	HMM-361
REA, WILLIAM	
READ, W.T. ("TED")	HMM-165
REAGAN, BOB	
REAMES, BILL	
REAMES, JOHN A.	HMM-165
REAP, TOM	
RECHONIE, ROBERT	VMO-6
REDMAN, CARROLL	VMO-6
REED, DAVID	HMH-463/ VMO-2
REED, K.R.	HMM-263
REED, VAN	
REESE, CLIFFORD E. ("CLIFF")	H&MS/VMO- 2/HML-367
REEVES, CHUCK	HMM-164
REEVES, WILL JR.	HMM-365
REICHERT, EL	VMO-2/ H&MS-16
REICK, JIM ("KING SNAKE")	HMM-161
REID, JOHN E.	HMH-463
REID, W. THOMAS ("DIRTBALL")	HMM-262
REIMAN, L.P.	HMM-265/364
REINER, ARNIE	HMM-265/164
REINHARDT, LEU	HMM-162
REINIKA, DOUGLAS E. ("DOUG")	HMM-161
REISTERER, DAVE	HMM-364
REMME, MIKE ("WNDR. WART HOG")	VMO-6
RENGEL, GREG	HMM-261/ 262/364
RENSCH, RON	
REPKO, WILLIAM A. ("BILL")	HMM-165
RERICK, BOB	
RESTIVO, JOHNNY DEAN	HMM-265/ FAC
RETTERER, RICHARD A.	HMM-164
REUSSER, KEN	MAG-36
REVER, W.H. JR. ("PAT")	HMM-161/ 262/H&MS
REYNOLDS, DWIGHT	HMM-262
REYNOLDS, FRANCIS T. ("FRANK")	H&MS-16
RHINE, DENNIS R.	HMM-161
RHODES, MGEN. J. ("TRADER JOHN")	HML-167/367
RIBBECK, AL	HMM-164
RIBECK, WALTER	HMM-262
RICE, DAVID	HMM-265

NAME	IN COUNTRY SQDN
RICHARDSON, DAVID E.	HMH-463
RICHARDSON, MIKE ("RED BARON")	HML-167
RICHARDSON, T.J. ("TERRY")	HMM-363/364
RICHART, JAMES	HML-167
RICK, EDWARD III ("HULK")	HMM-361
RIDDICK, WALT	HML-367
RIDGEWAY, LARRY	
RIDGWAY, ELMER	HMM-262
RIDGWAY, JACK ("J.J.")	HMM-263/163
RIEFFER, STEVE J. ("REEFER")	HMM-362/ 262/H&MS
RIERSGARD, DARYL	HMM-364
RIESTERER, DAVID K.	HMM-364
RIGGS, JAMES L. ("JIM")	HMM-163
RIGGS, SHELBY A. III	
RIGGS, STEVE	HMM-161
RING, BILL ("THE CHICKEN MAN")	HMM-364/165
RIPPEY, JOHN S. ("RIP")	HMM-265/165
RITCHIE, FRANK	
RITCHIE, JIM	HMM-161/364
RITORTO, ROBERT	VMO-6
RITTERMEYER, BILL	HMM-362
RIVAS, FRANK	HMM-263
RIVERS, RICK ("HAWK")	HMM-163
ROACH, JAMES W. ("JIM")	HML-367
ROAKE, MICHAEL ("HIPPIE")	HML-167
ROARK, DANIEL E. ("DANNY")	HMM-164
ROBBINS, H.R. ("RUSTY")	HMM-262
ROBBINS, M.G. ("ROBBY")	VMO-2
ROBERTS, GERRY	HMM-264
ROBERTS, HOWARD S. ("SCOTTY")	HMM-364/ 363/263
ROBERTS, JAMES A. ("JIM")	HMM-261
ROBERTS, JOSEPH T.	HMM-265
ROBILLARD, ALAN T. ("AL")	HMM-364
ROBILLARD, JACK ("SCARFACE 14")	HML-367
ROBINSON, BOB ("ROBBY")	HML-367
ROBINSON, DAVE	VMO-2
ROBINSON, JOHN C. ("J.C.")	HMM-163
ROBINSON, LARRY D.	HMM-164/161
ROBINSON, TED ("9 FINGERS")	HML-367
ROBISON, ERIC L. ("ROBBY")	HMM-364
ROBISON, PAUL B.	
ROBSON, JON R. ("LI'L JON")	VMO-2/HMM-165
ROCHE, BEN	HML-167
ROCHFORD, RAE [R. ROCHFORD]	HMM-263/363
ROCHFORD, TUSTIN [R. ROCHFORD]	HMM-263/363
RODGERS, PHIL	HMM-164/165
RODINSKY, CHARLES ("SKI")	H&MS-16
RODRIGUEZ, RONALD	HMM-165
ROEHRIG, TOM ("DUCKFOOT")	HMM-363
ROGERS, BOB	HMM-264
ROGERS, DEAN ("ROG")	HMM-164/ 165/265
ROGERS, JIM	HMM-164
ROGERS, ROBERT P. ("BUCK")	VMO-3/ HMM-165
ROGERS, TOM ("ROGERS #1")	HML-167
ROGERS, WAYNOR	HMM-361
ROHR, TERRY	HMM-265
ROLAND, JOE ("GATOR")	HMM-264/263
ROLFE, BRIAN W. ("MOOSE")	HMM-265
ROLLINS, PAUL R. ("ROLLO")	VMO-3/ HML-367
ROMAN, JOE	HMM-265
ROMAN, RON	VMO-2
ROMINE, E.	HMM-165
ROOK, RONN	HMM-263/ H&MS-16
ROOTS, JOHN C.	HMM-164/ FAC
ROPELEWSKI, LEE J. ("ROPE")	HMM-263
ROPELEWSKI, ROBERT	
ROSE, ALLAN ("AL")	HMM-364/ 165/265
ROSE, E.V.	HMM-362
ROSE, MIKE ("SALTY")	1st MAW
ROSS, B.J. ("BARNEY")	VMO-6
ROSS, DANIEL C. ("DAN")	HMM-361
ROSS, DAVE	HMM-163/ 369/VMO-2
ROSS, DAVID R. ("ROSSINI")	HMM-164
ROSS, GEORGE A.	HMM-163/164
ROSS, LEONA [ROBERT ROSS]	HMM-363
ROSS, THOMAS J. ("TOM")	HMM-361
ROSSER, RICHARD C. ("RAT")	VMO-6/ HMM-164
ROSTAD, DAVE	
ROTH, RICHARD ("RICK")	HMM-165/364
ROTHENBERGER, RAY	HMM-164
ROTHSCHILD, TONY	VMO-6
ROTHWEILER, HUGH D.	HMM-362
ROTTRUP, LOWELL ("MURPH")	HMM-164
ROUNSEVILLE, PETE ("PERFECT")	
ROUSSEAU, RICHARD F. ("TWIGGY")	H&MS-16/ HMM-364
ROWLAND, FRED ("SCREWS")	VMO-6
ROY, WILLIAM B. ("MOOSE")	HMR-361
RUBY, JESS	HMM-263
RUCK, BYRON ("B.J.")	HMM-363
RUDD, GEORGE T. ("G.T.")	H&MS-17
RUDOLFS, EDWARD	HMM-365
RUFFIN, BARRY L.	HMM-363
RUIZ, CECILIO L. JR.	H&MS-16
RULE, J.M. III ("JIM")	HMM-161
RULLI, JOHN	MAG-36
RUNDALL, AL	
RUNDLE, T.J.	HMM-263
RUNSVOLD, CHARLES A. ("CHARLIE")	HMM-364
RUNYAN, KEN	HMM-165
RUNYON, JIM	
RUSH, TOM	
RUSSANO, FRANK	HMM-262
RUSSELL, PHIL	
RUSSELL, RICHARD L. ("RUSS")	HMM-362/462
RUSSELL, STEVE ("DOC")	HMM-364/ 263/165
RUSSELL, TEDDY A. ("TED")	HMM-164
RUSZIN, J.W.	VMO-2/ HMM-262
RUTH, BILL ("BABY RUTH")	HMM-363
RUZICKSA, JOE	
RYAN, RICHARD L. ("DICK")	HMH-463
RYTI, BRAD	

— S —

NAME	IN COUNTRY SQDN
SAARELA, EDWARD	
SABATTUS, DONALD J.	HMM-362
SABIN, RON ("THE PRINCE")	HMM-361
SACHS, RUSTY ("GUNNY")	HMM-362
SADDLER, DAVE	HML-367/ HMM-161
SADOWSKI, JOE	HMH-463
SAKAOKA, FRANK T.	HMM-164
SALAU, ED	HMM-265
SALLAZ, JAN W. ("SAL")	HMM-361
SALLEE, GARY G.	HMM-265
SALMON, CHRIS ("SAKANASAN")	HMM-361
SALMON, NOEL	HMM-263
SALOMON, EDWARD A.	HMM-363
SALTER, GENE	HMM-363
SALTRY, ROBERT	HMM-164
SALUM, GEORGE D.	HMM-362
SAMPLE, E.J. ("ED")	VMO-
SANCHEZ, XAVIER	HMM-161
SANDELL, FRANK	VMO-2
SANDERS, CHARLES E. ("GENE")	HMM-265/161
SANDERS, HASKELL D.	HMM-365
SANDERS, JIMMY	HMM-263
SANDERS, W.L. ("BUDDY")	HMM-361
SANDOVAL, G. ("SANDY")	HML-167
SANDSTROM, GEORGE L.	HMM-263/364
SANDVOSS, BURT	
SANGER, EDWARD F.	
SANTINI, THE GREAT	In the rear with the gear
SANTOS, KENNETH	HMM-364
SARGENT, MICHAEL T.	HMM-164
SARLES, TREV	HMM-263/364
SASS, DAVE	1st MAW
SASS, JEFF	HMM-161
SAUCEDA, EFREN ("CHICO")	HMH-463
SAUNDERS, BOB	VMO-6
SAUNDERS, N.E. ("NICK")	HMM-263/ 161/463
SAWCZYN, PETER G. ("PETE")	HMM-262
SAWYER, MICK	HMM-165/265
SAYES, DAVE	HMM-165
SCAGLIONE, PETE	
SCANLON, MARTIN J. ("MARTY")	HMM-261/VMO-6
SCARBOROUGH, N.H.	HMM-265
SCHAEFER, JIM	HMH-462
SCHANEY, THOMAS E.	HMM-362/363
SCHECTER, LARRY ("ABE")	HMM-163/ HML-167
SCHEFFLER, BILL ("SCHEFF")	HMM-164
SCHERMERHORN, RICHARD C.	HMM-164
SCHERZINGER, JOHN ("SHITSLINGER")	HMM-165
SCHEWSTER, JOHN	HMM-361/ 161/462
SCHIFFBAUER, BOB ("SHIFTY")	VMO-3/2
SCHMIDT, JIM	HMM-265/164
SCHMIDT, PHIL	
SCHMITZ, DAVID K.	HMM-364
SCHMITZ, ROGER	MABS-16
SCHOENER, C.J. III ("CHIC")	HMM-364
SCHOENFELDER, ROGER G.	HMM-263
SCHOLLE, JOSEPH F. ("CRAZY JOE")	HMM-363/ H&MS-36
SCHRADER, CHARLES ("WHIP")	HML-367/ HMM-265
SCHRAMM, ALAN ("TANGO")	HMM-164
SCHREIBER, BOB	HMM-364
SCHREINER, DENNIS ("DENNY")	HMM-164/265
SCHRIBER, ED	HMM-364
SCHRUPP, RICHARD A. ("DICK")	HMM-163
SCHUH, RICK	HMM-161
SCHULDHEISZ, WM. ("LIVING SKULL")	HMM-365
SCHULTZE, BARRY	HMM-163
SCHUTT, HERB ("SPIKE")	HMH-463
SCHVIMMER, JOSEPH ("JOE")	VMO
SCHWANDA, BRUCE A.	VMO-3
SCHWARTZ, JIM	HMM-363
SCHWARZ, BILL	HMM-364
SCHWEITZER, KEN	VMO-2
SCHWEND, JIM ("MR. 'D'")	HMM-165
SCHWINDT, LOU ("SKIPPER")	HMM-163
SCIESKA, JERRY	HMM-164
SCOGGINS, WILLIAM	HMM-161
SCOTT, GARY L.	HMM-164
SCOTT, H. RAE ("SCOTTY")	HMH-463
SCOTT, KENNETH D.	H&MS-16/ HMM-163
SCOTT, ROLAND ("SCOTTY")	HML-367
SCOTT, WALTER T.	
SCOTT, WILLIAM F.	
SEAMAN, JERRY	
SEARING, ED	HMM-163/ 363/MAG-16
SEARS, JOHN	
SEBECK, THOMAS E.	HMM-361
SECKINGER, RALPH	HMM-165
SEEPS, PHIL	
SEGARRA, LEN	VMO-2
SEIBEL, ROBERT ("ROB")	HMM-165
SEIFERT, ED ("EASY ED")	HMM-263
SEITZ, FRED	
SELLERS, TREVOR	HMM-265
SELLERS, WILEY J.	HMM-164/ 262/H&MS
SENECAL, PAUL	HMR-262/362
SENKUS, VETE	HMM-262
SERGEANT, JERRY L.	H&MS-36
SERMAN, PAT	
SEVERSON, MIKE ("BROW")	HMM-261/362
SEXTON, DONALD K.	HMM-165
SEXTON, JAMES P.	HML-367
SEXTON, TONY W.	HMM-263
SHADRICK, BILL	HMM-262
SHAFFER, D.L. ("WHITEY")	HMM-164
SHALLCROSS, WALT	
SHALLENBERGER, WILLIAM ("BILL")	HMM-364
SHANAHAN, JAMES L. ("JIM")	MABS-15/ HMM-263
SHANNON, JERRY	
SHANNON, ROBERT C.	HMM-161
SHANTRY, ROBERT E. JR. ("BUZ")	HMM-363
SHAPIRO, BRUCE ("SHMIT OFFICER")	HMM-261
SHARKEY, CHRIS	HMM-163
SHARP, JERRY	
SHAUER, WALTER ("EASY WALT")	HMM-362/361
SHAW, THEODORE L. III	HMM-265
SHAW, WAYNE	HMM-265
SHEA, MICHAEL L. ("MIKE")	HMM-165
SHEA, TOM	HMM-265
SHEARY, TOM	HMM-262
SHEEHAN, BOB ("SCARFACE ONE")	HML-367
SHELDON, GARY	
SHELTON, JAMES L. ("JIM")	HMM-362/463
SHELTON, RAYMOND S.	
SHERER, LARRY	HMM-163/ 262/161
SHERMAN, WAYNE F. ("BRONCO")	HMM-364
SHIELDS, JOHN M. ("JOHNNY MIKE")	HMM-362/ VMO-6
SHIMER, RICHARD ("DICK")	HMM-361
SHINNICK, J. NELSON	VMO-2
SHIRK, W. BRUCE ("ONE-SHOT")	HMM-163
SHIVELY, BILL ("SPEEDBRAKES")	HMM-263
SHOCKLEY, DENNIS	HMH-462/ HMM-165
SHOUB, JOE	HMM-163
SHRIBER, ED	
SHRINER, WILLIAM J.	HMM-362
SHUBIN, DEREK A. ("QUAKER")	HMM-265/361
SHUNK, HOWARD W. ("BILL")	HMM-164
SHUSTER, JOHN ("POPS")	HMM-361/ HMH-462
SHUTER, BGEN. DAVE	HMM-162/ 364/361
SIEGAL, JEFF	HMM-262
SIEGEL, HARVEY D.	
SIFUENTES, ART	HMM-163
SIGLER, THATCHER ("RACER LEADER")	HMM-361
SIGMAN, BERLE JOHN III	HMM-362/ HMH-463
SILARD, CON D. JR. ("CONNIE")	HML-367/ H&MS-16
SILVA, HERB	HML-367
SILVER, DAN	HMH-463
SILVEY, DANA	HMM-164
SIMKO, LARRY	HMM-262
SIMMONS, CARLYLE L. ("BUCK")	HML-367
SIMMONS, ROGER	HMM-165/265
SIMMONS, WILLIAM J. ("BILL")	HMM-165
SIMONS, GARY F.	HMM-363
SIMPSON, A.C.	HMH-463
SIMPSON, GEORGE ("AJIT")	HMM-363/364
SIMPSON, JOAN [BOB SIMPSON]	HMM-161
SIMPSON, TONY	HMM-262
SIMS, EVERETT E. ("BUTCH")	HMM-261/ HML367/167
SIMS, RONALD ("RON")	HMM-362
SIMS, TOM ("MONK")	HMM-363
SINAGRA, CIRO	HMM-265
SINCLAIR, JEFF	HMM-165
SINNOTT, WILLIAM T. ("TOM")	HMM-163/362
SIRACUSA, PHIL	1st MAW
SIROIS, BYRON K.	HMR-363/162 [Korea]
SISSON, MIKE	VMO-6
SIZER, CHARLES L. ("CHUCK")	HMM-161
SKAGGS, RONNIE J.	HMM-164
SKALSKI, STANLEY A.	H&MS-16
SKIBINSKI, JERRY ("SKI 2")	HMM-364
SKINNER, B.F.	HMM-365
SKUSA, DAVE ("SKOOZ")	HML-367
SLACK, BAIN	HMM-165
SLAGLE, ARTHUR ("ART")	HMM-364/164
SLAGLE, LOUIS E. ("SKY KING")	HMM-364/ 164/H&MS
SLAWINSKI, BUCKY ("SKI")	HMM-164
SLOAN, VINCENT J. ("RUSTY")	HMM-165
SLUIS, KEN R.	
SMILANICH, W.E.	HMM-164
SMILEY, JERRY	HMM-163
SMITH, BGEN. LLOYD	
SMITH, BILL	HMM-164
SMITH, BILL ("BIG BILL")	HMM-261
SMITH, C.L. ("SMITTY")	HMM-163
SMITH, CHARLES H. ("PETE")	HMM-364
SMITH, CHARLES W. JR. ("C.W.")	HMM-161
SMITH, CRAWFORD ("C.W.")	HML-367
SMITH, DON ("SMITTY")	HMM-161
SMITH, FRANK ("FIREBASE")	HMM-164
SMITH, GEORGE	HMM-164
SMITH, GORDON	HMR-263/165
SMITH, H.L.	HMM-161
SMITH, HUGH	HMM-161/163
SMITH, JAMES J. ("J.J.")	1st MAW
SMITH, JERRY	VMO-2
SMITH, JIM	HMM-364
SMITH, JOE P.	
SMITH, LARRY	
SMITH, LARRY ("GREBE")	HMM-161
SMITH, MICHAEL A. ("MA")	HML-167
SMITH, P.R.	
SMITH, RAY F.	HMM-265/165
SMITH, RUSSELL	H&MS-36
SMITH, STEVE	HMM-161
SMITH, WALT ("WHISKEY WHISKEY")	HMM-362
SMITH, WILLIAM J. ("BILL")	HMM-164
SMITH, WILLIAM R. ("HOOK")	HMM-261
SNEAD, GARY	
SNEAR, CLAY	HMM-262
SNIDENBACH, JOHN	VMO-2
SNIDER, NEIL A. ("HOSS")	HMM-161/163
SNIFFEN, PAUL	HMM-263
SNYDER, DARRELL	VMO-2
SOHM, MARC T.	HMM-362
SOKOLOWSKI, MARK	HMM-261/263
SOLANO, SAM ("DA SARGE")	1st MAW
SOLIDAY, TED ("HOSTAGE SUGAR")	VMO-2/ HML-367
SOLOMON, CHARLES G.	VMO-6
SOMERS, MYLES	HMM-265
SOMMERS, JOHN	HMM-361
SOMMERS, TOM	HML-367
SONGER, CHARLES R. ("CHUCK")	HMM-161
SOPER, D.W.	HMM-263
SORENSEN, RON	HMM-362/ 363/367
SOUTH, MARTY	HMM-261
SOUTHWICK, CURT	
SOUTHWORTH, E. G. ("McSOUTH")	HMM-362
SOWELL, TOM	VMO-2/ HML-367
SOWERS, D. ERIC ("RICK")	H&MS-36
SOWIK, WILLIAM	MAG-16
SPARRY, H.L. ("SKIP")	HMM-361
SPATARO, JAMES R.	HMM-165
SPECK, JERRY	
SPECHINGER, TERRENCE ("SPIKE")	HMM-361
SPENCE, JIM	VMO-1
SPENCER, DAVID L. ("DAVE")	HMH-463
SPERONI, VICTOR J.	HMM-362
SPEVACEK, GARY D.	HMM-265
SPIARS, EARLY W.	VMO-2/ HMM-362
SPICER, BILL	HMM-362
SPICKARD, BRUCE	HMM-362/261
SPINK, SHEPARD C. ("SHEP")	VMO-2
SPINNER, DAVID	HML-167
SPOHN, DICK	HMM-263/264
SPRADLEY, RUSSELL F.	HMM-165
SPROULE, AL	HMM-165
SPROULE, WILLIAM D. ("WILLIE D")	HMM-362
SPURLOCK, DAVID A. ("DAVE")	HMM-263/ H&MS-36
SPURRIER, ROYAL J.	HMH-461/462
SRAMEK, JAMES B. ("JIM")	HMM-165
ST. CLAIR, LOWELL T. ("SAINT")	HMM-164
STACK, RICH	HMM-365
STACK, VAUGHN	HMM-261/364
STACKHOUSE, GARY E.	
STAHL, CHARLES H.	HMM-261/ HML-167
STAHL, JUDD	
STAHL, ROY E.	MAG-16/39
STAKA, EDWARD A.	HMM-164/ 165/364
STAKES, RUSSELL J.	HML-167
STAMM, MELVYN R. ("MEL")	HMM-262
STANKO, BERNARD J. ("BERNIE")	HMM-363
STAPLES, JOHN N. III	HMM-364
STARN, PETER	HMH-463
STARNER, DEL R.	VMO-2
STARR, LARRY	HMM-261
STAUTZENBACH, LYN ("STATZ")	HMM-361
STEADMAN, HANK	HMM-361/ 364/MAG-16
STECHER, RON	HMM-263
STECKBAUER, JIM	HMM-161/263
STEELE, CHUCK	HML-167
STEELE, DAVID L. ("GATOR")	HMM-263/ 362/VMO-6
STEELE, GREG	
STEELE, JAMES R.	
STEELE, RICHARD	HMH-463
STEEN, ORRISON	HMR-161
STEFAN, GERALD	HMM-164
STEFAN, LOU ("BUF")	HMM-362
STEFFAN, WILLIAM J. ("BILL")	HMH-463/ H&MS-16
STEGICH, STEVE	HMM-262
STEINBERG, BOB ("SUPER JEW")	HMM-364
STEINBERG, MEL	HMM-262
STENEMAN, ROBERT ("BOB")	HMM-263
STEPHAN, GERALD F.	HMM-164
STEPHENS, LUTHER C. ("STEVE")	HMM-261/265
STEVEMER,	
STEVENS, LUTHER	HMM-265
STEVENS, STEVE	
STEWARD, ROGER	HMM-261
STEWART, DAVID L.	HMM-165
STEWART, JIM	HMM-364
STEWART, JOHN K. ("STEW")	HMM-261
STEWART, MIKE ("FESTER")	HMM-263
STIEGMAN, DON ("STIGGY")	HMM-363
STIERS, JAMES C.	HMM-165
STIGER, JIM ("STIGS")	HMM-262
STILWAGEN, WILLIAM W. ("DINGER")	HMM-364/ MABS-16
STITH, FRED ("THE CROONER")	HMM-164
STOCKER, NORMAN R. ("BOB")	HMM-364
STOEHR, CARL II	HMM-364
STOEL, CLARENCE	VMO-2
STOFFERAN, A.G. ("GUNNY")	HMM-163/ 363/367
STOFFEY, BOB	VMO-2
STONE, JAMES	
STONEKING, DONALD B. ("STONEY")	HMM-263/265
STOOPS, RICKY L.	HMM-164
STORMS, ALAN D. ("STORMY")	VMO-6
STORY, CHUCK ("C/L MAX")	HMM-364/262
STOUT, HENRY A.	HMM-262
STOVER, ("SMOKEY")	HMM-261
STRAIN, THOMAS S.	VMO-2/ HML-367
STRATTON, JERRY R. ("PACK")	HMM-362
STRAUSBAUGH, EDWARD S.	HMM-164
STRAUSBORGER, RICHARD L. ("DICK")	VMO-2
STREET, JAMES H. ("JIM")	HMM-161/ 263/362
STREET, JOHN G.	HML-369
STREETER, JOHN ("SHAKEY")	HMM-363
STRICKLAND, RICHARD F. ("DICK")	HMH-463
STROFF, M.	1st MAW
STROMPOLIS, MICHAEL ("THE GREEK")	HMM-361
STRONG, JERRY	HMM-162
STROUT, CLIFFORD	MAG-36
STUBBE, RAY W.	1st MAW
STUBER, KEN	HMM-164
STUPP, CLAY	HMM-164
STURGIS, JACK	H&MS-13
STURKEY, MARION F. ("STURK")	HMM-265
STUTEVILLE, B.J.	HMM-263
STYLES, ED	HMM-161
SUDICK, TODD ("CADILLAC")	1st MAW
SUEDES, ROBERT E. ("BOB")	HMR-163 [Korea]
SUIDZANSKI, JERRY	HMM-164
SULLIVAN, DICK ("RINGER")	HMM-363/265
SULLIVAN, GENE	
SULLIVAN, JOHN J.	VMO-6
SULLIVAN, MIKE	HMM-164/262
SULLIVAN, THOMAS F. ("SULLY")	HMM-164
SULLIVAN, TOM ("SULLY")	HMM-161/364
SUMP, JERRY	HMM-265
SUPLITA, JOSEPH	HMM-365
SUTTER, GILBERT	MAG-16
SUTTER, WAYNE L.	HMM-165/362
SVENDSEN, EVERETT ("SWEENEY")	HMM-363
SWAILES, ROBERT ("BOB")	H&MS-16/ HMM-263
SWAIM, STEVE ("SWAMI")	HMM-164/165
SWANBERG, BRUCE ("SWANNY")	HMM-265/ 165/VMO-6
SWANEY, KENNETH	HMM-164
SWANSTORM, DEXTER	HMM-165
SWARTZ, STEWART	VMO-2
SWEPPY, PERRY JR.	HMX-1
SWETE, BOB	HMM-163/162
SWICKARD, DEANE	VMO-2/ HML-367
SWINBURN, CHARLIE	VMO-6
SWOSZOWSKI, RICHARD H. ("SKI")	H&MS-16/ HMH-361
SYC, ("DUKE")	HMM-164
SYSLO, JOE	HMM-362/ 363/H&MS

— T —

NAME	IN COUNTRY SQDN
TACKITT, MERLE	HMM-163
TAFOYA, ANTHONY R. ("SONNY")	HMM-164/ AIR AMERICA
TAGGART, ANDY	
TALENT, ROBERT M. ("4-STORY")	HMM-163
TALMADGE, R.W. ("BILL")	HMM-363/364

NAME	IN COUNTRY SQDN
TATE, ROGER E.	
TATROW, GERARD N. ("GERRY")	HMM-165
TATUM, JOHN C.	HMM-165
TAVERNIER, CLAUDE E. JR. ("TAV")	H&MS-36/ HMM-165
TAYLOR, A.J.	HMM-362
TAYLOR, BILL	HML-167/ FAC
TAYLOR, DAVID E.	HMH-462
TAYLOR, DOUGLAS ("CRAIG")	HMM-164
TAYLOR, DUDLEY	HMM-261
TAYLOR, JIM	
TAYLOR, MGEN. LARRY ("KIM CHI")	AIR AMERICA
TAYLOR, R.D. ("DAD")	HMM-165
TEAGARDEN, JOSEPH W.	
TEELE, JOHN	HMM-263
TELLES, JOHN	HML-367
TENNENT, MIKE ("FESTUS")	HMM-364/263
TESDAHL, LAUREN ("BIG T")	HMM-364/165
TETZLOFF, JIM ("TETZ")	HMH-462
TEWES, MAURICE P.	HMM-364
THARP, JOHN J.	HMM-165
THERRIAULT, TONY	HMM-262
THIESSE, J.C.	
THIRY, GARY E. ("GAR")	HMM-363/ 367/VMO-2
THOMAS, AL ("BRAVO")	HMM-361/362
THOMAS, BUD	
THOMAS, ED	HMM-265
THOMAS, JAMES A.	VMO-6
THOMAS, JOHN C.	HML-167
THOMAS, JOHN R. ("JACK")	HMM-162
THOMAS, JOSEPH A.	VMO-6
THOMAS, LOUIS ("MAGNET ASS")	HMM-364
THOMPSON, CHARLES	HMM-161
THOMPSON, JACK	HMM-164/263
THOMPSON, JOHN	
THOMPSON, LARRY	HMM-262/364
THOMPSON, MARVIN	HMM-165
THOMPSON, RICHARD	HMM-162
THORESON, DICK	
THRASH,	
THRASHER,	
THURBER, TOM ("THUMPER")	HMM-362/263
THURMOND, RUSS	
THURNER, ARTHUR C. ("ART")	HML-367
TIERNEY, TOM	HMM-363/362
TIFT, TED	HMM-164
TILLY, ROBERT	
TIMBLIN, JOSEPH W.	HMM-165/161
TIMMERMAN, LYLE	HMM-161/263
TIMMONS, TIM	HMM-164
TINSLEY, BILL	
TINSLEY, DALE L. ("PAPA-BEAR")	HMM-265
TISDALE, DAN	
TISRON, JIM	H&MS
TIVNAN, J.M.	HMM-265
TOBEN, TED	HMM-262/263
TODD, JAMES W.	HMM-262
TODD, SHIRL L.	HMM-163/ H&MS-16
TOETTCHER, RICHARD P. ("TOUCH")	HMM-161/361
TOLLS, GENE	HMM-262
TOMAIKO, J.G.	
TOMARO, GREGORY J.	HMM-364
TOMLIN, ROBIN	HMM-262
TOMLINSON, JOE	VMO-2/ HML-167
TONERO, LOUIS V.	HMM-161
TONGE, FRED W.	
TOOHEY, JAMES A. ("JIM")	HMM-361
TOPE, LYLE	HMM-161
TOPP, DENNIS	
TORRES, FRANK	HMM-262/263
TORRES, ROBERT ("R.T.")	HMM-163/361
TOWER, ROBERT B.	MAG-16
TOY, JOHN	
TRAINOR, JOHN K. ("GUNNY")	HMM-361/163
TRAUT, EARL	
TREMONT, FRED K.	HMM-161/363
TRENKLE, RANDY	
TRIMBLE, HENRY L. ("HANK")	VMO-2
TRINER, MIKE ("MA")	HMM-164
TRIPP, MIKE ("TRIPPER")	HMM-363
TRIVETT, MARION C. JR. ("RED")	VMO-6
TROJAN, DAVE	
TROPE, JAMES G. ("JIM")	HMM-262
TROTMAN, ERNEST	HMM-265
TROTTER, CHARLES D.	
TROWER, DANIEL R.	HMM-164
TROYER, DEAN	
TRUNDY, MGEN. RICHARD T. ("DICK")	HMR-162/ HMM-164
TRUNFIO, VINCENT R.	H&MS-36
TRUPE, JAMES L.	HMH-463
TUBESING, GORDON E. ("TUBE")	HMM-265
TUCKER, GEORGE E.	HMM-364/263
TUCKER, ROBERT E.	VMO-2/ HML-367
TUCKER, STEVE ("SAM")	HMM-364
TUELL, HANK	
TUGGLE, DAVID M.	HMM-361
TURANO, JIM ("CRAZYHORSE")	HMH-462
TUREK, BILL	HMM-165
TURLEY, GLEN	HMM-365
TURNER, HENRY III	
TURNER, JIM	HMM-364
TURNER, JOE	VMO-6
TURNER, LARRY ("FLUBBER")	HMM-362
TUSHKOWSKI, RON ("SKI")	HMM-363
TVEIT, GENE	HMM-163
TWARDZIK, GEORGE F. ("POLOCK")	HMM-163/ HMH-463
TWEED, McDONALD D. ("MAC")	HMM-361

— U —

NAME	IN COUNTRY SQDN
UBEL, JACK	HMR-162/ HML-167
UDELL, FARRELL	HMM-165
ULBRICHT, WALT	HMH-462
UNDERWOOD, DAVID F. ("DOG")	HMM-163
UNDERWOOD, JOE ("UNDERDOG")	HMM-361/ HML-367
UNRUH, PERRY J.	
UNSER, AL J.	HMM-262
UNTHANK, JACK	VMO-6
UPRIGHT, TOM	HMM-161/263
UPSHAW, CHARLES R. ("CHARLIE")	HMM-364/362
UPSHAW, DICK	HMM-265
UPTHEGROVE, JOHN ("U-GROVE")	HML-367/167
URBANSKI, WALTER	HMM-265
USHER, JIM	

— V —

NAME	IN COUNTRY SQDN
VALDEZ, ED	HML-367
VALDEZ, GEORGE R.	HMM-362/164
VALDEZ, RAYMOND	HMM-262
VALENTINE, WALTER ("VAL")	HML-367
VALENTINO, BILL ("S. STUDHORSE")	HMM-263
VALINSKY, MATT	HMM-161
VALLIERE, AL ("BUTCH")	HMM-363
VALLUZZI, ROCCO ("ROCK")	HMM-263/ HMH-463
VAN GORDER, JIM	HMM-263
VAN HOUTEN, ED ("VAN")	HMM-165
VAN LEEUWEN, N.R.	HMM-263
VAN LIEW, DENNIS S.	HMM-362/263
VAN NORTWICK, JOHN	HMM-363/ 263/463
VAN VOORHIS, THOMAS ("BARON")	HMM-363/364
VAN ZWALUWENBURG, JACK ("VAN Z")	HMM-363
VANDAVEER, LEE	
VANEK, KEN ("P.V.")	HMM-361
VANN, LARRY Q. ("EL QUE")	HMM-361/363
VARDEN, ARNOLD W.	VMO-2
VARELA, SENICO	HMM-364
VARELLI, JACK	HMM-363/ HML-167
VAS DIAS, RICHARD ("VAS")	HMM-263
VASTERLING, AL	
VAUGHN, DON	HMM-161
VEAZEY, DAVID J. ("EASY")	HMM-163/ 363/VMO-6
VECCHITTO, WAYNE A. ("VIC")	HMM-265/262
VELLEUX, MICHAEL ("GRIMEY")	HMM-364
VENTRIS, KEN	VMO-2
VERBAEL, RUSS	HMM-165/ 164/VMO-2
VERBECK, A.J. ("GENE THE MACHINE")	HMM-161/ 364/VMO-2
VERGARA, EMILIO D. ("SONNY")	HMM-164
VERNIER, DARCY	HMM-263
VERSAGGI, JOSEPH A.	HMH-462
VIANELLO, CLAUDIO	HMM-163
VIANO, PAUL JR.	HMM-362
VICE, NANCY [D. SPOTTSWOOD]	HMH-463
VICKERS, JACK	HMM-163
VIECELLI, WILLIAM J.	HMM-364
VIELHAUR, G.C. JR.	HMM-263
VIGLIONE, JOHN	HMM-263
VIGNERE, JOEL G.	HMM-362
VILLAREAL, J. ("FRIENDLY GUNNER")	HMM-362
VINSON, DENNIS L. ("MULEY")	HMM-163
VISCONTI, JAN [FRANK VISCONTI]	HMM-261/362
VITI, AL	HML-167
VOEGELE, ALLEN E.	VMO-6
VOGEL, PETER J. ("JERRY")	HMM-163

— W —

NAME	IN COUNTRY SQDN
WADDELL, BILL	HMM-162
WADE, KENNETH R. ("KEN")	HMM-364
WADE, L.R. JR. ("BOB")	HMM-363
WAGENER, PAUL ("MIGHTY MOUSE")	HMM-361
WAGNER, JACK	
WAGNER, JOHN	HMM-264/364
WAGONER, DON ("WAGS")	HMM-361
WALKER, DAVE	HMM-263/161
WALKER, DON	HMM-163
WALKER, HAROLD G. ("HAL")	HMM-262
WALKER, JERRY	VMO-6
WALKER, MARTIN R.	HMM-164/ 165/262
WALKER, MICHAEL G.	HML-367
WALL, TOM	HMM-164
WALLACE, J. BRIAN	MASS-2/ HMM-263
WALLACE, L.C. ("RED")	HMM-261
WALLACE, WILLIAM	HMM-265
WALLER, FRANK L.	HMM-165
WALLIN, DAVE	HMM-363/364
WALLIS, THOMAS E. ("TOM")	HMM-263
WALLIS, WILLIAM P. ("BILL")	HMM-362/ 361/H&MS
WALSH, HUEY C.	HMM-265/164
WALTER, TIM	
WALTERS, JON ("COWBOY")	HMH-463/462
WALTRIP, STEVE	VMO-2
WALZ, NICK	HMM-261
WANDRIE, DAVID F.	HMM-165
WARD, BGEN. JERRY E. ("COACH")	VMO-6
WARD, JOEL	
WARE, SAMUEL J. ("SILVERWARE")	HMM-364/165
WARGO, JOHN	HMM-164
WARLEY, CHARLES ("CHARLIE")	VMO-6
WARNER, JIM ("WHITE RAT")	
WARNER, R.W. ("DEAK")	HMM-362
WARNING, TOM ("FREE")	HMM-361/362
WARREN, RICHARD C.	HMM-265
WASCO, MIKE	
WASHAM, FRANK	HMH-462
WASSON, LOREN	
WATERBURY, WILLIAM F. ("BILL")	HMM-265/262
WATERS, BILL ("MUDDY")	HMM-163/ 362/364
WATERS, K.D. ("JAWS")	VMO-6/2
WATKINS, CARY	HML-167
WATKINS, DAVE	H&MS-36/ HMM-362
WATKINS, ROBERT	HMM-263
WATKINS, STEVE	HMM-161
WATKINS, T.C. ("TERRY")	
WATKINS, TOM	HML-367
WATO, ED SR.	HMM-261/362
WATSON, EDWARD L.	HML-367
WATSON, JAMES R. ("JIM")	HMH-361/ HMM-764

NAME	IN COUNTRY SQDN
WATSON, JOHN L.	VMO-6
WATSON, JONATHAN ("JOE")	HMM-361
WATSON, WARREN ("DAD")	HMM-164
WATT, LEWIS ("JUNKMAN 3")	H&MS-16
WATTS, JOHN R. ("J.R.")	HMM-161
WAUNCH, DON ("PAUNCH")	HMM-163/262
WEATHERS, DENNIS D. ("DRIPPY")	VMO-2/ HML-167
WEAVER, BILLY	HMM-164
WEAVER, BOBBY	HMM-164
WEBB, MICHAEL C.	HMH-463/461/ 163
WEBER, DICK	VMO-6
WEBER, HARRY R. ("BOB")	HMM-262
WEBER, JACK	VMO-1
WEBER, JACK E.	HMM-162
WEDDING, MIKE	
WEED, GEORGE C.	VMO-6
WEEKLY, BYRON ("SPADE")	HMR(L)-363
WEEKS, HARRY E. ("HIGH EXPLOSIVE")	HMM-163/ 263
WEGENER, J.B. II ("WHEELS")	H&HS-1/ HMM-165
WEHRLY, JAMES J.	H&MS-16
WEIG, DANIEL J. ("DAN")	H&MS-16
WEIGOLD, TED	HMH-463
WEISEL, BOB	HMM-363/362
WEISS, ED ("WEASEL")	HMM-163
WEISS, JOSEPH JR.	HMM-163
WELCH, ROBERT C. ("CHUCK")	HMM-165
WELCH, WILLIAM H. ("BILL")	VMO-6/ H&MS-36
WELDING, AL	
WELDON, BOBBY R.	VMO-2
WELLS, ANDY ("AIRSPEED")	VMO-3
WELLS, GREGORY D.	HMM-363
WELSH, ("FLIP")	HMM-261
WELSH, JERRY ("RAQUEL")	HMM-364
WELSH, JOHN	VMO-6
WELTY, GEORGE W.	VMO-6
WEMHEUER, ROBERT F. ("BULL")	HMH-463
WENDLANDT, JERRY ("BIGFOOT")	HMM-263
WENK, JOHN A.	HMM-263
WENTZ, TERRY L.	HMM-165
WERNLI, JOE	HMM-163/ 262/VMO-6
WERT, LARRY ("STOGGIE")	HMH-463
WERVE, NICK	HMM-365
WESLER, HARRY ("WES")	HMM-265/165
WESOLOWSKI, GENE ("SKI")	HMM-263
WESSELL, JIM	HMM-261
WEST, BRIAN ("WILD-WILD")	HMM-363
WESTERGREN, GARY	HMM-261
WESTMORELAND, C.W. ("WES")	HMM-162/ 261/361
WESTON, BOB	
WESTON, JAMES A.	HMM-361/ 161/363
WHALEN, SCOTT ("SCOOCH")	HMM-363
WHALEY, B. ("SEAWORTHY WHISKEY")	VMO-6/ HMM-364
WHARLEY, CHARLES	VMO-6
WHARTON, GEORGE	HMM-161
WHEATLEY, HENRY ("BUTCH")	VMO-6
WHEELER, CARL	HMM-362
WHEELER, DICK	
WHEELER, LEONARD ("LEO")	HMR(M)-461
WHELPLEY, STEVEN H.	HMM-164
WHIDDEN, HARRY	HMM-165
WHITBECK, NORMAN F. ("WHIZ")	HMM-361
WHITCOMB, JIM	HMM-363
WHITE, ALEX	HMM-161
WHITE, BRENDA [DICK DODD]	HML-167
WHITE, DAVE	HMM-262
WHITE, JAMES ("SNEAKY")	HMM-262
WHITE, LTGEN. W.J.	
WHITE, RICHARD	
WHITE, RONALD B.	HMM-365
WHITE, STEVE	HMM-364
WHITE, WADE S.	VMO-3/ HML-367
WHITEHORN, WALT ("HOOT")	HMM-165
WHITEHOUSE, RAY	HMM-165
WHITEHURST, NEIL ("SCARFACE 38")	HML-367
WHITEHURST, WILLIAM D. ("WHITEY")	HMH-463
WHITFIELD, HOWARD M. ("HOWIE")	HMM-361/262
WHITLOW, JOHN	HMM-364
WHITMORE, DAN	
WHITTACRE, RONALD	HMM-165
WHITTINGHAM, DAVID K.	HMM-165
WHITWORTH, JOHN ("WHIT")	HMM-262
WICKERSHAM, FRANK ("WHISKEY 3")	HMM-363/ 367/VMO-2
WIEGAND, ROBERT ("RAZOR")	HMM-162/364
WIGNALL, D. ("NITEWINDS SHADOW")	HMM-165
WILBUR, HAYDEN M. ("HONDO")	HMM-364
WILCOX, KENNETH	H&MS-16
WILCOX, TERRY LEE ("PIRATE")	HMM-263/365
WILEY, BO	
WILEY, CLIFF ("DWARF")	VMO-2/HMM- 364/367
WILEY, OLIVER	
WILKIE, JIM ("WENDELL")	HMM-262
WILKINSON, BOBBY R.	HMM-161/ HML-367
WILKISON, BILLY G. ("BILL")	MABS-16/ HMM-362
WILKISON, CHARLIE ("WILKIE")	VMO-2/6
WILLEUMIER, ROBERT	HMM-362
WILLEY, BILL	HMM-362
WILLIAMS, BENJAMIN ("GENTLE BEN")	HMM-364
WILLIAMS, BGEN. MIKE ("SEABAGS")	HML-367
WILLIAMS, DAVID	HMM-263
WILLIAMS, GARY	HMM-163
WILLIAMS, J.C.	HMM-362/ 162/163/164
WILLIAMS, PAUL L. ("SCARFACE 49")	VMO-2/ HML-367
WILLIAMS, PETER D.	HMM-165
WILLIAMS, ROBERT A.	HMM-361
WILLIAMS, ROBERT W. ("WILLIE")	H&MS/ HMM-361/363
WILLIAMS, STEVEN R.	HMM-165

NAME	IN COUNTRY SQDN
WILLIAMS, THOMAS L. JR.	H&MS-16/ VMO-2
WILLIS, BUD	VMO-2
WILLMAN, ROBERT	HMM-263
WILMOT, DONALD W.	HMM-361
WILSON, AL	HMM-161
WILSON, D.M.	
WILSON, DON R. ("THE SHADOW")	HMM-163
WILSON, DONALD	HMM-164
WILSON, GENE ("WILLIE")	HMM-164/ 364/H&MS
WILSON, HUGH A. ("BEAR")	HMM-263
WILSON, J.J.	HMM-265
WILSON, JOHN	
WILSON, P.A.	HMM-161
WILSON, RICK	HMM-363/364
WILSON, STEVE	VMO-6
WILSON, T.C.	HMM-363/362
WILSON, WENDALL	VMO-3
WIMMLER, CHARLES A. ("WIMPY")	HMM-362/361
WINDHAM, AL ("WINDY")	
WINKLEPECK, RANDY	
WINTER, RON	HMM-161/ 164/364
WIRE, SYD	HMM-265
WISE, R.K.	HMM-365
WISE, WALTER R. ("WORM")	HMM-364
WISHARD, JOHN W. ("PADRE")	MAG-16
WISSMAR, JAMES L. ("WIZZY")	HMM-165
WISTRAND, STEVE	
WITHERS, M.H. ("MIKE")	HMR-163 [Korea]
WITHROW, JAMES C.	HMM-161
WITLOW, JOHN	HMM-364
WITSELL, JOHN A. JR.	HML-367
WITTE, RICHARD ("SLUSH")	HMM-262
WOEDSTRA, CHRIS	HMM-164
WOIDYLA, BILL ("CRAZED EX W.O.")	HMM-361
WOLF, NORMAN	HMM-161
WOLFE, JOHN C. ("WOLFMAN")	HMM-265
WOLFSEN, JIM ("DOC")	HMM-164
WOLSON, ABRAHAM	HMM-265
WOLTER, MIKE	HMM-165
WOLTZ, JAMES D.	HMM-265/462
WOMACK, BILLY D.	HML-367
WOOD, DAVID S.	HMM-165
WOOD, MARION G.	
WOODMAN, CARL	MAG-16
WOODRUFF, LARRY L. ("WOODY")	HMM-263
WORDEHOFF, GERRY	HMM-264/265
WORDEN, PETER R.	HMM-365/ HMH-463
WORLEY, ALIN	HML-167
WOUDSTRA, CHRIS R.	HMM-164
WRIGHT, ALEX ("BUZZARD")	HMM-164
WRIGHT, DONALD L. ("LARRY")	VMO-6
WRIGHT, ROBERT O.	
WRITESEL, GERALD L. ("GARY")	HMM-165
WURTHMAN, PATERSON C. ("GUNS")	HMM-363
WYDNER, C. ("CHARLIE CIGAR")	HMR-161
WYSER, BILL ("THE WIZARD")	HMM-165/ H&MS-16
WZOREK, JIM ("THE COUNT")	

— Y —

NAME	IN COUNTRY SQDN
YANCHIS, ROBERT T. ("BOB")	HMM-162/262
YASKOVIC, ROBERT A. ("YAZZ")	HML-367
YATES, CHUCK ("ROWDY")	HMM-261
YAUNERIDGE, BILL	
YEAGER, BOB ("YOGI")	HMM-262
YEEND, BOB	
YOHE, FRANCIS L.	HMM-362
YONKER, PARK	HML-267/ VMO-1
YOST, JOHN	
YOUNG, BENNY	HMM-363/ 364/161
YOUNG, BILL	HMM-164/161
YOUNG, ERNIE	HMM-163/161
YOUNG, GARY L.	HMM-164/ 265/163
YOUNG, JOHN	HML-367
YOUNG, MIKE	HMM-161/263
YOUNG, WILLIAM	HMM-265
YULE, GRANT	HMM-364/361
YUNG, CARL ("CY")	HMM-362

— Z —

NAME	IN COUNTRY SQDN
ZABINSKI, DAVE	
ZACEK, KARL G. ("ACE WINGTIPS")	HMM-262/265
ZACKER, MIKE ("ZACK")	HMM-362/ HMH-463
ZACZEK, RONALD ("ZACK")	VMO-3
ZALESKI, ALAN J.	HMM-165
ZAMORA, DON	HMM-164
ZAMORA, HAL ("Z")	HMM-364
ZAPATA, ROBERTO G. ("Z")	HMM-165
ZAPPARDINO, RONALD R. ("RON")	VMO-2/FAC
ZEHMS, RICHARD W. ("ZOOMIE")	HMM-361/ 363/362
ZEIGLER, LEWIS	HMM-363
ZELINSKY, ROGER	HMM-263
ZELLER, DON	HMM-263
ZELLICH, JACK ("SMILEY")	HMM-162/ 167/VMO-2
ZIBROFSKI, TOM ("SKI")	HMM-261
ZIMMERMAN, C. ("MONGOOSE")	HMM-363
ZIMMERMAN, ROY ("SWAMP RAT")	HMM-363
ZIPSER, LARRY	
ZOBENICA, PETE	HMM-164
ZOK, LARRY ("LZ")	HMM-262/ 263/MWHG
ZOLLER, JOHN C. ("ZIPPER")	HMM-163
ZUCH, HAROLD A. ("BUTCH")	
ZULLER, DON	HMM-263
ZUMPANO, RICHARD J. ("RICH")	H&MS-16/ HMM-264
ZUPPKE, TOM ("HOOPER")	HMM-261/ 263/461/462
ZWAAGSTRA, BERT	HMM-165

Index

— A —
A Shau Valley 12, 14, 42, 43, 48
Aardema, Cpl 31
Alford, TSgt 28
Allgood, Frankie 29
An Hoa 15, 17, 27, 31, 48
Angle, Pete 36
App, John 37
Armstrong, Charles B. 10
Armstrong, Greg 20
Atkinson Jr., James O. 14

— B —
Baker, Jon 31
Barber, J. D. 6
Barbour, Alan 18
Barnes 8
Batangan Peninsula 15
Beasley, Cpl 31
Becker, John B. 14
Ben Hai River 14
Berry, Gene 18
Binh Dinh 12
Bolton, Jim 30, 108
Bosler, Ed 11

— C —
Cam Lo 12
Carey, Mick 28
Carley, Mike 40
Carlson, Rod 46
Cascio, Ben 20
Catawba Falls 17
Charlie Ridge 15, 17
China Beach 46
Chu Lai 11, 12, 13, 14
Clapp, Archie J. 4, 6, 10
Clark, Joe 26
Clausen, R.M. 17
Coady, E.J. 12
Con Thien 18, 35, 45, 50
Condon, John 4
Cook, Gary 46
Cook, Lemuel 16
Crawford, Danny 6
Culver 8
Cummings, Dave 17
Cutri, J.A. 19

— D —
Da Krong Valley 15
Da Nang 10, 11, 13, 15, 19, 24, 26, 29, 30, 33, 46
Daniels 8
Diaz 8
Dong Ha 13, 14, 28, 39, 41, 42, 45
Dorfeld, Robert 31
Duda 8
Duke, Sergeant Major 45
Dyer, John T. 6

— E —
Eisensen, Major 45
Elder 8
Elwood 8

— F —
Flores, Samuel Jr. 17, 35
Fox 38, 39

— G —
Garrettson, Frank 11
Glenn, John 31
Gordon 8
Gregorie, Paul 36
Grizz 17

— H —
Hachtel, Robert 11, 13, 15, 17, 20
Hai Lang Forest 28
Haiphong Harbor 18
Hall, Richard H. 21
Hammock, Roy D. 43
Hansen 8
Hardesty 8
Hendrie, R.E. 24, 109
Henry, Roger 17
Herman, Roger A. 5, 6, 8, 18, 46
Herron, Buck 31
Hiep Duc 11, 16
Hill 845 17
Hippert, Jim 40
HMA-369 18
HMH-361 17
HMH-462 14, 16, 18
HMH-463 13, 15, 17, 18, 32, 37, 50
HML-167 13, 15, 18
HML-367 13, 14, 16, 18
HML-369 18
HMM-161 11, 12, 14, 15, 16, 17, 28, 30, 31, 35, 37, 54
HMM-162 10, 11
HMM-163 10, 11, 12
HMM-164 12, 15, 16, 18
HMM-165 10, 13, 15, 17, 18
HMM-261 10, 11
HMM-262 14, 15, 16, 17, 18, 32, 35, 45
HMM-263 14, 15, 16, 17, 31
HMM-265 14, 15, 16
HMM-361 10, 11, 14, 46
HMM-362 4, 6, 8, 10, 11, 13, 14, 15, 20, 40
HMM-363 12, 13, 14, 26, 28, 29, 30, 39, 41
HMM-364 2, 10, 12, 15, 17, 23
HMM-365 10, 11, 15
Hoffman 17
Hoi An 12, 24, 26
Houser, J.J. 48
Hue 13, 22

— I —
Imperial Lake 17
Interrante 8

— K —
Keller, "Red Dog" 41
Kew, George 11
Key, John Henry 14
Khe Sanh 10, 11, 13, 14, 17, 18, 28, 37, 42
Kimmel, Gene 14
Koral 8
Ky Ha 13
Kyllo, Kellan 6, 27

— L —
La Croix 8
Lach Huyen 18
LaFountaine, Norm 6
Leahy, A.M. 6, 14, 19, 20, 21, 22, 23, 26, 28, 30, 108, 111
Ledbetter, Walt 17
Leftwich, William 42
Leonard, Corporal 43
Lorisey 8

— M —
MAG-11 16
MAG-16 11, 12, 13, 14, 15, 16, 18
MAG-36 11, 12, 13, 14, 15, 16, 46
Marble Mountain 13, 14, 15, 26, 32, 39, 44, 46
Marquey 8
Massey, Gene 35
Mazon, Len 31
McAmis, R.W. 45
McCallum 8
McClellan, Robert A. 37
McDade 8
McElwee, Michael M. 35
Medina 8
Mekong Delta 10
Meland, Q.R. "Goose" 44
Miller, Huey P.L. 36
Mix, Tom 32
Moore, D. 17
Murphy 36
Mutter's Ridge 30, 31

— N —
Nam Dong 10
Nang, Da 33
Nazareth 8
Nebel, R.M. 42
Niesen, Paul W. 30, 35, 36
Norcott, "Bobby" 32
Norman, John F. 39
Novak, Joe 18
Nui Loc Son 40

— O —
Odom, Bob 30
Operation Bold Mariner 15
Operation Dewey Canyon 15
Operation Double Eagle 12
Operation Durham Peak 16
Operation Endsweep 18
Operation Essex 28, 29
Operation Foster 29
Operation Frequent Wind 18
Operation Harvest Moon 12, 16
Operation Hastings 13
Operation Hoang Dieu 17
Operation Maui Peak 15
Operation Meade River 15
Operation Medina 28
Operation Nightingale 10
Operation Pipestone Canyon 15
Operation Prairie 13
Operation Shu-Fly 10
Operation Starlight 11, 15
Operation Stone 13
Operation Taylor Common 15
Operation Upshur Stream 17

— P —
Penn, David 16
Petteys, D. M. 42
Phouc Valley 12
Phu Bai 11, 12, 14, 16, 18, 23, 43, 48
Phu Cat 12
Pilck, V. 17
Pinkerton, Walter E., Jr. 44
Pittman, Chuck 17
Pless, Steve 13
Poleski, Martin L. 13, 42
Powell, Terry 35
Puckett, Sgt 48
Puller, "Chesty" 41
Pultorak, Joseph 10

— Q —
Quang Nam 17
Quang Ngai 12
Quang Tin Province
Quang Tri 10, 11, 14, 16, 18, 22, 28, 30, 32, 35
Que Son Mountains 16
Que Son Valley 11
Qui Nhon 11, 36

— R —
Read, Ted 18
Ring, Bill 14, 45
Rivers, Rick 45
Rockpile 32, 42
Royer, Mike 11
Ryan 8

— S —
Sabin, Ron 46
Sachs, Ernest P. 40
Sadowski, Lt. Col. 37
Salter, M.E. 2, 18, 26, 30, 39
Sawczyn, Peter G. 16, 33, 34
Scott, H.R. 37
Shanahan, J.L. 2, 6, 18, 44
Sharp, Scott 36
Sigman, John 8
Silva 8
Smith, Bill 18
Smith, Tim 35
Soc Trang 10
Southworth, E.C. 33
St. Pierre 8
Stiegman, D.L. 41
Stockburger, Arthur L. 14
Stoffey, Bob 12, 18
Stragal 8
Sugar Bear 33
Swoszowski, Richard H. 33
Syslo, Joseph A. 15

— T —
Talen, R.M. 38
Tam Ky 10
Task Force Delta 12
Thuong Duc 10, 15
Thurman 8
Tobin, Captain 41
Tonkin Gulf 18
Twardzik, George 48
Tweed, McDonald D. 10, 12

— U —
USS *Princeton* 11, 35, 42
USS *Tripoli* 16, 17

— V —
Valentine, F. William 42
Van Nortwick, John 2, 6, 12, 15, 18
Van Tuong Peninsula 11
Vieira, LCpl 6, 29, 30
VMO-2 11, 12, 13, 14, 15, 16, 28, 29, 30
VMO-6 11, 13, 14, 15, 16

— W —
Walker, Mickey 31
Walsh, Huey C. 41
Warford, Zin D. 10
Warner, Jack 45
Warring, Tom 36
Wasco, Mike 17
Wegener, J.B., III 16, 41
Wesolowski, Gene 12
Wickersham, F.G. III 33
Williams, Michael E. 14
Wilson, T.D. 40
Wolter, Mike 17

— Y —
Yaskovic, Robert A. 32